nternational.urban.ideas.
competition.for.the.new.multi-functional.administrative.city.
in.the.republic.of.korea

韩国首尔迁都规划竞赛作品集

[韩] 复合型行政中心城市建设推进委员会
复合型行政中心城市建设厅 编

武凤文 傅博 冯辽 译

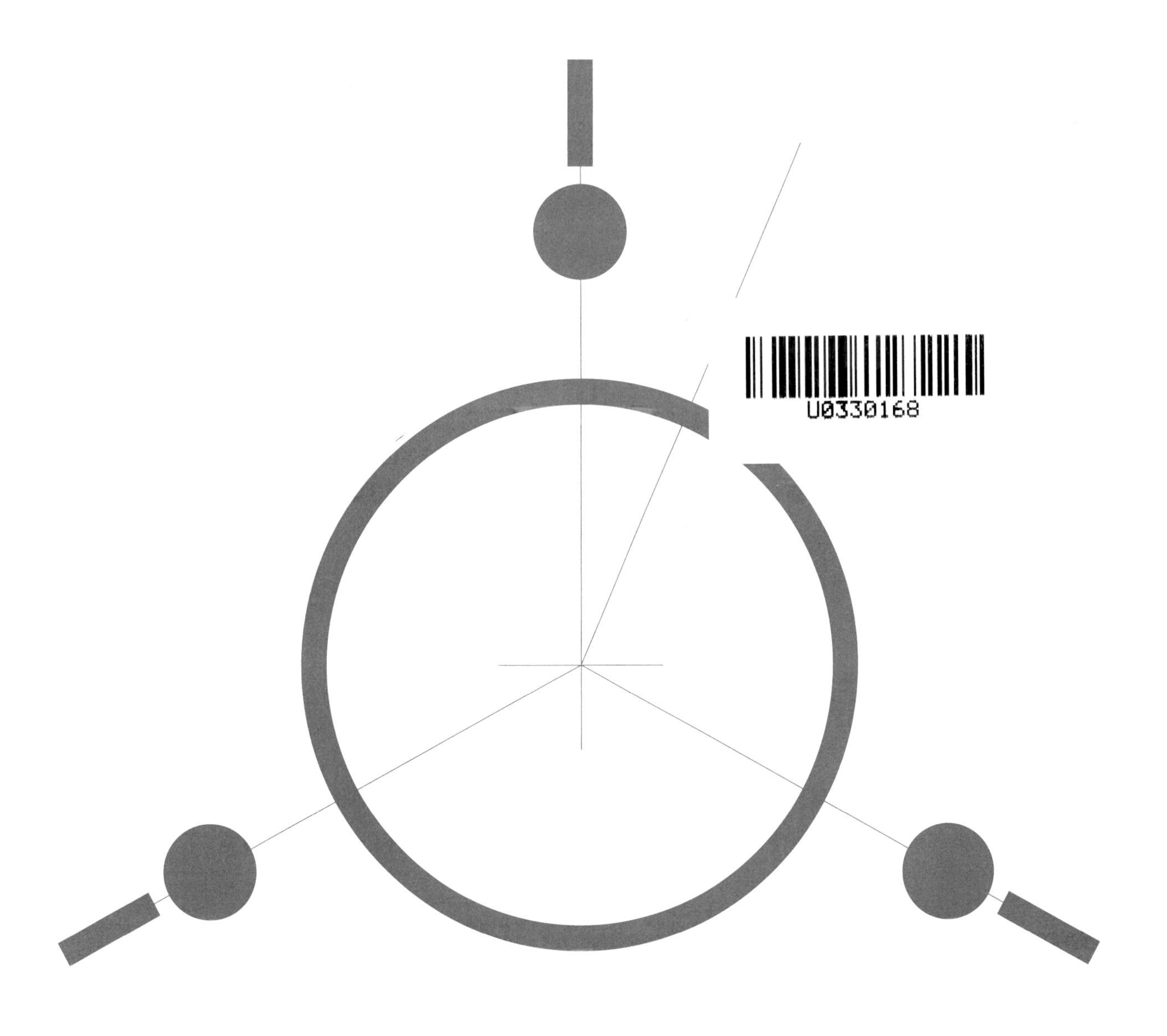

中国建筑工业出版社

目 录

C/o/n/t/e/n/t/s

前 言

Lee, Hae Chan / 李海瓒

我们十分感谢此次对复合型行政中心城市建设工程关注的每个人，同时对那些参与国际城市概念性竞赛并为之提供了许多有创新性和原创性观点的参赛者表示诚挚的谢意。正如大家所期待的那样，韩国政府承诺：新城将会被建设成为一座广为称颂的模范城。

自从20世纪60年代时起，韩国飞快的经济增长速度震惊了整个世界。这种飞速增长的结果使得韩国在2005年的GDP总值位列世界第11位。在这个时期，首尔地区集中的人口和经济活动都急速发展，从而加剧了区域发展的不平衡性。韩国政府希望通过此次竞赛来缓解过度集中的首都地区，从而获得区域发展的平衡。这作为政府急需解决的头等大事，使得本次竞赛意义重大。通过这次竞赛，首都地区和韩国的其他地区将会呈现出一派繁荣的景象，从而为国家经济的增长提供新的动力。到那时韩国国内所有的地区都将会平等地发展成为宜居地。虽然迄今为止还存有许多疑惑和困难，但本次竞赛作为一个已成事实是不会改变的。

设计和建造一座新城要有一整套表现城市未来发展的图集。在这次高水准的竞赛中，那些关于建筑物和城市组成要素的创新性观点对此作出了贡献，同时韩国政府组织的这个竞赛就是这种尝试的一部分。本次竞赛共有351个小组，来自40个不同的国家，其中26个国家的121位参赛者在竞赛中提交了他们各自的作品。

韩国政府组织出版了这次国际竞赛中提交的所有作品。这些作品中包含不同的理念，它们不仅被看作是把新城建设成为世界一流城市的基本原则，同时从历史的角度上看也是具有记录价值的。我们衷心地感谢本次竞赛的所有参赛者，感谢他们卓越的才华；同时我们也深深地感谢那些使本书得以出版的编者们。

复合型行政中心城市建设委员会会长

李海瓒　崔秉宣

2006年3月

FOREWORD

Choe, Byung Sun / 崔秉宣

We would like to thank everyone who takes an active interest in the Multi-functional Administrative City Construction Project and those who participated in the International Urban Ideas Competition and provided many creative and original ideas for the project. To follow up all your expectations, the Korean government gives its assurance that the new city will be built as an exemplary city that will be praiseworthy in the world.

Since the 1960s, the Republic of Korea has accomplished a rapid economic growth that surprised the whole world. As its result, Korea's GDP ranked the 11th largest in the world in 2005. During this time, centralization of the population and economic activities in Seoul metropolitan area (capital region) has been accelerated, therefore the regional disparity has deepened in Korea. The Korean government promoted Construction of the Multi-functional Administrative City as the most effective and fundamental solution for easing concentration of the capital region and actualizing balanced regional growths. Construction of the Multi-functional Administrative City is hugely meaningful as a leading business that the capital region and the other regions throughout the whole country make common prosperity building up a new engine for national economic growth and then all regions in this country will equally develope to be attractive places to live. Even though there has been many criticisms and complications so far, the Multi-functional Administrative City is now a historical reality that cannot be reversed.

Planning and building a new city requires the entire picture of it which explains the desired layout of that future city. Highly creative ideas on the structure and composition of the city would certainly contribute to upgrading the quality of the Multi-functional Administrative City and the International Urban Ideas Competition was held by the Korean government as a part of this effort. Total of 351 teams from 40 different countries registered and 121 participants from 26 countries submitted their pieces of work in the competition.

The government has published 『The Portfolio of International Urban Ideas Competition』 by collecting all your pieces of work submitted to the international competition. Various ideas in the portfolio are serving as fundamental basis for making the Multi-functional Administrative City as a top world-class city and also that would be worthy documentaries in a view of our history as well. We would like to deeply re-appreciate all the entrants to the competition for their excellent ideas and those involved in publishing of this portfolio.

March 2006

Co-chairman

Presidential Committee on Multi-functional Administrative City Construction

Hae-Chan Lee & Byung-Sun Choi

贺 词

Lee, Choon Hee / 李春熙

举办这次国际城市概念性竞赛，是为了给复合型行政中心新城的建设集思广益，现在随着这本作品集的出版，使得本次竞赛终于接近尾声。

作为关于韩国城市发展的第一个国际竞赛，开始时我们充满了期待和担忧，然而随着竞赛的顺利进行，我们的忧虑很快就消除了，并最终在令我们惊喜的状态下圆满结束。

提交的作品来自 26 个国家共 121 件，所有的作品都是那么优秀以至于我们很难抉择。最后，我们选出了 5 个一等奖和 5 个优秀奖。我们相信本次竞赛不管从质量上还是数量上都是成功的。

在 2005 年 5 月竞赛通告公布的时候，我们就宣布获奖作品将被用作新城建设的总体规划中。既然现在我们已经选出了获奖作品，那么在新城的总体规划设计中，它将为我们提供其卓越的设计理念。

自从 2005 年 11 月 15 日公布获奖者名单后，城市规划师们举办各种论坛和会议来探讨怎样将那些获奖作品中的理念应用到新城的总体规划中。国土研究院和其他相关研究所的研究员们组成了一个专家组，他们亲自调研了新城的基地，并试图通过场地的测量和技术评估来预测实际开发建设中可能存在的问题。此外，我们还提出了 15 个专业方面的议题，包括运输、通信、景观、文化和福利。从 2006 年 1 月开始每周都举办研讨会，以便可以广泛吸取有关专家和学者们的真知灼见。

通过采取这些创新性的举措，我们确信新城的总体规划不是一个拘泥于传统的、简单的规划蓝图，而是一个富有远见、切实可行的方案。据此，这个规划方案就成为了一个包含有不同价值观的新城建设的范例。这里每个要素都起着重要的作用，同时这也暗合了以知识和信息为主的 21 世纪对速度和多变性的追求。

我们计划在将来为新城的综合和详细发展规划举办其他的设计竞赛。同时我们也期待着你们的关心和支持，因为这样可以公正地评审、选择和使用竞赛中那些有创意的观点和设计，从而促进其进一步发展。这本作品集不仅包括获奖作品和其他提交的作品，还包括与竞赛有关的一些信息，如从开始到结束的所有程序。我们希望本书的出版，可以给规划师和建筑师，当然也包括给举行类似设计竞赛的政府或非政府组织提供有用的参考。同时我们也希望本书，对于那些对新城建设保有热情的人，可以成为一个有趣的素材。

我们衷心地感谢每个参与和支持竞赛的人。我们期待着你们在新城建设中的继续支持，这些都将成为其他城市建设和规划的范例。

复合型行政中心城市建设厅长

李春熙

2006 年 3 月

GREETINGS

The International Urban Ideas Competition, held to collect ideas to build the Multi-functional Administrative City as a city that is creative and innovative is now drawing to close with the publication of this collection.

As it was the first international competitions in a case of new city development in Korea, we had to start with much expectations as well as much concerns. However, our concerns were soon dispelled as the competition went well and finally ended up with great success so now we are pleased with that.

Total of 121 entries from 26 countries were submitted and all the submitted works were excellent so that it was hard to choose a winner and ended up awarding 5 upper tier winners and 5 honorable mentions. And it is believed as a huge success on that competition for its quality of the works as well as its quantity.

When the announcement for the international competition was made in May 2005, it was also made that the winning ideas would be used as the basis for the master plan. Now that we have chosen the winning entries, it remains for us to work out those outstanding ideas into the master plan of the new city.

Since the announcement of prize winners on November 15, 2005, forums and meetings attended by urban planners were held to discuss the incorporation of the winning ideas into the master plan. A team of experts composed of researchers from Korea Research Institute for Human Settlements and other relevant institutes have visited the site of the new city to conduct site survey and technical review to determine potential problems in practical application of the ideas. In addition, we identified 15 professional research topics including transportation, information & communication, landscape, and culture & welfare and have been holding open seminars every week since January 2006 to pool the knowledge of experts and researchers in those fields.

By taking these innovative planning steps, we are ensuring that the master plan is not just a simple blueprint that is tied down to formality, but it turns out to be a practical plan containing the urban vision and feasible implementation methods. In this way, the master plan will present a model of the new city that places importance on diverse values, where all members of the community play leading roles, and that suits 21st century's era of knowledge and information which calls for fast and flexible responses.

We plan to hold other design competitions in the future for the new city's general and detailed development plans. We would like to ask for your interest and support so that such competitions where creative ideas and designs can be fairly judged, chosen, and used can be further promoted.

This portfolio not only contains the winning entries and other submitted works, but also has information on the competition itself including all the procedures from the beginning to the end. Our hope in publishing this portfolio is that it can be utilized as useful reference for urban planners and architects as well as for other governmental/non-governmental organizations that are planning similar design competitions. And we hope that it will be an interesting material for all those showing keen interest in the Multi-functional Administration City Construction Project.

We would like to thank everyone who participated in the competition and supported, We look forward to your continued support in building the Multi-functional Administrative City that could be as a model in other city construction and urban planning projects.

March 2006

Administrator,

Multi-functional Administrative City Construction Agency

Choon-Hee Lee

国际城市概念竞赛概要

新复合型行政中心城市背景介绍

1. 新城建设的背景

在过去的40年中，首尔大都市圈（首尔、仁川、京畿道）不断增加的人口和工业促进了韩国经济的发展。政府推行各种政策来调控首尔大都市圈的增长以促进其他地区的发展，但都不成功。

首尔大都市圈集中了韩国48%的人口，相比日本（32%）、英国（12%）和法国（19%），韩国人口的集中是目前最为严重的，需要迫切解决这个问题。这种过度的集中是区域平衡发展的一个很大障碍。此外，首尔大都市圈还面临着许多社会经济学的问题，诸如高额的居住成本和土地价格、拥堵的交通和环境的污染。

韩国政府在1982年颁布了首尔大都市圈调整规划法，并使用各种方法譬如国土综合规划，以促进国土平衡发展，但效果甚微。

20世纪70年代后半期，朴正熙总统设想在忠清地区建设一个具有行政职能的临时首都，以解决首都地区过度集中的人口问题，同时加强国家的防御能力。最初，朴正熙总统在忠清地区挑了3块候选基地，并建议建设一个面积为85km²，人口为500000的新首都。然而，这个计划随着总统的去世而宣告流产。一直到20世纪90年代中期，作为分散中央行政机构政策的一部分，在大田广域市新建了第三政府办公厅，从首尔大都市圈搬迁来的10个中央政府机构落户在这里。虽然政府为此付出很多努力，但首尔大都市圈人口过度集中的现状仍就没有任何改观，作为一个基本问题就如过度拥挤和中央集权一样仍然存在。

这种情况下，在2002年的总统大选中，候选人卢武铉再次提出要在忠清地区修建一个新的行政首府。在他的竞选宣言中即强调：拥有500000人口的新行政首府将在2007年破土建设。

因此，参与式政府在2004年1月29日颁布了作为国家在2020年的国家发展战略新国土设想，其中把国土的均衡发展选定为一个核心推广策略。为了推广这个策略，一个综合一揽子政策正在推行并试图与权力下放和东北亚地区的商业枢纽工程紧密结合起来。

首先，每个区域在信息的生产和决策中都扮演不同的角色，在政策的制订上首尔拥有垄断权。通过新城的建设，依靠转换国家管理范例，即将集中于首都地区的权力下放到各个区域中去（建设新行政首都的特别法令），通过跨区域的合作和互补，把现存的构筑物转换成建筑物。同时，也可以通过区域和战略性产业，来加强自己的独立能力，

通过各行业、机构、研究中心和产业群的合作，全面推进原市属国有土地的均衡发展（国家均衡发展的特殊法令）；此外，也可以通过发展区域文化、旅游和经济欠发达地区来消除区域间的不平衡性。通过在实践中应用这种方法，全面提升了地方自治政府的能力和其自主权力（权力下放的特别法令），中央政府也将通过实行自我管理机制——比如教育和市政警察制度，把权力下放到各个区域。

到目前为止，参与式政府一直致力于新城的发展，并把土地的分散作为其首要任务，且认为这种方式有助于均衡发展战略。这个战略就是在新城和其他区域之间形成一个网络，即在2～3小时之内均可从新城到达其他的大城市。

然而，由于宪法法院否决了2004年10月21日的这个提议，有关新城的计划遭到了全面的禁止。因此，为了国土的均衡发展，就出现了一些符合政策要求的替代方案，并密切关注宪法法院的决议。与此同时，无论是国民议会的执政党还是反对党都认为应该兴建的是新复合型行政城市而不是新的行政首都。并且颁布法令以确保其施行（关于新复合型行政城市建设的特殊法令）。

根据新复合型行政城市建设的特殊法令，总统和这六个部（统一、 外交、国防、司法、政府管理和家庭事务以及妇女）将继续留在首尔都会区。中央行政机构：包括12个部、4个处、2个厅将被转移到燕岐和公州地区。

Summary of the International Urban Ideas Competition

Background of the New Multi-functional Administrative City

1. Background of the New Multi-functional Administrative City

For the past four decades, the concentration of population and industries in the Seoul Metropolitan Area (Seoul, Incheon, and Gyeonggi) has continuously increased as a result of the economic growth in Korea. The government implemented a variety of policies to regulate the growth of the Seoul Metropolitan Area as well as to promote the development of other regions, but in vain.

Forty-eight percentage of Korea's population is concentrated in the Seoul Metropolitan Area. Compared to the population of other countries -Japan(32%), Great Britain(12%), and France (19%)- Korea's population concentration is by far the severest and indicates the urgency of addressing the issue of population concentration in the Seoul Metropolitan Area. This heavy concentration has been a great barrier to achieving balanced regional development and national reconciliation by widening the gap between the Seoul Metropolitan Area and other regions. In addition, the Seoul Metropolitan Area is faced with many socioeconomic problems, such as high housing costs, land prices, traffic congestion, and environmental pollution.

The Korean government enacted the Seoul Metropolitan Area Readjustment Planning Act in 1982, and implemented various measures, such as the National Territorial Plan, in order to promote regional development, but in vain.

In the second half of the 1970s, President Park Chung-Hee envisioned a temporary administrative capital in Chungcheong area (Chungcheongbuk-do, Chungcheongnam-do and Daejeon) to resolve the concentration problem in the capital and to strengthen national defense. Originally, the Park Administration selected three candidate sites in the Chungcheong area and proposed a new capital city with a population of 500,000 and 85km^2 of land. However, the plan was aborted abruptly following his death. In the mid 1990's, as a part of the policy aimed at dispersing central administrative institutions, the Third Government Complex was built in Dunsan in the city of Daejeon, where ten central government agencies in the Seoul Metropolitan Area were relocated. In spite of such efforts by the government, concentration in the Seoul Metropolitan Area has not shown any signs of relief as fundamental problems such as overcrowding and centralized governmental power still persist.

In these circumstances, during the presidential election campaign in 2002, the idea of constructing a New Administrative Capital in the Chungcheong area was announced by the presidential candidate at that time, Roh Moo-Hyun, as one of his campaign pledges: A New Administrative Capital with a population of 500,000, whose groundbreaking ceremony would take place in 2007.

In this regard, the Participatory government established 「New National Land Design (Jan. 29, 2004)」 as one of the national development strategies aiming at the year 2020, and chose the balanced development of the national land as a core promotion strategy; in order to promote this strategy, a synthetic package policy has been promoted while being closely combined with decentralization and Northeast Asia Business Hub Project.

First of all, each region will play a role in information production and decision making in policy that have been monopolized by Seoul, by means of transforming the paradigm of national governance from the focus on the Metropolitan area to decentralization through the establishment of the New Multi-functional Administrative City (Special Act on the Construction of the New Administrative Capital). In doing so, one can transform the existing structure into the structure with inter-regional cooperation and supplement. In the meantime, one can

国际城市概念竞赛概要

新复合型行政中心城市背景介绍

2. 复合型行政中心城市建设的组织机构

复合型行政中心城市建设推进委员会

该委员会是总统下属的一个机构。30名成员中既有中央相关部门的部长也有普通民众的代表。并由总理和民众代表联合进行具体运作。正如一个临时办公室一样，该委员会已经开始运作了，并已履行其职能，商榷了新城建设的主要政策。

复合型行政中心城市推进团及建设厅的成立

建设推进团是建设委员会的一个下属机构，其主要职责是：制订土地的开发规划和土地补偿政策等。在2006年1月1日新城正式开始建设的时候，建设推进团被解散，取而代之成立了建设厅。

3. 新复合型行政中心城市建设历程表

新城的规划包括以下四个阶段：筹备、规划、建设和重新安置。2005年作为一个前期的准备阶段，主要任务包括：颁布特殊法令，巩固推进系统，确定新城边界、安排工程的工作人员，以及建立一个重新安置的计划。

于此同时，从2005年开始一直到2007年属于规划设计阶段，这一时期主要工作内容是：制订总体规划和开发规划，进行环境评估，研究环境问题和交通压力，评估土地的价值，以及建设厅的成立和特殊预算的出炉。

从2007年开始进入建设阶段，基地的准备工作和政府建筑将开始建设，有关新城选址的法律也有望被颁布。从2012年开始，那些行政机关和居民们将按照计划逐步外迁。

3.1 前期准备阶段

制订了行政首都的后续对策，为了复合型行政中心城市在燕岐和公州地区的建设，在2005年颁布了特别法令。这个特别法令要求有一些相关的规划，其中包括：基地，周边地区，将要迁移的政府机关，项目的负责人，总体规划，开发规划以及区域规划。

推广体系的巩固（2005年3月~12月）

在总统的领导下将成立一个推进委员会和一个推广办事处，每一个涉及的领域都将成立一个单独委员会。为了项目的实施还将成立一个“建设厅”，并于2006年1月开始编写有关被安排和补充的人事和组织规章。

选定基地、周边地区和项目负责人（2005年5月）

这块购买来用作搬迁行政机构和新建行政城市的场地被指定为选中的基地。该区域的相邻地块，即那些可以防止城市无限扩张并促进其规划管理的区域被指定为周边地区，同时，还指定了一个该项目的负责人。

3.2 规划设计阶段

制订总体规划和开发规划（2005年5月~2006年12月）

制订新城的总体规划是对国际城市概念性竞赛成果的回应。这个总规包括：新城建设项目的概述、人口的安排、土地的利用、迁移部门的部署、交通、景观、环境、教育、文化和福利设施、基础设施、融资方式以及对其他规划的指导作用。

总体规划的下一个阶段是开发规划。这个发展性规划包括：人口容纳规划、中央行政机构搬迁规划、基础设施规划、施工进度和分阶段实施规划以及第一部分地区单位规划。

建立实施规划（2005年9月~2007年3月）

关于区域建设的执行计划，在第一期将开始准备。该计划包括：关于图纸和计划的详细说明手册，被保留的建筑物和设施以及第一部分地区单位规划。

关于土地的使用补偿计划（2005年12月）

基于相关的法律，这个项目的负责人组织成立了一个委员会，这个委员会主要负责土地的使用和有效

Summary of the International Urban Ideas Competition

Background of the New Multi-functional Administrative City

strengthen one's independent capacity through regional and strategic industry, joint cooperation among industries, institutions, research centers, and industry clusters by comprehensively promoting the balanced development of the national land led by municipalities for independent decentralization (Special Act on Balanced National Development); in addition, one corrects the unbalanced development among regions through the development of regional culture, tourism, and underdeveloped areas. By putting the realistic effects of decentralization into practice while raising the capacity of local self-governing bodies and the uplift of their autonomous power in a comprehensive manner (Special Act on Decentralization), the power and authority of the central government will be transferred to each region through the introduction of the self-governing bodies for education and municipal police system.

So far, the Participatory government has been promoting the new city as a leading task for the decentralized formation of land and it has been supposed to contribute to the balanced development strategy based on the city network between the new city and other areas in the country by enabling (any major cities to be reached) within two to three hours from the new city.

However, the project concerning the new city was comprehensively barred due to the Constitutional Court's decision that disfavored the project dated on Oct. 21, 2004. Accordingly, there were some alternative plans for the continuous operation of the policy of the balanced development of the national land while observing the decision of the Constitutional Court. Meanwhile, both the governing party and opposition party in the National Assembly compromised over the establishment of New Multi-functional Administrative City instead of building a new administrative capital city, and a <Special Act on the New Multi-functional Administrative City Construction> was enacted in order to ensure its promotion.

According to Special Act on the New Multi-functional Administrative City Construction, the president and six Ministries (Ministries of Unification, Foreign Affairs, Defense, Justice, and Government Administration and Home Affairs, and Gender Equality) among the Administrative organizations will stay in the Seoul Metropolitan area and the central administrative organizations including 12 Ministries, 4 Agencies, and 2 State-run Organizations will be transferred to Yeongi and Gongju area.

2. Organizations in Charge of Planning the New Multi-functional Administrative City

Presidential Committee on Multi-functional Administrative City Construction

The committee has been under the supervision of the president, composed of 30 members of the ministers concerned and non-official civilians, and operated in the joint chairperson system of the Prime Minister and Non-Official Civilian. The committee has been operating as an emergency office and has been performing the function of deliberating the main policies.

Multi-functional Administrative City Construction Task Force and Establishment of the Construction Agency

The business promotion has been proceeded with the aid of the construction promotion office under the supervision of Promotion Committee. Furthermore, the construction promotion office was dismissed on Jan. 1, 2006 -when the construction business such as the establishment of development and the compensation of land are in their full swing- and the Construction Agency was established.

3. Timetable for the New Multi-functional Administrative City

国际城市概念竞赛概要
新复合型行政中心城市背景介绍

赔偿问题，并由居民、相关机构、当地政府和专家们组成。

建设厅的成立（2006 年 1 月）

成立了负责项目实施的建设厅，同时组织了具体的实施人员和制定了机构的规章制度。

特殊账号的设立（2005 年 3 月 ~12 月）

从 2006 年开始，将进入用于基础设施建设的特殊账号的准备期，且保持其与总体规划和开发性规划的一致性。

3.3 建设阶段

新城和政府办公建筑建设期（2007 年末～ 2030 年）

这个时期基础设施将开始建设。对基地的测量也将在 2007 年末或 2008 年的上半年开始。此外，基础设施、政府办公建筑、房屋也都同期投入建设。

3.4 重新安置阶段

政府机构和居民重新安置阶段（始于 2012 年）

从 2012 年开始，中央政府部门将进入重新安置阶段，民众也有望迁入到新城中去。

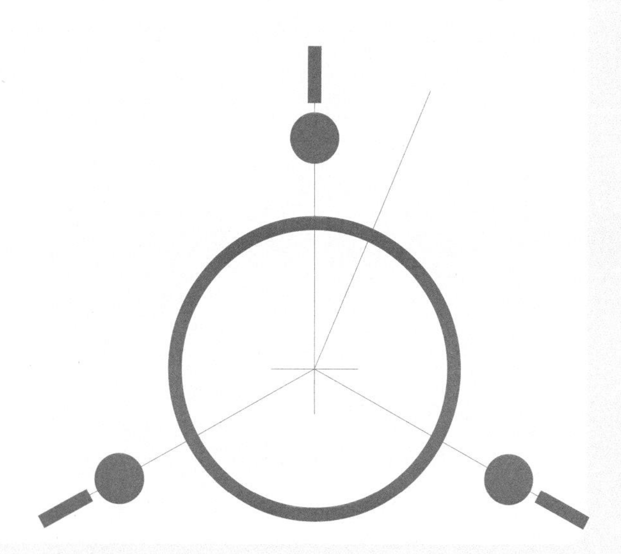

The New Multi-functional Administrative City Plan consists of four stages: preparation, planning, construction, and relocation. In 2005, as the preparation stage, a special act has been enacted, a promotion system was consolidated, a territory boundary and a project operator were designated, and a relocation plan was established.

At the same time, from 2005 which is the first year of the planning stage to 2007, establishment of master plan and development plans, environmental review, environmental crisis and traffic impact study, purchase of a site, and establishment of a Construction Agency and special accounts have been carried out.

From 2007, in the construction stage, site preparation and government building construction will be commenced, and a law related to the position of the administrative city is supposed to be enacted. From 2012, administrative agencies and residents will be moved in step by step.

3.1 Preparation Stage

Special Act for the Establishment of the New Multi-functional Administrative City in Yeongi & Gongju area as a successive countermeasure for the new administrative capital was enacted on 2005

The special act calls for related plans including a site, a peripheral area, government offices to be relocated, a project operator, master plan, development plan, and a regional plan.

Consolidation of the Promotion System (March to December, 2005)

A presidential committee and a promotion office under the president were created and individual committees for each field were organized. Establishment of a 'Construction Agency' responsible for project implementation beginning on January 2006 was arranged and supplement of personnel and organizational regulations was prepared.

Summary of the International Urban Ideas Competition

Background of the New Multi-functional Administrative City

Designation of site, peripheral area and project operator (May, 2005)

The area, which is purchased in order to relocate administrative organizations and construct an administrative city, was designated as the site. The area neighboring the site, which needs to prevent urban sprawl and facilitate a planned management, was designated as a peripheral area. Also, a project operator was designated.

3.2 Planning Stage

Establishment of the Master and Development Plan (May, 2005 to December, 2006)

Reflecting the results of International Urban Ideas Competition, a master plan will be established. The Master plan includes an outline of city construction project, population arrangement, land use, disposition of transferred offices, transportation, landscape, environment, educational, cultural and welfare facilities, infrastructure, financing method, and a guideline for additional plan.

Reflecting the results of master plan, a development plan will be established. The development plan includes a population plan, a plan for accommodating central administrative organizations, a infrastructure plan, the construction period and phased implementation plans, an enforcement plan and the guidelines for the first district unit plan.

Establishment of the Enforcement Plan (September, 2005 to March, 2007)

Enforcement plan on districts constructed during phase 1 will be prepared. The plan includes detailed manuals on drawing and program, a plan for maintained buildings and facilities, and the first district unit plan.

Commencing a compensation service by land taking. (December, 2005)

Based on related laws, the project operator organizes a council for land taking and execute compensation. The council consists of residents, related organizations, local governments, and professionals.

Establishment of the Construction Agency (January, 2006)

A Construction Agency responsible for project implementation was established and necessary officials and agency regulations were arranged.

Preparing for the Establishment of the Special Accounts (March to December, 2005)

From 2006, Special accounts for infrastructure construction are prepared, in accordance with the master plan and development plan.

3.3 Construction Stage

Construction of the City and Government Buildings (late 2007~ 2030)

Construction on infrastructure will begin. Leveling on site in late 2007 or 2008 will start. In addition construction on, infrastructure, government buildings, and housing will begin.

3.4 Relocation Stage

Relocation of Government Organizations and People (beginning 2012)

From 2012, Ministries of Central government will be relocated in stages and people are expected to move into the New Multi-functional Administrative City.

国际城市概念竞赛概要

韩国首尔迁都规划竞赛基本情况

■ 主办方

复合型行政中心城市建设推进委员会

■ 竞赛类别

开放的国际城市概念性竞赛

■ 工程选址和尺度

- 选址：在韩国忠清南道公州和燕岐市周围
- 基地面积：73.14km^2（7314hm^2）
- 规划人口： 500000

■ 城市功能

- 主要功能：国家行政中心
- 辅助功能：国家政策研究中心，国际文化交流中心，教育中心（大学水准），商业和科技中心，旅游中心，休闲及其他基本功能（诸如商业、办公、居住等）

■ 竞赛细则

- 竞赛公布日期：2005年5月27日
- 报名参赛时间：2005年6月1日～2005年7月11日
- 对参赛者介绍竞赛的详细情况：2005年7月12日
- 疑问解答期：2005年7月12～29日
- 参赛作品提交期：2005年10月18～25日
- 评审期：2005年11月11～14日
- 公布获奖作品：2005年11月15日

■ 提交的作品

- 提交的作品共分为121组，其中韩国59组，国外62组（德国10个，美国9个，荷兰5个，日本、意大利、澳大利亚、西班牙各4个，法国、瑞士、英国、中国香港各2个，印度尼西亚、以色列、马来西亚、新加坡、泰国、阿拉伯联合酋长国、阿根廷、巴西、智利、亚美尼亚、奥地利、芬兰、葡萄牙、塞尔维亚和黑山共和国各1个）

■ 评委会成员

- 大卫·哈维（David Harrvey，美国，联合会主席），纳德（Nader Tehrani，美国，联合会主席），矶崎新（Arata Isozaki，日本），韦尼·马斯（Winy Maas，荷兰），Min Hyun Sik、Park Sam Ock、Ohn Yeong Te、Yoo Kerl（韩国）

■ 详细计划

- 总体规划的制定：2005年5月至2006年7月
- 开发规划的制定：2005年8月至2006年11月
- 实施规划的制定：2005年9月至2007年6月
- 广域城市规划的制定：2005年7月至2007年6月
- 破土动工时间：2007年7月
- 第一阶段建设完成并可以入住期：2012年
- 工程结束期：2030年

■ 政府部门的重新部署

- 政府部门的重新部署分为三个阶段（从2012年～2014年）
- 12个部门：财政经济部、教育部、科学技术部、文化观光部、农林部、产业资源部、信息通信部、健康福利部、环境部、劳动部、建设交通部、海洋水产部
- 4个处：规划预算处、法制处、国家广告处、爱国者和退伍军人事务处
- 2个厅：国家税务厅和国家突发事件处理厅

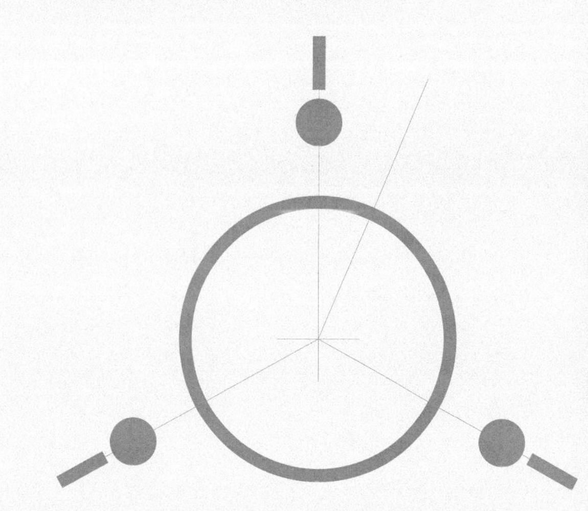

Summary of the International Urban Ideas Competition

Summary of the International Urban Ideas Competition

Summary of the International Urban Ideas Competition for the New Multi-functional Administrative City in the Republic of Korea

■ Host

Presidential Committee on Multi-functional Administrative City Construction

■ Competition Type

Open international urban ideas competition on one stage

■ Project Site and Dimensions

- Location : Around Yeongi-Gun and Gongju-City, Chungcheongnam-Province, South Korea
- Site Area : 73.14km^2 (7,314ha)
- Target Population : 500,000

■ City Functions

- Main function : National administration
- Subsidiary functions : research of national policies, international cultural exchange, education(up to university-level), businesses involved in innovative technology, tourism, lei-sure and other basic urban functions. (neighborhood commercial, office, residential, etc.)

■ Competition Details

- Competition announcement : May 27, 2005
- Registration : June 1 ~ July 11, 2005
- Distribution of detailed information to participants : July 12, 2005
- Questions & Answers : July 12 ~ July 29, 2005
- Submission of Entries : October 18 ~ October 25, 2005
- Jury's Deliberation : November 11 ~ November 14, 2005
- Public Announcement of Winning Entries : November 15, 2005

■ Submitted Entries

- 59 Korean teams, 62 foreign teams, total 121 teams.

  (59 teams from Korea, 10 teams from Germany, 9 teams from USA, 5 teams from Netherlands, 4 teams each from Japan, Italy, Australia and Spain, 2 teams each from France, Switzer-land, England, and Hong kong, 1 teams each from Indonesia, Israel, Malaysia, Singapore, Thailand, UAE, Argentina, Brasil, Chile, Armenia, Austria, Finland, Portugal, and Serbia and Montenegro)

■ Jury Members

- David Harvey(USA, Co-chairman), Nader Tehrani(USA, Co-chairman), Arata Isozaki(Japan), Winy Maas(Netherlands), Min Hyun Sik, Park Sam Ock, Ohn Yeong Te, Yoo Kerl (Korea)

■ Further Schedule

- Establishment of the master plan : May, 2005 ~ July, 2006
- Establishment of the development plan : August, 2005 ~ November, 2006
- Establishment of the enforcement plan : September, 2005 ~ June, 2007
- Establishment of the regional plan : July, 2005 ~ June, 2007
- Construction begins : July, 2007
- Completion of the 1st phase construction and resident : 2012
- Project Completion : 2030

■ Agencies to be Relocated

- Relocation in 3 phases from the year 2012 to 2014
- 12 Ministries : Ministry of Finance and Economy, Ministry of Education & Human Resources Development, Ministry of Science and Technology, Ministry of Culture & Tourism, Ministry of Agriculture and Forestry, Ministry of Commerce, Industry and Energy, Ministry of Information and Communi-cation, Ministry of Health and Welfare, Ministry of Environment, Ministry of Labor, Ministry of Construction and Transportation, and Ministry of Maritime Affairs and Fisheries.
- 4 Agencies : Ministry of Planning and Budget, Ministry of Government Legislation, Government Information Agency, The Ministry of Patriots and Veterans Affairs
- 2 State-run Organizations : National Tax Service, and National Emergency Management Agency

国际城市概念竞赛概要 / 评委会成员和专业顾问简介

Summary of the International Urban Ideas Competition / Jury Members & Professional Advisor Profile

评委会成员

大卫 · 哈维（评审委员会联合主席 美国）
- 纽约研究所城市大学著名人类学教授
- 上海同济大学城市规划学院客座教授
- 《后现代性的状况》（1989 年）等著作的作者

David Harvey (Co-chairman, USA)
- Professor of Anthropology at the Graduate Center in the CUNY
- Advisory Professor in the Department of Urban Planning in Tonjing University in Shanghai
- Author of The Condition of Postmodernity (1989) and others

纳德 · tehrani（评审委员会联合主席 美国）
- DA 事务所的负责人
- 哈佛大学设计学院建筑学副教授
- DA 事务所最近荣获原生态住居竞赛一等奖，以及获得过其他的各类奖项

Nader Tehrani (Co-chairman, USA)
- Principal of Office dA
- Adjunct Associate Professor of Architecture at the Harvard Graduate School of Design
- Office dA recently won first place in the Elemental Housing Competition, and other awards

矶崎新（日本）
- 1954 年毕业于日本东京大学工学部建筑专业
- 1963 年创建了矶崎新设计室
- 曾任日本横滨国际港口设计竞赛评委
- 1986 年荣获英国皇家建筑师学会的金质奖章和其他奖项

Arata Isozaki (Japan)
- Graduate from Architectural Faculty of University of Tokyo(1954)
- Establish Arata Isozaki & Associates (1963)
- Juror of Yokohama International Port Terminal Design Competition, Yokohama, Japan
- RIBA's Royal Gold Medal for Architect (England), and others (1986)

韦尼 · 马斯（荷兰）
- MVRDV 建筑设计事务所的创始人之一和主要负责人
- 设计作品有：2000 年汉诺威世界博览会荷兰馆，荷兰埃茵霍温的“飞行论坛”工业园区，荷兰阿姆斯特丹 Silodam 办公住宅综合体

Winy Maas (Netherlands)
- Co founder and co director MVRDV
- Dutch Pavilion for the World EXPO 2000 in Hannover, an innovative business park 'Flight Forum' in Eindhoven, the Silodam Housing complex in Amsterdam

Min, Hyun Sik（韩国）
- 韩国国家文科艺术综合大学建筑学教授
- 寄傲轩建筑师事务所的顾问
- 威尼斯国际建筑双年展
- 光州，亚洲文化枢纽城市概念性设计（2004 年）以及其他作品

Min, Hyun Sik (Republic of Korea)
- Professor of Architecture, Korean National University of Arts
- Advisor, KIOHUN Architects & Associates
- Venice Biennale's International Architecture Exhibition
- Schematic design of A Culture Hub City of Asia, Gwangju (2004), and others

Park sam ock（韩国）

- 首尔国立大学社会科学学院地理系教授
- 第 16 届 PRSC 的主席（1997 ~ 1999 年）
- 美国区域科学论文集，国际杂志编辑
- 亚洲太平洋沿岸地区和全球化（1995 年）等

Park, Sam Ock (Republic of Korea)

- Professor of Department of Geography, College of Social Sciences, Seoul National University
- Chairman of the 16th PRSC (1997～1999)
- Papers in Regional Science (USA), International Journal Editor
- 'The Asian Pacific Rim and Globalization' (1995) and others

Ohn, Yeong Te（韩国）

- 从 1992 年至今，任庆熙大学建筑学院教授
- 汾镇新的城市发展总体规划和城市设计（1989 ~ 1991 年）
- 河内新镇开发项目的可行性研究（1996 ~ 2003 年）

Ohn, Yeong Te (Republic of Korea)

- Professor of Graduate School of Architecture, Kyunghee Univer-sity (1992 ～present)
- Bundang Newtown Development Master Plan and Urban Design (1989～1991)
- Feasibility Study on the Hanoi Newtown Development Project (1996～2003)

Yoo, kerl（韩国）

- IARC 公司创始合伙人 （1995 年至今）
- 2000 年荣获美国建筑师联合会荣誉奖
- 一山千禧年社区服务中心（2005 年）以及其他设计

Yoo, Kerl (Republic of Korea)

- Founding Partner, IARC Inc. (1995～present)
- AIA Honor Award for Built Architecture (2000)
- Ilsan Millennium Community Center (2005), and others

专业顾问

Ahn kun hyuck 博士（韩国）

- 首尔国立大学环境和地理系统工程学院教授
- 复合型行政中心城市研究组联合主任

Dr. Ahn Kun Hyuck (Republic of Korea)

- Professor, School of Civil, Urban and Geo-Systems Engineering, Seoul National University
- Co-director of Multi-functional Administrative City Research Team

技术委员会

- **柳重锡** 中央大学 城市工程系 教授
- **林钟华** 仁荷大学 建筑系 教授
- **权宁相** 国土研究院 研究员主任
- **郑春容** 国土研究院 研究委员
- **裴炯敏** 首尔市立大学 建筑系 教授
- **李泰勇** 建国大学建筑学院 教授
- **李旺建** 国土研究院 研究委员

评审委员会的报告

评审委员会关于韩国首尔迁都规划竞赛的报告

评审委员会由大卫·哈维、纳德·Tehrani(评审委员会联合主席)、矶崎新、韦尼·马斯、Min、Hyunsik、Park,Sam Ock、Ohn, Yeong Te、Yoo,Kerl 组成。2005年11月11日评审委员会成员集中起来，11月12日委员们调研了基地，11月12日和13日开会讨论并对参赛作品进行了评审。

本次竞赛意义十分重大，旨在建立新复合型行政中心城市。评委们深刻认识到了此次竞赛项目的复杂性——规模大、涉及范围广以及应当满足多层次的需要，他们也为自己能够参与竞赛的评审工作感到非常荣幸。正如我们所愿，本次竞赛的首要目标是在首都首尔地区建立一种新的城市发展模式，以此缓解城市中功能及活动过度集中带来的压力；这将有助于扭转韩国城市中存在的地区性发展不平衡的状态。在这一过程中，我们将建立一种人性化、环境优美、和谐创新的新型城市，它将成为一支催化剂，为韩国、亚洲乃至全世界的城市化进程提供范本。这些目标在项目组委会提出的项目的四个现状和对城市远景的四个展望中明确得以体现。

在我们收到的121件参赛作品中，蕴含着很多非常好的想法，评委们从中筛选出一些对新城建设有益的建议。最终，5项作品获得了一等奖，这些作品之间既有相通之处，也存在不同；一项作品获得了二等奖。这些作品中有价值的观点将被整合并运用于项目建设中。城市规划是一个动态发展的过程，而并非是一个静止固定的模式，因此，我们希望此次竞赛能够为建设推进委员会的研究决策提供帮助。

评委在评审过程中，指出以下这些问题需要认真权衡：

1. 政府计划将12个部、4个处、2个厅迁移到新城中去。围绕行政中心规划城市好像是合理的，但在评委们看来，这似乎犯了一个深层错误。显然，行政机构是吸引其他职能部门和机构的“诱饵”，例如研究与开发机构、高等院校、文化机构、旅游和服务业等。因此，必须考虑行政机构对其他城市生活的吸引作用，并据此来规划城市。这就意味着，城市规划不能以行政机构为中心，即使它们不可或缺、地位显赫。参赛者也必须有基本的概念，例如，对政府部门采取分散布置，相对集中或高度集中的布置手法，这些取决于采用何种城市设计手法。

2. 政府部门职能需要通过人的工作来实现，最初的雇佣制度毫无疑问是不公平的。权利掌握在受过良好教育的技术界、科学界、经济领域的精英手里，但广大的平民阶层养活着他们。问题在于，找到一种方法，改善这种社会结构，为城市增加更多的功能和兴趣点。因此，城市设计应该是开放的，以容纳从根本上不同的活动（诸如文化、手工、甚至农业），从而对那些占优势的产业起到一个潜在的缓冲作用。

3. 任何一个想要名副其实的城市必须具备一种特性、构建一种形象，并找到一种方法来表达这种独特性。一个城市的形象很重要，当然这是历史长期演变的结果，而城市也具此证明自己的区域和民族性。那么，如果新城设计成功的话，在世界上将形成其独特的形

Jury Report

Jury Report of the International Competition for the New Multi-Functional Administrative City in the Republic of Korea

The members of the jury were David Harvey and Nader Tehrani (Co-chairman); Arata Isozaki, Winy Mass; Min, Hyun Sik; Park, Sam Ock; Ohn, Yeong Te; Yoo, Kerl; The jury convened on November 11th 2005, visited the site of the proposed city on November 12th and met to evaluate the submissions to the competition on the 12th and 13th of November.

The project to construct this new multi-functional administrative city is of outstanding significance and importance. The jurors were both impressed with the scale, range and multiple objectives of the project and felt privileged to have the opportunity to influence its outcome. As we understood it, the immediate aim is to create an alternative pole of urban development -a grand attractor for national and international development- to counteract the overloaded concentration of activities and functions in the Seoul metropolitan region. The aim is to re-equilibrate the uneven geographical development within the Korean urban system. In the process, it became possible to think of constructing a new kind of city (humane, environmentally friendly, innovative) that could function as a catalyst in human affairs and thus be a model for the twenty first century urbanization not only for Korea and Asia but also for the rest of the world. These aims were spelled out most directly in the four 'project objectives' and the four 'urban visions' set out by the Presidential Committee.

There was a fecundity of excellent ideas within the 121 submissions that we examined. The jury therefore, concentrated on sifting through the submissions to select a bundle of good ideas that might contribute to the design of this new city. To this end, we arrived at five 'first tier' prize-winners to be considered in relation (and not, for the most part, in competition) to each other and a 'second tier' of prize-winners that might contribute interesting ideas to be incorporated into the final project. Planring is a process rather than a fixed plan and we hope to have contributed creatively to the furtherance of the Presidential Committee's reflections, deliberations and research program.

The jury, in the course of its deliberations, defined a series of issues that needed careful evaluation :

1. The government intends to relocate 12 of its ministries to this new city. The temptation is to build the city around the ministries but that, in the jury's view, would be a profound mistake. The ministries are plainly intended as 'bait' to draw in all manners of other activities and functions (R&D, universities, cultural industries, tourism and services). It is, therefore, imperative to design the city in such a way that the ministries act as attractors to other activities. This means that they must not dominate, even thought they may have an important founding and iconic role. A basic choice needs to be made, for example, as to whether the ministries should be dispersed, clustered or centralized in relation to whatever overall design for the city is chosen.

2. The initial employment structure, as a consequence of the devolution of ministerial functions, will undoubtedly be unbalanced. It will rest primarily on the educated, technical, scientific and polio economic elite backed by an extensive service class. The problem is to find ways to transcend this structure and to attract other functions and interests to the city. The urban design has, therefore, to be open to accommodate

评审委员会的报告

象。因此，本次设计的首要任务是使城市清晰易识别，同时表现力也是不能被忽视的问题。

4. 基础设施也是评委们关注的一个重要方面。水和废水处理就是经常被提及的方面，比如，水的循环利用和垃圾处理，其他的如：运输和通信系统方面，设计者们则希望打破陈规有所创新。医院、学校、托儿所和市政服务处在日常的生活中都扮演着一个至关重要的角色，在多数的迁徙运动中它们维系了社会的稳定性。但如何整合这些基础设施并利用它们使它们作为一种引导发展的手段，是一个很有意思的问题。

5. 环境问题也具有重大的意义。其具有双重特性。一方面，城市设计必须尝试将水、能源和其他资源投入的消耗量降到最低，并认真思考再循环和再利用这种方式。现在如何让人们最大限度地减少汽车的使用量就是一个很重要的问题。因此城市生活中的新技术和新的组织形式需要被探索，并纳入到我们的规划设计中去。另一方面涉及具体的问题，如怎样处理指定地区已有的标记和现存的景观形式，如流动的河流、自然分开的地貌(城乡关系的象征)和绿树覆盖的小山。在这方面，文化的制约是很重要的。譬如不能兴建超过山脉顶端的建筑物，让河流径流自由和扩大洪泛区的范围等。也许这些做法才更具吸引力。

6. 日常生活的质量，如户外活动、邻里关系、工作和消费场所、娱乐和休闲等在我们的生活中变得越来越重要。如果新城可以扮演好这样一个引人瞩目的角色，那么就可以去超越有着强烈吸引力的首尔。

7. 最后是有关政治方面的问题。政治决策的制定往往是组织城市内部空间框架的依据。在大众的民主政治中，城市公民身份是个非常重要的问题。基于这一点，选择城市空间形态迄今为止都难以是中立的。

总之，我们关注的主要是这五个方面的问题：我们与自然的关系，我们与社会关系，日常生活的质量，我们的世界观以及现知的科学技术。

兴建城市时我们首先考虑的是如何规划和建造的问题。城市的空间形态（总平面图）可以帮助解决和回答一些这样的问题。 圆点、圆圈、网格、网络和群岛都是一些潜在的、有趣的形式，它们都试图在过程和形式之间建立一种关联。以上的这些探讨可以直观地解释评审团对一等奖和二等奖作品推荐的原因。

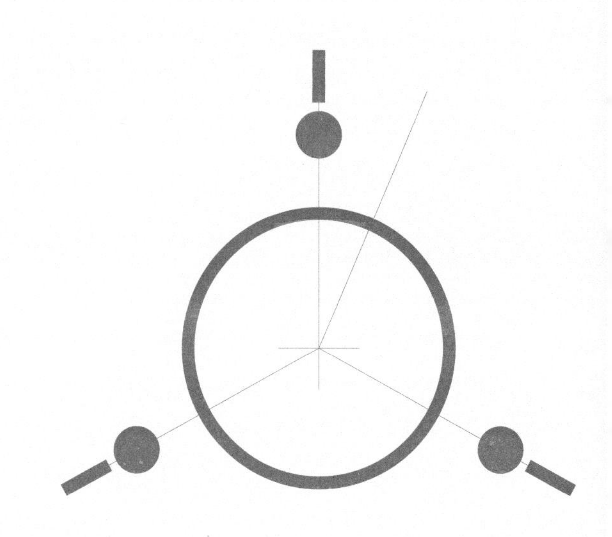

radically different activities (cultural, artisanal, even agricultural) to those that will otherwise dominate with a potentially deadening effect.

3. Any city worthy of its name must define an identity, construct an image and find a way to represent its special qualities to the world. Its iconography is important and this implies the evolution of symbolic meanings. The city will thus speak to its region and to the nation, and if it is successful, project a distinctive image around the world. The mundane task of making the city internally legible as well as expressive cannot be ignored either.

4. The interrogation of infrastructural provision played a major role in the jury's considerations. Water and waste water treatment were frequently mentioned in submissions, for example, as were recycling schemes and garbage disposal. Others went out their way to attempt something innovative with respect to transport and communication systems. Hospitals, schools, day-care centers, and municipal services play a vital role in daily life and provide for rootedness and social stability in the midst of much migratory movement. How to integrate this infrastructure and to use them as a means to channel development is an interesting question.

5. The environmental issue was also considered to be of major significance. It has a dual character. On the one hand, urban design must attempt to minimize consumption of water, energy and other inputs and to take seriously the idea of recycling and of reuse. The major issue here is how to minimize automobile usage without denying the advantages that car ownership can confer. New technologies and organizational forms of urban living need to be explored and incorporated into design schemas. The opportunity to experiment and test out innovations in this sphere must not be lost. There is, secondly, the specific question of how to deal with the existing landscape forms already strongly marked in the designated territory, such as the flow of the river, the nature of the rice-paddy fields (emblematic of the rural-urban relation) and the tree-covered hills. In this regard, the cultural constraint of not building over the tops of mountains is important, while the possibility of letting the river run free and of broadening the flood plain has major attractions.

6. The qualities of daily life as lived out in homes, neighborhoods, work places and places of consumption, recreation and creative leisure will play an important role, if this new city is to act as an attractor against the strong magnet of Seoul.

7. Finally, there is the issue of political form, of how political decision making will be organized within the territorial framework of the city and on what basis. In any democracy, the question of urban citizenship is of great importance and on this point the choice of spatial form is far from neutral.

There were, in summary, five very broad questions that ordered our deliberations: that of our relation to nature, our social relations, the qualities of our daily life, the mental conceptions we hold of the world, and the technological mixes that get proposed and deployed.

The general problem is how to imagine, let alone construct, a spatial form (a master plan) for the city that can help unpack and answer some of these questions. Dots, circles, grids, networks and archipelagos are all potentially interesting forms to try to establish a coherence between processes and forms. These topics will be taken up directly in explaining the jury's decision on the first and second tier recommendations.

评审委员会的报告

评审委员会对设计方案的评论

评审委员会从121个设计方案中最终选出了10个候选方案，并对它们作了评论。然而出乎预料的是：在最佳方案的评选时，评委会却无法达成共识。这个决定所获得的一致认可并连同一种责任感促使我们作出了一项富有建设性和严谨性的答复，以此作为我们提交的大纲的一部分。

首要的是，评审委员会明白韩国面对21世纪的背景下举办此次竞赛所肩负的重要使命和在极其有限的时间内组织机构执行其中一个或多个方案的决心。最终，我们将10个候选方案以5个一组的形式分成了两组：第一组由名次较高的方案共同分享了优胜奖，第二组的5个候选方案共享了荣誉奖。

一等奖（优胜奖）

虽然有“5个”候选方案获得了一等奖，但它们是经过特定的评选以创造一种有助于最终可能被付诸实施的连贯规划的一致和较为统一的口径。保留的两个候选方案是不同城市规划“流派”所提议的一部分，并被选作原有三个方案的重要补遗；换句话说，他们三个一组作出了概念性的表达，方案却没有被认定为拥有独立的自主参赛权。而评审委员会的目的是分析和汲取不同方案的变化，形成能够提高方案的潜力和发掘某些隐含于这些方案中的潜在可能性的组织机构。

评审委员会在讨论形态、组织和意义之间的关系时，强烈意识到新的行政中心需要包含既是作为韩国的新建城市，又是立足于21世纪的国际化城市这种背景下的浓厚象征意义。有基于此，评审委员会认为城市形态须达到强烈的“象征性”特点，以此作为体现现存于韩国的图式化或符号化的一条途径。从这一结论范围的另一方面来讲，规划方案需要建立一种同样强烈的“综合象征性”逻辑关系以支撑他们的象征形态；并使街区布局、基础设施组织、交通网络和其他特征一起将必要的外形实体化成为可能。在所有提议中，候选方案三人组巧合地具有同一特征：区域尺度级别的一个巨大放射状圆形。

二等奖（优秀奖）

二等奖方案形成了虽富意义但更为零散的提案。它们引发了评委们的争论，或者说是更为尖锐的反驳和反对的声音。尽管如此，它们却有助于一等奖的形成标准以及在许多方面为我们的研究和思考提供更多的线索。

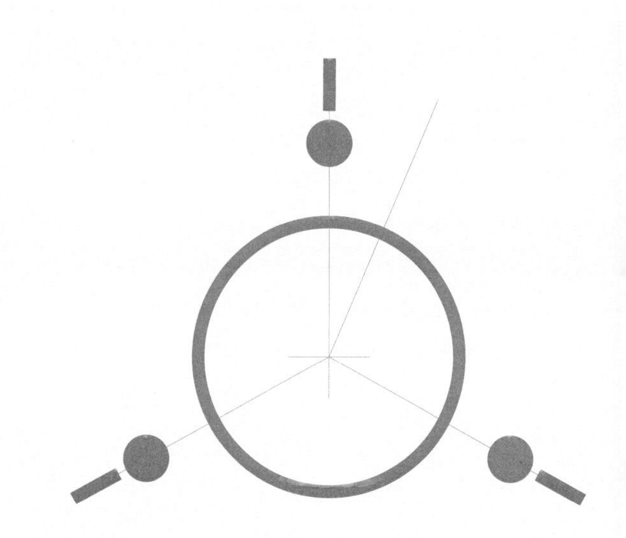

Jury Report

JURY COMMENTS on the projects

From a total of 121 schemes, the jury was able to shortlist ten schemes on which to comment and make recommendations. Contrary to original expectations, however, the jury was unable to arrive at one winning scheme. This decision was reached unanimously, and was coupled by a sense of responsibility to make a structured and calibrated response as part of the brief we are hereby submitting.

First and foremost, the jury understood the significant mission of this competition in the context of the 21st century Korea, and the will of the organizing body to implement a scheme -some scheme- on a very restricted schedule. To this end, we have divided the shortlist into two groups of five submissions: one, an upper tier shares the winning prize and a second, another tier of five shares the honorable mention status.

THE UPPER TIER (First Tier Prize)

Though the winning schemes are 'five', they were selected specifically to create a single and more unified dialogue towards a coherent plan -a plan that may be eventually implemented. The remaining two schemes are part of a different 'genre' of urban planning proposals, and were selected as a critical addendum to the original three; in other words, they speak conceptually to the triad, but were not selected to stand alone as schemes in their own right. In this way, the jury's intention is to have the organizing body examine and draw variations of these plans to further their potentials and expose certain latent possibilities lurking within these schemes and strategies.

In discussing the relationship between form, organization, and meaning, the jury felt strongly that the new administrative city would embody a strong symbolic presence within the context of new cities in Korea, indeed within the new 21st century global city. To this end, it was felt that the form of the city should attain a strong 'figurative' quality as a way of giving iconic or symbolic presence within Korea. On the other end of the spectrum, it was felt that the plans need to establish an equally strong 'configurative' logic to support their symbolic form; this came in the block layouts, infrastructural organization, networks, among other features were needed to substantiate the figure. Of all of the submissions, the short-listed triad coincidentally shared one trait: the circle, a colossal radial figure rendered at the scale of the region.

THE SECOND TIER (Honorable Mention)

The second tier of projects form a significant but more fragmented set of proposals. They contributed to the debate of the jury, and maybe more pointedly are the results of contradictory and dissenting voices. Nonetheless, the helped form criteria for the first tier, and in many ways offer more clues for research and speculation.

参赛作品名单 / List of Submitted Entries

一等奖作品 First Tier Prizes

AP - 22411 西班牙 — 千城之城 …… 36
Perea Ortega, Andrés / EuroEstudios, S. L .

OO - 88888 瑞士 — 轨道交通 …… 46
Duerig, Jean Pierre

LM - 10678 韩国 — 水城 …… 54
Shong, Bok Shub / Petit, Olivier / Brochet-Lajus-Pueyo / D&B Architecture Design group

YO - 62102 韩国 — 双面城 …… 62
Kim, Young Joon / Zaera Polo, Alejandro

DO - 17888 意大利 — 城市语法 …… 72
Aureli, Pier Vittorio
Geers, Kersten / Tattara, Martino / Van Severen, David

二等奖作品 Honorable Mentions

YJ-16103 德国 — 流动之城 …… 82
Kunzemann, Juergen / Jacob, Dirk / Kunzemann, Yumiko Izuta

MT-60012 奥地利 — 阴阳 …… 88
Pucher, Thomas / Heidrun, Steinhauser / Norbert, Adam / Martin, Mathy / Roland, Muller

CK-70980 韩国 — 场地的回归 …… 92
Choi, Hyun Kyu / Kang, Mi Kyung

FZ-28583 日本 — 千岛城 …… 98
Sumiya, Mamoru / Hayashida, Toshiko

UD-21278 智利 — 孕育新都市风格 …… 104
Undurraga, Cristian / Allard, Pablo / Lopez, Pablo / Taller Undurraga Deves Arquitectos

注：在评审委员会的报告中列举了一等奖和二等奖获得者的名单

First tier prize winners and honorable mentions are listed in Jury Report's order

入围作品
Entry Works

AX-96961
日本
无题 …… 112
Kurokawa, Kisho / Kumazawa, Akira / Hashimoto, Kenichi / Jin, Guanyian / Koike, Daisuke

AK-10009
美国
快速发展的城市 …… 114
Arbanas, Magraret / Krarmuk, Uenal / Kuo, Jeannette

KN-78462
德国
无题 …… 116
Schenk, Leonhard / Lehan Drei, Architects+Urban Planners Feketics Schenk Schuster / Muller, Felix / Flury, Christina / Witulski, Stephanie

OK-54321
韩国
平和的生态政策 …… 118
Ock, Han Suk / Seo, Tae Yeol / Lee, Sang Yong / Huh, Tai Yong / Lee, Sang Kyoo

HM-15688
韩国
开园 …… 120
Lee, In Won / Mooyoung Architects&Engineers / Michell, Anthony / Kim, Tae Kyung

KM-13579
韩国
无题 …… 122
Lee, Houng Chul / Eric, Strauss / Shin, Kyungsik Irene / KunHwa Engineering CO., LTD. / Urban and Regional Plannning Program Michiegan State University

MM-02030
韩国
混合巨尺城市 …… 124
Space Group / Lee, Sang Leem / Purini, Franco

AR-65211
韩国
无题 …… 126
Pak, Hun Young / A. rum Architects / Post Media / Go, Eun Tae / Kim, Han Jun

HH-10101
韩国
数码城堡 …… 128
Hwang, Doo Jin / Hong, Soon Jae

GI-56765
韩国
无题 …… 130
Yi, Eun Young

CA-72800
法国
无题 …… 132
Treuttel, Jerome / Treuttel, Jean Jacques / Garcias, Jean Claude / Pourrier, Stephane / Fichou Torres, Laurent

入围作品
Entry Works

MB-60657 美国 — 无题 ······ 134
Jonathan D. Solomon / Kutan A. Ayata / Aleksandr Y. Mergold

EY-13557 韩国 — 无题 ······ 136
Choi, John / Choi, Ropiha / Mcgregor and Partners / Garbutt, Michael

NY-11217 美国 — 花园城市 ······ 138
Stienon, Christopher / Kim, Eung Soo / Huang, Shin Yau

KM-14444 印度尼西亚 — 衍生城市 ······ 140
BaritoAdi Bulden Raya Ganda Rito
Ilya Fadjar, Maharika / Revianto Budi, Santosa /
Ariadi, Susanto / Prihatmaji, Yulianto Purwono

OB-21010 德国 — 无题 ······ 142
Grosch, Leonard / Wessendorf, Joerg

UC-58679 韩国 — 无题 ······ 144
Kim, Kyung Hwan

OS-05090 德国 — 韩国金城 ······ 145
O.S.A Ochs. Schmidhuber. Architects
Heiner, Luz

GS-23577 中国香港 — 绿色通道 ······ 146
Chang, Chung Kul / Yamane, Cintia / Hsieh, Alice / Kim, Sang Ok

PC-12489 韩国 — 人类的绿色之城 ······ 147
Cho, Sung Eun / Kim, Jeong Min / Back, Seung Mok

SM-77777 塞尔维亚和黑山共和国 — 无题 ······ 148
Milaca, Bajic Brkoric / Archi 5 team

JC-12046 泰国 — 忠清新城 ······ 149
Jumsai, Sumet / Michael, Sorkin /
SJA 3D Company Limited / Architects 49 Limited / Buro, Happold

YC-85130 韩国 — 新都市主义 ······ 150
Lee, Young Chun

PY-11701 韩国 — RID ······ 151
Park, Won Woo / Yoon, Jae Woong

入围作品
Entry Works

HS-12121
日本
无题 ······ 152
Kohki Hiranuma

OZ-12030
韩国
混合单元体 ······ 153
Min, Bum Kee / Sir, Min+D&A Associates INC. /
Kim, Sei Yong / Lee, Jae June / Park, Byung Hun

DM-53982
韩国
齿状城市 ······ 154
Nam, Soo Hyoun / Leem, Jea Eun / Choe, Sun Mi

YK-00408
韩国
无题 ······ 155
Yoon, Woong Won / Kim, Jeong Joo /
Yoon, Seung Hyun

IR-31215
韩国
幸福之城 ······ 156
Lew, Deok Hyun / Yeom, Yoon Sook

CL-01278
韩国
重叠之城 ······ 157
Lee, Dong Shin / Lee, Sung Geun / Choi, Min Ah

DR-40713
韩国
市民心中的城市 ······ 158
Lee, Seok Woo / Kim, Woong Tae / Yoo, Young Mo /
Lee, Seung Bae / DongRim P&D, Co.

CH-35109
韩国
模糊之城 ······ 159
Park, Young Woo / Kim, Yong Kyun / Ji, Hyun Ae /
Kim, Sung Kyum / Kim, Yun Jung

PD-02468
韩国
无题 ······ 160
Lee, Pill Soo / Lee, Jong Ho

HR-03410
韩国
山水画 ······ 161
Jeong, Young Kyoon / Kim, Don Yun / Chung, Jae Yong /
Han, Gwang Ya / Heerim Architects&Planners Co., Ltd.

WP-10101
韩国
生态型轻巧城市 ······ 162
Chung, Gu Yon / Sanin, Francisco / Magerand, Jean /
Mortamais, Elizabeth

YM-19753
韩国
多形态城市 ······ 163
Lee, Hee Chung / Kim, Yong Sung / Nam, Seung Kyun /
Ko, Kwang Sung / Myoung In Architects & Engineers

入围作品
Entry Works

HJ-33399
韩国
新世纪的不朽之城 ··· 164
Park, Hyun Jin / Toshiken Korea co., Ltd

HS-91084
韩国
芙蓉城 ··· 165
Lee, Byung Jin / Yu, Jung Hyuk / Hu, Sung Kyun / Cho, Sung Gil / Cho, Young Gyu

JY-18055
韩国
活力之城 ··· 166
Ahn, Young Su / Shin, Min Joong / Joo, Sung Hak / Kim, Hyeon A

UN-08025
韩国
融入自然 ··· 167
Kim, Uk / Kang, Chul Hee / Lee, Sang Ho / Kang, Jun Mo / Bahn, Sang Chul

WK-20049
韩国
林中筑巢 ··· 168
Kim, Young Sub / Waro Kishi+K. Associates Architects

AS-08605
韩国
进化中的都市体系 ··· 169
Kim, Woo Sung / Lim, Jong Ah / Archiplan INC

MS-11111
韩国
梦幻之城 ··· 170
Shin, Min Jeong / Park, Sung Nam / Kim, Hee Sung

KJ-26334
韩国
人类的活动与城市 ··· 171
Kim, Nam Jung / Kang, Joo Yeon / Kim, Jin Hee

LK-12356
韩国
未来之城 ··· 172
Lee, Jong Ho / Kim, Jong Sik

NA-03582
韩国
永恒之城 ··· 173
Lee, Jae Won / Jang, Seong Il / Shin, Sang Hyun / Choi, Yun Jung / Kwag, Ji Hee

CL-99923
韩国
未来的主要模式 ··· 174
Choi, Won Seok / Lee, Jun Hyung / Lee, Jee Young / Kim, Min Ho

HS-17317
韩国
无题 ··· 175
Lee, Ho Joon / Sung, In Jung

10164090
韩国
无题 ··· 176
Han, Ju Hyung

入围作品
Entry Works

KJ-05111 韩国 — 无题 …… 177
Sim, Kil Je / No, Tae Young

HU-21309 韩国 — 无题 …… 178
Eom, Su Ryu / Jung, Hee Joo / Lee, Eun Hee / Lee, Jae Woo

KA-09796 韩国 — 无题 …… 179
Eun, Jin Pyo / Kim, Dong Young / Sung Hwan

JK-59147 韩国 — 幸福之城 …… 180
Kim, Jin Young

ON-64925 韩国 — 无题 …… 181
Nam, Young Hyun / Jim, Young Sup / Lee, Myung Hee / Jin, Yoon

ET-77777 韩国 — 无题 …… 182
Shin, Ye Kyeong / Lee, Jeong Houn

SA-01023 韩国 — 无题 …… 183
Kim, Tai Young / Seon Architects & Engineers Group / School of Architecture and Engineering, Cheongju University

GE-23681 韩国 — 无题 …… 184
Lee, Jang Gun / Koh, Dong Wook / Jang, Yong Hun

LA-50403 韩国 — 进化中的城市 …… 185
Lee, Sang Hyun

JG-32373 韩国 — 仿生态城市 …… 186
Yun, Hee Jin / Ji, Jang Hun / Han, Geang Nam / Lee, Yu Mi / Braun, Adrea

AL-25271 澳大利亚 — 无题 …… 187
Landstrom, Roger

BX-83800 中国香港 — 无题 …… 188
Karakiewicz, Justyna / Kvan, Thoms / Hwang, Se Young Iris / Zhai, Binging / Rotmeyer, Juliana

VC-20046 澳大利亚 — 共生城市 …… 189
Yu, Eric / Lee, Louise

入围作品 Entry Works

KO-10000
以色列
无题 …… 190
Moshe, Salomon

DS-19125
意大利
无题 …… 191
Balsimelli, Andrea / Schineano, Sallydagnaiz / Cappoqi, Pino / Belloccisessa, Ilaria

KA-22222
韩国
银杏之城 …… 192
Kahng, Jang Wan

DO-74408
德国
无题 …… 193
Marx, Christian / Team 408

SC-73781
德国
玲珑之城 …… 194
Haimerl, Peter / Peter Haimerl Studio Fur Architecture

GC-55133
德国
无题 …… 195
Bomerski, Martin / Bomerski, Artur / Bomerski, Albert

MN-20059
澳大利亚
大型旅游城市 …… 196
Yip, Ting Hin Michael / Fung, Si On / Njo, Victor / Horsting, Bas

DG-16205
德国
无题 …… 197
Witthinvich, Jochch / Stelmach, Mathias / Braig, Monika

IN-02090
韩国
无题 …… 198
Cho, Seong Ju

AN-23822
韩国
无题 …… 199
Ahn, Jung Gil / Inhousing Co., Ltd / Kang, Sung Joong / Haein Architects & Engineers Co. Ltd

FL-0220
日本
自我—连接—网络 …… 200
Furuya, Nobuaki / Lee, Doo Yeol

LB-28000
新加坡
无题 …… 201
Butler Sierra, Luis

SL-75110
韩国
Daul …… 202
Lee, Su Man

入围作品
Entry Works

LM-39054
美国
紧凑城市 203
Mundwiler, Stephan / Lee Mundwiler Architects / Lee, Cara / Santos, Gustavo / Watanabe, Hiroyuki

MN-27277
亚美尼亚
巨型城市 204
Petrosyan, Noure / Zoroyan, Suartin

AD-22601
法国
间隔地带 205
Haddad, Albert

BB-32123
阿拉伯联合酋长国
无题 206
Joyanovic, Bratislav

XS-00018
荷兰
无题 207
Schaap, Ton / Venhoeven CS Architecten / Van Geissen, Cees / Kuiken, Rene

DN-05001
荷兰
无题 208
Tooruerol, Paul

UU-60395
英国
无题. 209
Vladin, Petrov / Tsocaiev, Georgi

UB-99999
意大利
UBI离散城 210
Conti, Anna

BK-58825
巴西
杨梅城 211
Dias, Carlos / Sales, Pedro M.R. / Yamato Newton, Massafumi / Fehr, Lucas / Ursini, Marcelo L.

JO-71685
荷兰
绿核城市 212
Spijker, wan't Jaakko / Bultstra, Henk / Deuten Bert, Karel / Steinarsson, Orri

MB-81713
美国
无题 213
Arellanes II, Michael

BV-72459
西班牙
江南平原城 214
Hostench Ruiz, Oriol / Bernat Quinquer, Sergi / Caba Roset, Joan / Crespo Solana, Rafel / Gonzalez Benedi, Sergi

入围作品
Entry Works

MW-56141 德国 — 无题 …… 215
Schelbert, Hannes / Hofer, Florian / Bucher, Sigfried

DJ-15054 澳大利亚 — 文城 …… 216
Jones Evans, Dale / Dculys Landscape Architects / Dale Jones Evans Pty Ltd. Architecture / Maki, Yamaji

AV-67825 阿根廷 — 无题 …… 217
Varas, Alberto / Estudio Alberto Varas Yasoc, Arqs / Montorfano, Hugo

RC-02049 瑞士 — 彩虹之城 …… 218
Pasqualini, Isabella / Kuhnholz, Olof / Gran, Carianne

FF-25787 德国 — 15个核心的城市 …… 219
Fritzen, Andreas / Feiling, Oliver / Rung, Hannelore

MU-72165 葡萄牙 — 无题 …… 220
Barbedo, Jose / Sottomayor, Nuno / Pinheiro Torres, Miguel / Jigueiroa, Joao Luis / Alves Pinho, Albino

DP-26366 马来西亚 — 无题 …… 221
Kamal, Zaharin

TN-04125 美国 — 链状城市 …… 222
Ngai, Ted / Schenk, Beat / Feng, Alice / Kim, Chae Won / Feevajee, Ali

KN-45810 韩国 — 无题 …… 223
Kim, Hee Seok

16062005 芬兰 — 无题. …… 224
Katainen, Juhani / Erra, Jyrki / Korpela, Pekka / Koivisto, Hannu / Haikio, Juho

UP-59604 韩国 — 无题 …… 225
Park, Sam Ho / Chang, Hyun Jung / Lee, Jung Min / Park, Chul Woon

EH-31687 韩国 — 无题 …… 226
Lee, Jae Yeol / Noh, Gyeo Sun / Koo, Min Cheol

入围作品 Entry Works

YE-33388 韩国
活力韩国　活力城市 …… 227
Eun, Min Kyun / Yoon, Ki Byung

DJ-88206 美国
南部明星城市 …… 228
Garratt, Dale A./ Garratt, Joy I. / Garratt, Daejo S.

GL-52947 意大利
无题 …… 229
Lorenzi Gianni / Untersvlzner, Johannes

AB-24786 美国
大都市测量控制 …… 230
Lynch, Catherine / Metropolitan Planning Collaborative / Todd, Lieberman / Benjamin de la Pena

ST-19123 英国
无题 …… 231
Luck, Robert / Cvrtis, James / Pattni, Krishan

PS-25773 荷兰
无题 …… 232
Boender, Arnest

HS-96989 韩国
无题 …… 233
Yang, Sung Goo / Bae, Hyoung Du / Oh, Hyun Il

MC-07061 韩国
无题 …… 234
Min, Yong Il

MB-06054 荷兰
无题 …… 235
Mattuiss, Bouw / NL Architects / Matton, Ton / Malkit, Shoshan

TE-01873 西班牙
无题 …… 236
Canovas, Andres / Nicolas, Maruri / Amann, Atxu / Lopez Fernandez, Ana / Bravo, Mauro

MA-22012 美国
无题 …… 237
Sommer, Richard / Miller, Laura / Kwon, Mee Hae / Modesitt, Adam / Dan, Adams

GC-50437 西班牙
无题 …… 238
Solid Arquitecnra S.L. / Maroto Raros, Frsncisw Javies / Soto Aguirre, Aluaro

韩国首尔迁都规划竞赛作品集／获奖作品

International Urban Ideas Competition for the New Multi-functional Administrative City in the Republic of Korea
Winning Works

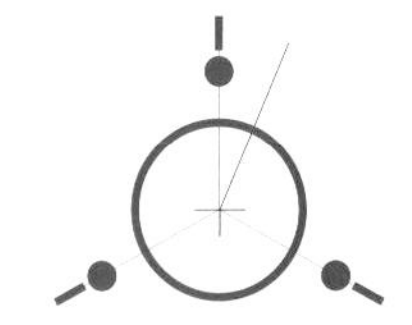

一等奖作品
First Tier Prizes

千城之城
The City of the Thousand Cities
Perea Ortega, Andrés_ 西班牙

轨道交通
The Orbital Road
Duerig, Jean Pierre_ 瑞士

水城
Thirty Bridges City
Shong, Bok Shub_ 韩国

双面城
Dichotomous City
Kim, Young Joon_ 韩国

城市语法
A Grammar for the City
Aureli, Pier Vittorio_ 意大利

千城之城

Perea Ortega, Andrés (西班牙)
EuroEstudios,S.L.

Perea Ortega, Andrés

- Perea Ortega, Andres是西班牙最优秀的建筑师之一，拥有超过30年的专业实践经验。在他的职业生涯中，我们了解到：他在马德里大学担任了多年资深讲师，在全西班牙和许多国家担任客座教授。
- 2005年首尔表演艺术中心概念性设计国际竞赛等奖的获得者。

- Perea Ortega, Andres is one of the most outstanding architect from the Spanish scenario with over 30 years of professional practice. Amongst his built works, we could outline the fact that he has been a senior lecturer in the University of Madrid for many years, and visiting lecturer all over Spain and several countries.
- 2005 International Ideas Competition for the design of The Seoul Performing Art Center : 1st Pri-ze winner

团队
EuroEstudios, S.L.

■ 评委意见

“千城之城”或许是最成熟的参赛作品，它展现给我们的不仅是一件很好的规划作品，还包括有大量的分析图表：定位、循环、景观、密度、交通、可持续与优先发展分析图等。这件作品最大的特点就是城市中的空地。这片空地位于市中心，在这里可以看到群山和稻田，它成了很好的公共活动预留场所。评委们一致认为，这块奇妙的空地构成了一种新型公共场所，在设计不够完善时可以随时调整，并在将场所定位为城市还是乡村这个问题上发挥着重要的作用。作品利用放射状网格来界定城市形态，评委们对利用网格结合地形与现状条件而使设计更富灵活性的设计方法赞赏有加。在这里，作品采用了一种可以任意变形的智能化网格。如果说网格的优点在于其灵活性，那么它也因为同样的原因而被质疑：即更多的考虑文脉而削弱了自身特色。

■ Jury Comments

'The City of a Thousand Cities' was maybe the most developed project, displaying not only a strong plan, but also various analytical diagrams showing layers of information: phasing, circulation, landscape, density, transportation, sustainability, and priorities. The most salient feature of this project is the void. It leaves in the center of the city, framing the mountains and the rice terraces as a new domesticated ground, ready to be appropriated for public use. This enigmatic void was seen by the jury as a new public ground, ready for transformation if undeveloped in the scheme and played an important role in negotiating the identity of the site, between its rural and urban status. The urban morphology of this scheme is planned on a radial grid; more importantly, the jury appreciated the way in which the grid negotiates the topography and existing conditions as if elastic. Here, the grid is not adopted off the shelf, but instead displays an intelligence to expand, contract, distort, and accommodate. If the strength of the grid comes in its flexibility, then it is also critiqued for the same reason: its accommodation of the context is the very feature that weakens its figure.

■ 摘要

没有界限的城市，不断变化的城市，自由发展的城市。在这里，理想化的城市和建筑成为可能；在这里，环境、社会和谐发展；在这里，美好的乌托邦不再是幻想。因为我们已经拥有了这一切：如画的风景，田园般的地貌……

■ Summary

City without limits. Constant city. City which contemplates itself through nature. The desirable city and architecture are possible. The citizen in harmony with his social and natural surroundings is possible. There is no need to appeal to utopias to build the city of peace and democracy. Reality provides all that is necessary: the landscape, the environmental conditions....

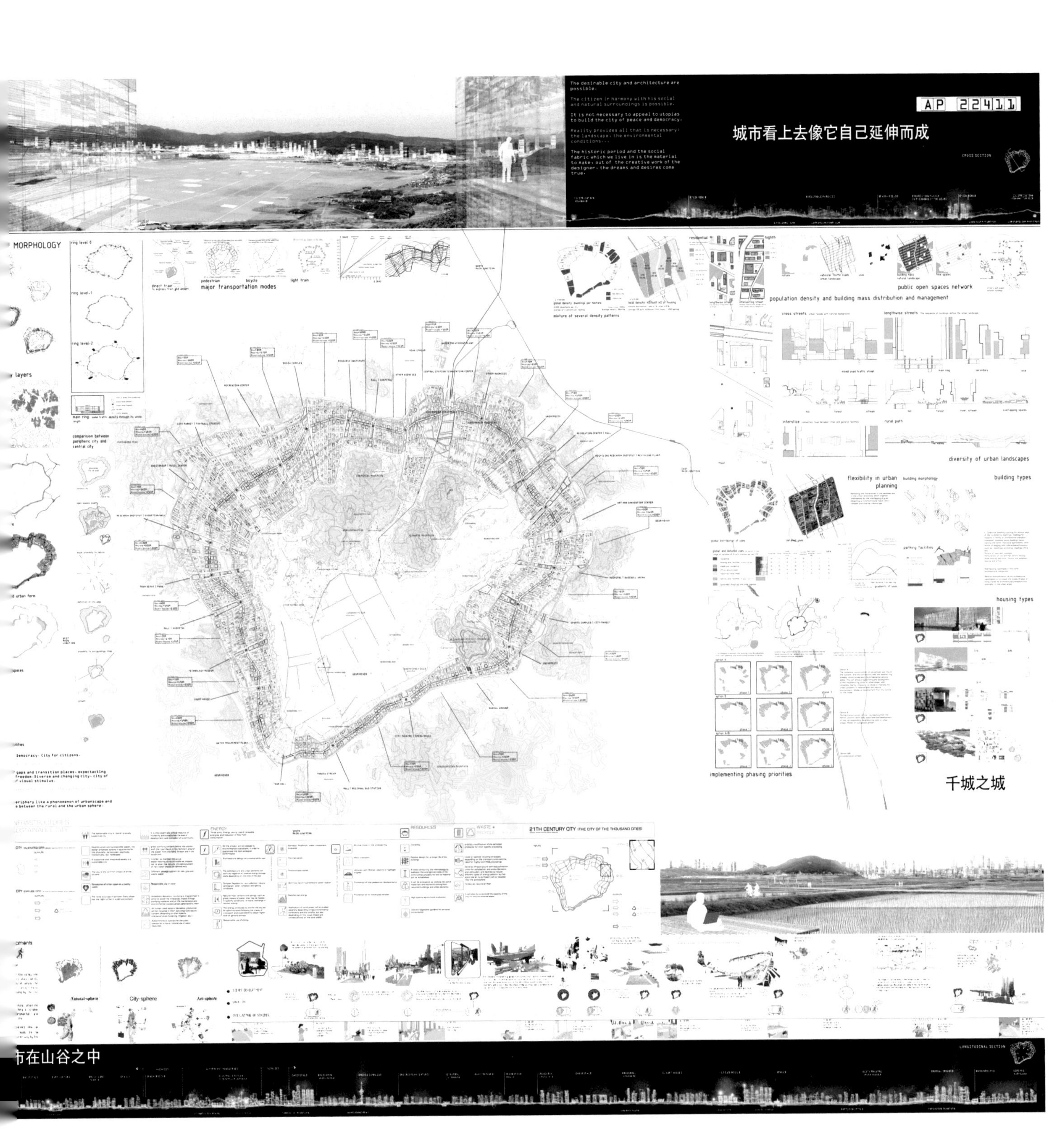

■ 千城之城

没有界限的城市。不断变化的城市。多中心城市。母城由 25 座子城之间的市政设施连接而成，每座子城拥有 2 万居民，子城中建有公共机构、市政设施、公共空间和居住区等。城市如同一个有机的生命体，是一项规模大、持续久、集体性的建设项目。城市作为使用对象而非象征符号，应当被管理、被规划，而不能仅停留于纸上。居民是城市活动的主体，也是城市中公共空间的主人，我们除了要懂得欣赏城市外，更要学会如何使用城市。

■ The city of the thousand cities

City without limits. Constant city. City of cities. City of 25 cities with 20,000 dwellers. Cities with social organizations, facilities, activities and dwellings.

Multi-city formed by 25 cities bonded together by the interstitial general facilities.

The city as a trans-generational project.

The city as a collective construction project extensive and constant.

The city as an object to use rather than a place for symbols.

City of management, city of program rather than a city of schemes.

The citizen as an active main character and owner of the public space.

A city to make use of rather than a city to contemplate.

■ 城市形态与结构 / Structure and urban form of the city

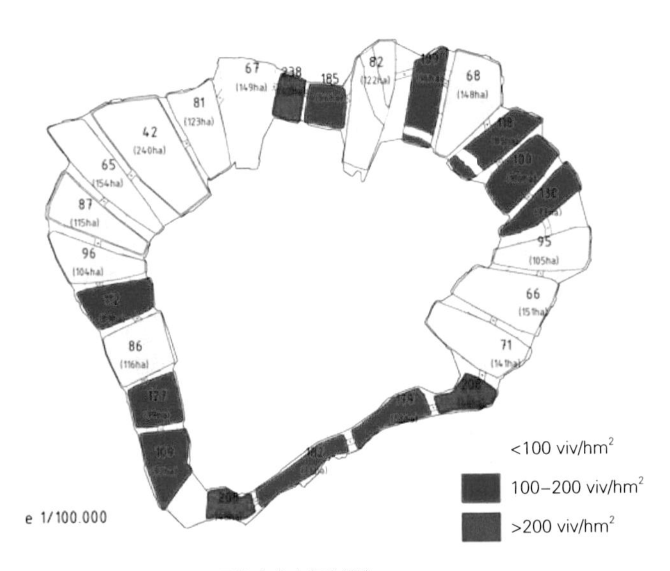

居住密度（每公顷）

自然环境

城市交通系统

功能分区

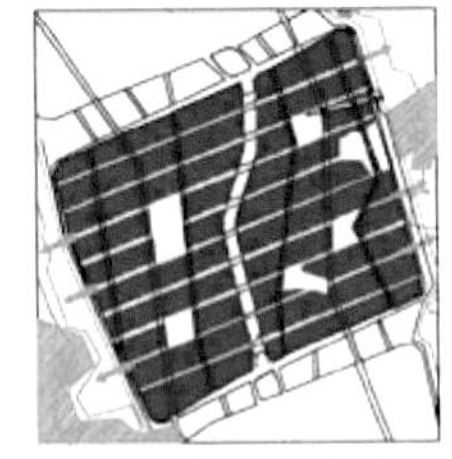

建筑物与自然景观

自由空间

每座小城拥有 2 万人口

每个居住单位平均 2 人

城市面积：$2585hm^2$

平均密度：$98viv/hm^2$

排除人口密度与人口活动的影响，将这些网格重叠在一起，我们可以得到一个综合性和多样性的城市。

By removing the hierarchies in the densities and in the urban activities, which organize themselves by the overlapping of grids, we can obtain a multi-functional result as well as a complex and diverse urban scape.

- **复合型城市的结构层次**

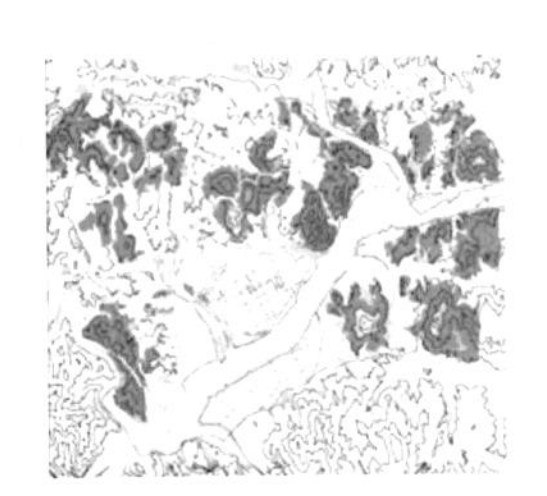

自然环境

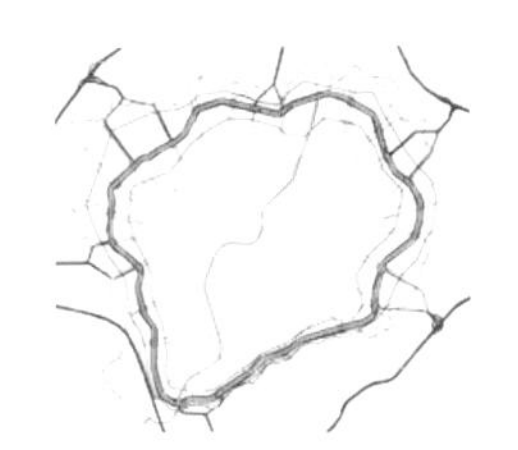

基础设施

城市结构和肌理

复合型城市市政设施

重叠空间

乡村圈

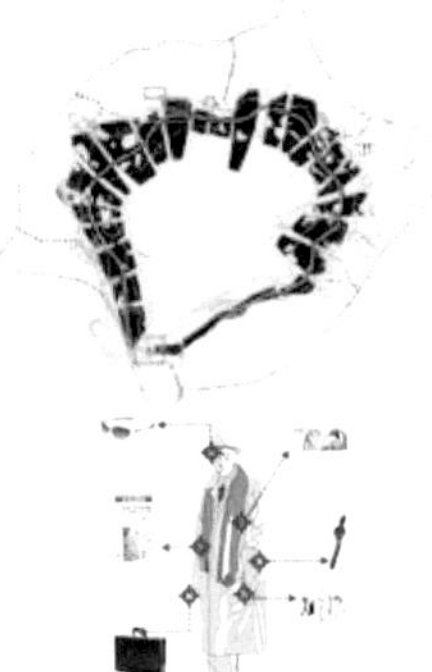

城市圈

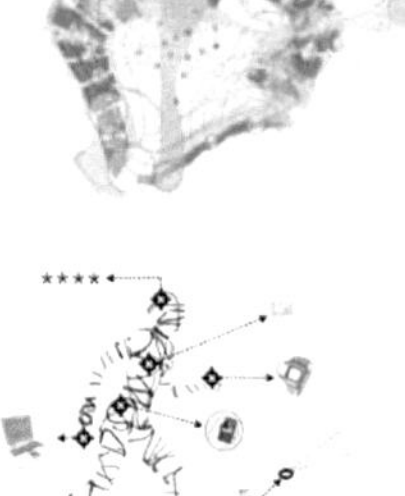

网络圈

山谷及其自然景观构成了城市边界，在城市中，乡村、城市和网络三者共存。在城市边沿散步，就仿佛行走在海边。

The urban limits with the valley, and with the landscape are places where the three spheres-rural, urban, and net-overlap. To walk through these limits will emulate the experience of walking by a sea shore.

■ 城市结构形态与自然环境的关系 / Relationship among the urban structure and form of the city and the environmental surroundings

• **城市中的童话——山谷**

为了实现城市与自然间的对话，在城市范围内建筑与环境的协调统一是极为重要。山谷作为城市文化遗产将被整体保护起来，而稻田作为自然景观则既能保证城市的水系平衡，又是生活方式与历史文化的见证。

• **The valley like urban utopia**

To overcome the dialogue between nature and the city, it is important to find a relationship between architecture and garden at an urban scale. The total preservation of the valley as cultural heritage. Keeping the rice fields like landscape as an element of river balance, as a way of life and as culture.

• **外围城市与中心城市的比较** / Comparision between peripheric city and central city

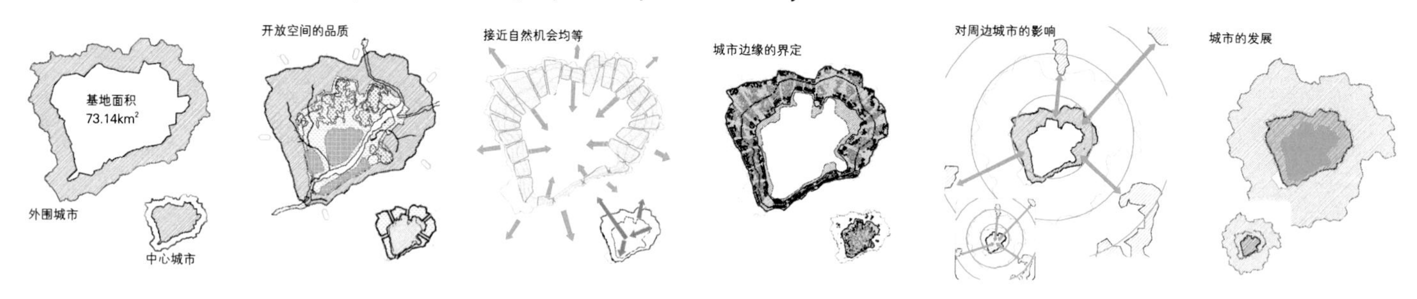

■ 城市全景 / Description of the overall landscape image of the city

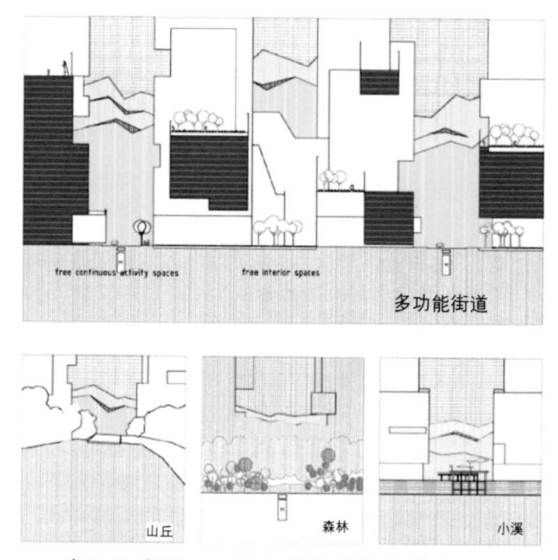

交叉的街道——以自然为背景的城市立面

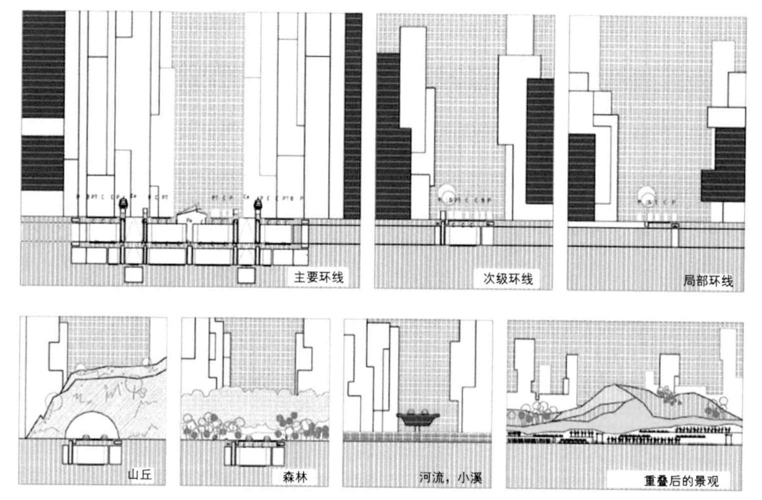

纵向街道 – 建筑序列界定城市景观

• **建筑物高度**

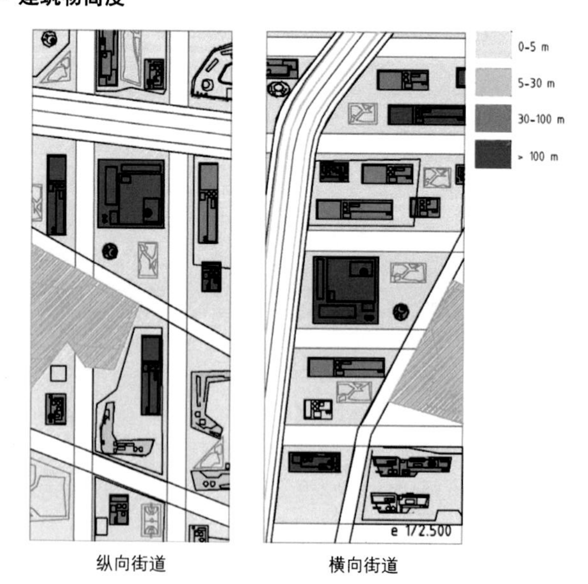

纵向街道　　横向街道

城市中有两种方向性的景观：人工景观沿纵向街道布置，横向街道两侧则以自然景观为主。

Dichotomy of two-directional scenarios: in the longitudinal direction of the urban flows, the scenery is private and architectural; in the traverse direction, to both sides (at a simple turn of the head), the scenery is always natural.

我们在25座子城之间都设计有网络终端，居民们通过网络不但能享受到硬件与软件的技术支持服务，还可以随时获取整个城市的实时图像、现状及历史背景资料。

We expect the creation of 'input-output' points for the net sphere in the itinerary between the 25 urban areas, where apart from the citizen support services of hardware and software, we propose interfaces that will allow to interact with the net sphere and obtain an 'on-line' vision of the whole city in real time and of its immediate and historical background.

■ 功能分区及人口密度

/ Distribution and management of the building mass and the population density

• 赋予变化的城市规划 / Flexibility in urban planning

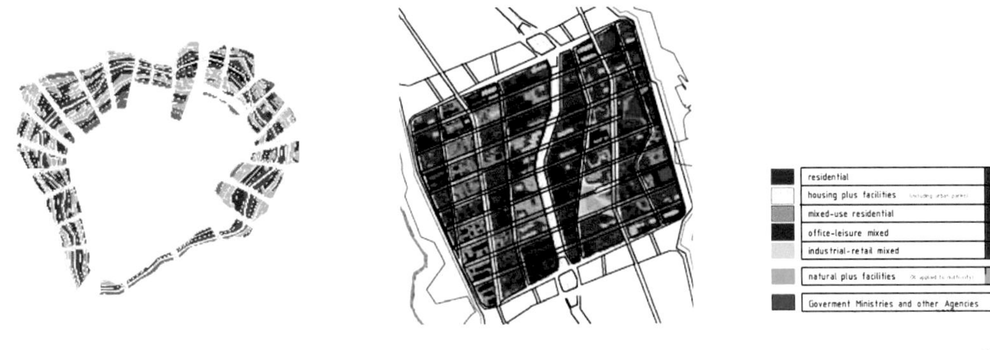

总体功能分区　　详细分区

	residential	office	mixed residential hotel	retail	facilities equipment	light indus workshop R&D	leisure	
residential	50	5	10	10	10	5	5	5
housing plus facilities	35	5	10	10	25	5	5	5
mixed-use residential	35	15	15	10	5	5	5	10
office-leisure mixed	15	25	15	10	5	5	15	10
industrial-retail mixed	20	10	15	10	5	25	5	10
natural plus facilities	-	-	5	1	5	-	1	5
Goverment Ministries and other Agencies								

总体功能分区和详细分区

■ 居住类型 / Housing types for residents

我们根据不同的生活方式设计了多种集合式住宅：多代同居住宅、外来人口住宅、独立式住宅（无土地产权）、单身公寓及其他形式的住宅，如住宅—工作室、住宅—办公室等复合形式。

Collective dwelling typology for various ways of life: multi-family dwellings, dwellings for citizens in family or professional transition (nomads), isolated family dwellings (never owning land), individual apartments, other forms of lodging with differentiated functions, such as dwellings-workshop, dwellings-office, and etc.

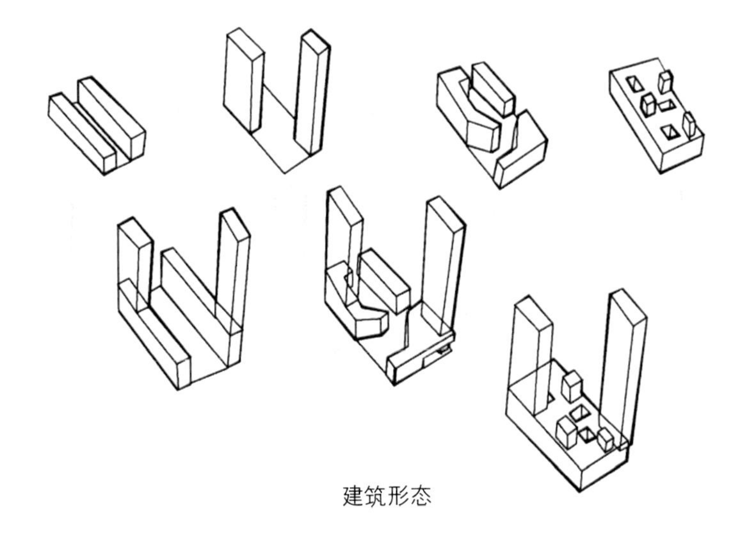

建筑形态

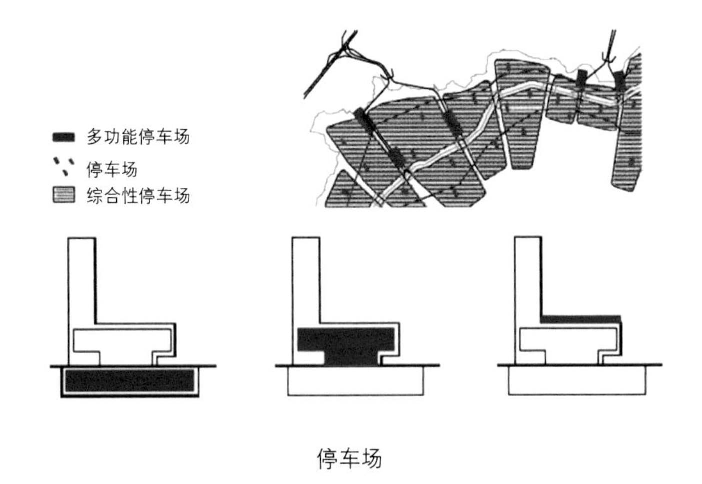

停车场

一个建筑综合体中的多种建筑类型。多功能型建筑和多功能型邻里单位等多样化的建筑类型可为城市营造丰富的生活方式。

Overlapping building types in one same architectural compound.

Mix of construction types on one same neighbouring area.

Radical diversification of the building types to increase the supply of ways of living within architectural boundaries, and specially in the urban areas.

• 居住密度：m^2 / 建筑房屋面积

比例统计差值在 ±5% 以内

50%(45–55)

35%(30–40)

20%(15–25)

15%(10–20)

混合使用功能的住宅

15%(10–20)

10%(5–15)

5%(1–10)

e 1/15.000

■ 日常交通 / Communication among the residents

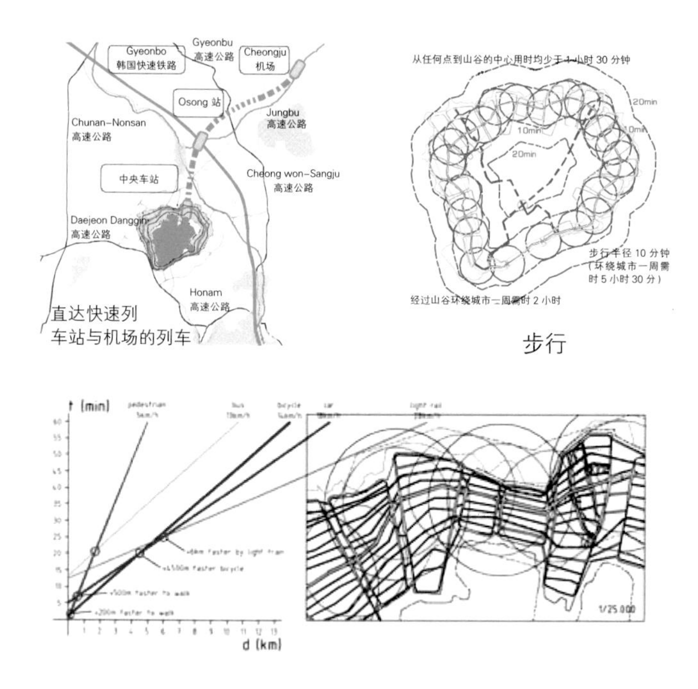

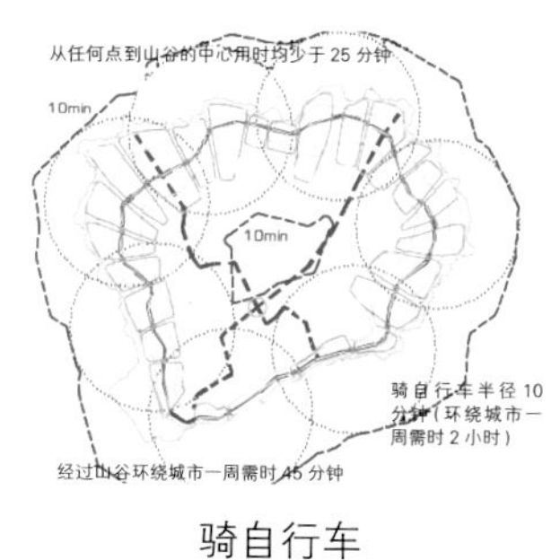

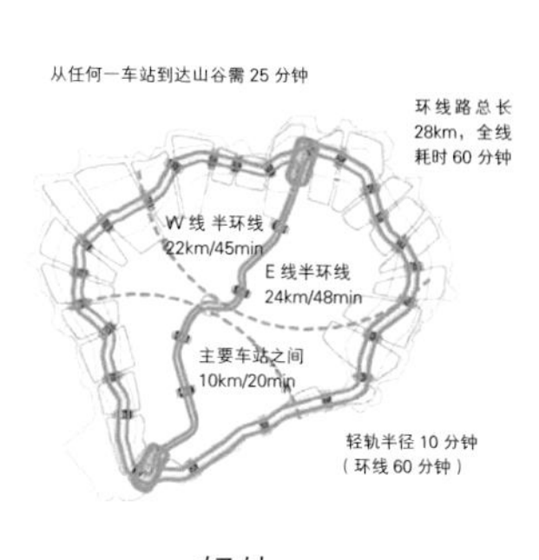

我们保留了山谷中现有的只供当地居民使用的道路网。这样就可以保证 25 座城市之间的交通联系，步行和骑车比乘车更为便利。

We keep the valleys existing road network with a restricted use only for its actual dwellers and their supplement.

The communication between the 25 urban areas or units will occur subordinating the vehicles to the pedestrians an the cyclists.

■ 主要交通方式 / Major transportation modes

地面层环线

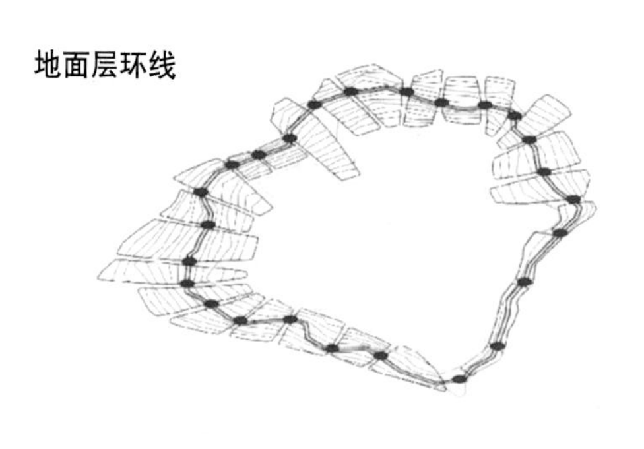

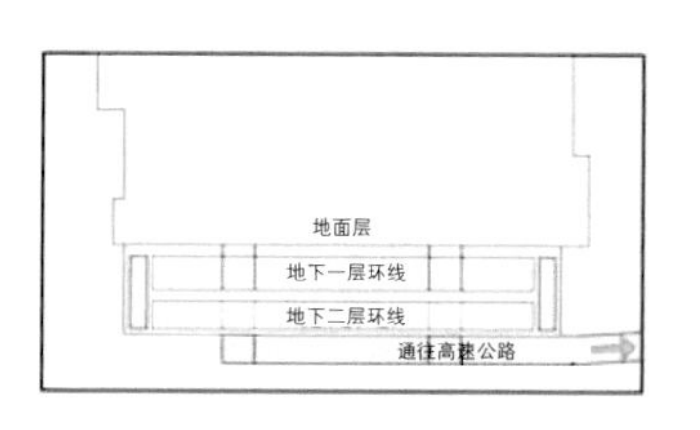

主环线：整个环线上的交通流量密度相等

地下一层环线

地下二层环线

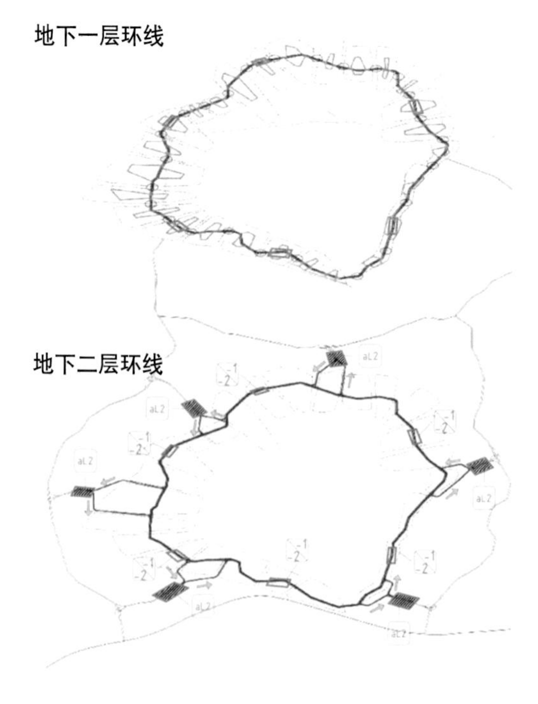

同轴体系的多层式环形道路组成了城市交通系统，交通系统构成了设计的主体框架。

所有车行道互不交叉，并与中心环线呈放射状相连。

小型车辆和私家车（小轿车、轻型货车和小型公共汽车）在上层环路上（第一层）行驶；重型车辆（卡车、公共汽车）在下层环路上（第二层）行驶。第一层与第二层道路之间有 6 处位于外环线入口附近的出入口。

The transportation modes are organized by coaxial systems overlaped to the main frame of the design.

All the different types of vehicular transportation are incorporated into the central ring, from which individual roads heading to separate directions disperse without crossings.

The upper or first level is intended for small public or private vehicles (cars, small vans, and microbus) while the lower or second level is for heavy vehicles (trucks, and bus). The heavy vehicles access the first level through six access points located near the entries from the exterior ring.

■ 公共开放空间网络 / Public open spaces network

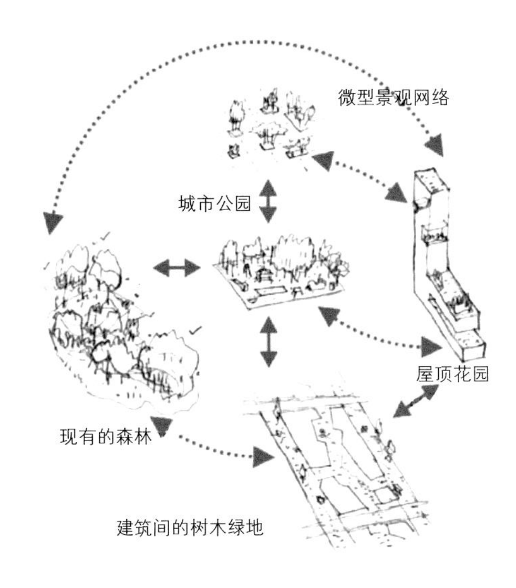

所有的空地均为公共用地。在传统的公共空间中，行为被限定在不同的建筑中，而作品则试图超越已有的公共空间形式：公共空间由多个开敞式公园组成。

Non built areas are all public.

This design is proposed over a globally public space, in which the activities accommodated in the different buildings are inserted.

This design intends for the public space to consist of a great variety of parks, on which the city settles as if on its tiptoes.

■ 可持续发展城市的基础 / Basic infrastructures for a sustainable city

19 世纪的城市（离散型城市）

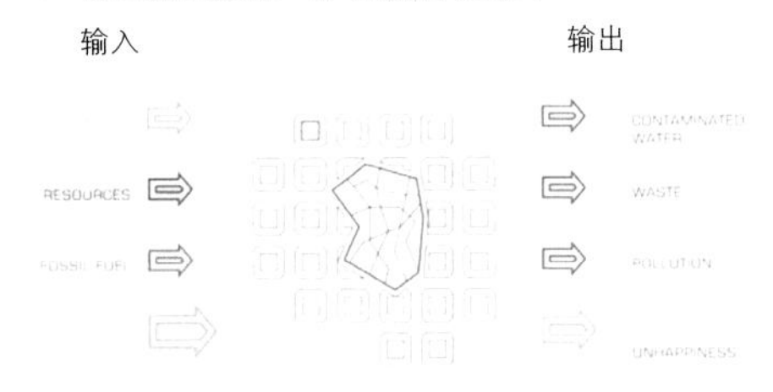

20 世纪的城市（弥漫型城市）

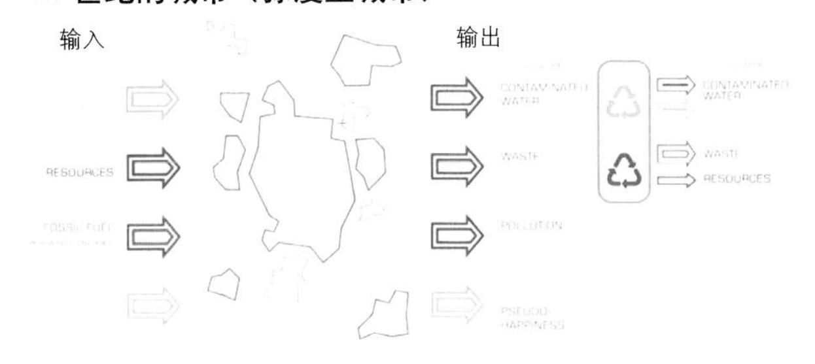

21 世纪的城市（千城之城）

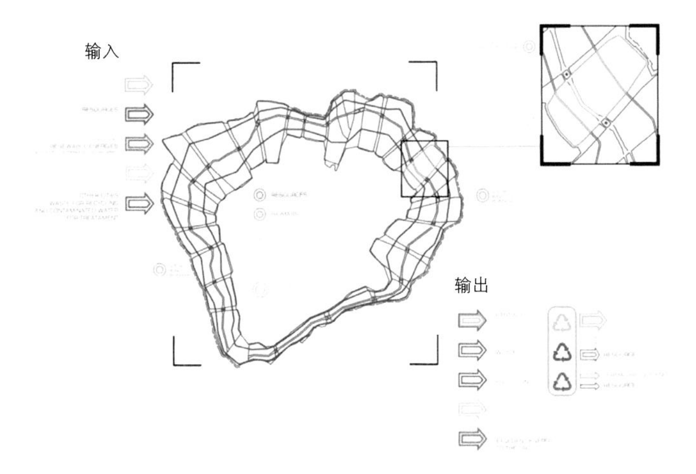

- **社会**
  可持续发展的城市是各方面和谐发展的城市。
- **水**
  人的根本品性决定了公平、协作和社会文明（社会进步和文化）的程度。
- **能源**
  3 个目标：节能、可再生能源的利用和减少化石类燃料的使用。
- **资源**
  采用一套方法，对城市基础设施建造、建筑施工以及市政设施建设过程中的能耗、物耗和由此造成的生态影响进行评估。
- **垃圾处理与循环利用**
  在城市基础设施的设计中考虑到了对城市生活产生的垃圾处理。

- **Socialization**
  The sustainable city is, overall, a socially supported city.
- **Water**
  It is the essentially ethical resource of the humanity that establishes the level of justice, cooperation, and culture (development and civilization) of a community.
- **Energy**
  Three objectives: Energy saving, Use of renewable energies, and Reduction of the consumption of fossil fuels.
- **Resources**
  A code containing the methodology to evaluate the energy costs and the ecological consequences of the construction process of the urban infrastructures, buildings, facilities, and its materials will be established.
- **Waste and Recycle**
  The design of the infrastructures will propose the process of urban activities related to waste disposal.

■ 文化策略 / Strategies regarding the cultural facilities

该作品将文化理解为源于市民而非是强加于市民。社会文化的形成是一个渐进的过程，期间既要保持文化的本土性，又要使之具有全球化的视角，这样才能形成被广泛认同的文化。网络的介入有助于这一目标的实现。

This proposal understands culture as the culture of the citizens and not as culture for the citizens.

This proposal conceives the space for the cultural expression of society as a continuous system, upon which to accentuate its most specialized version.

We intend for all the area of the project to be, in itself, a global cultural facility in its own natural environment. As such is the understood urban environment. The insertion of the Net sphere over the rural and urban spheres corresponds to this intention.

■地下道路设施 / About the underground infrastructure

地下基础设施不允许靠近那些建成的城市项目纵向轴线的环行通道。

这些道路是私家车和公交车均可使用的车行环线。设计者在认真研究地形后，根据功能需要，调整了它的水平高度。

The underground infrastructure is limited to the annular corridor -collector that structures the longitudinal axis of the urban project.

This infrastructure is essentially the ring for vehicular circulation, both private and collective.

This infrastructural ring has been designed by optimizing the topography, and adjusting it to the adequate levels for its functionality.

■分阶段建设 / Priorities in the implementation by phases

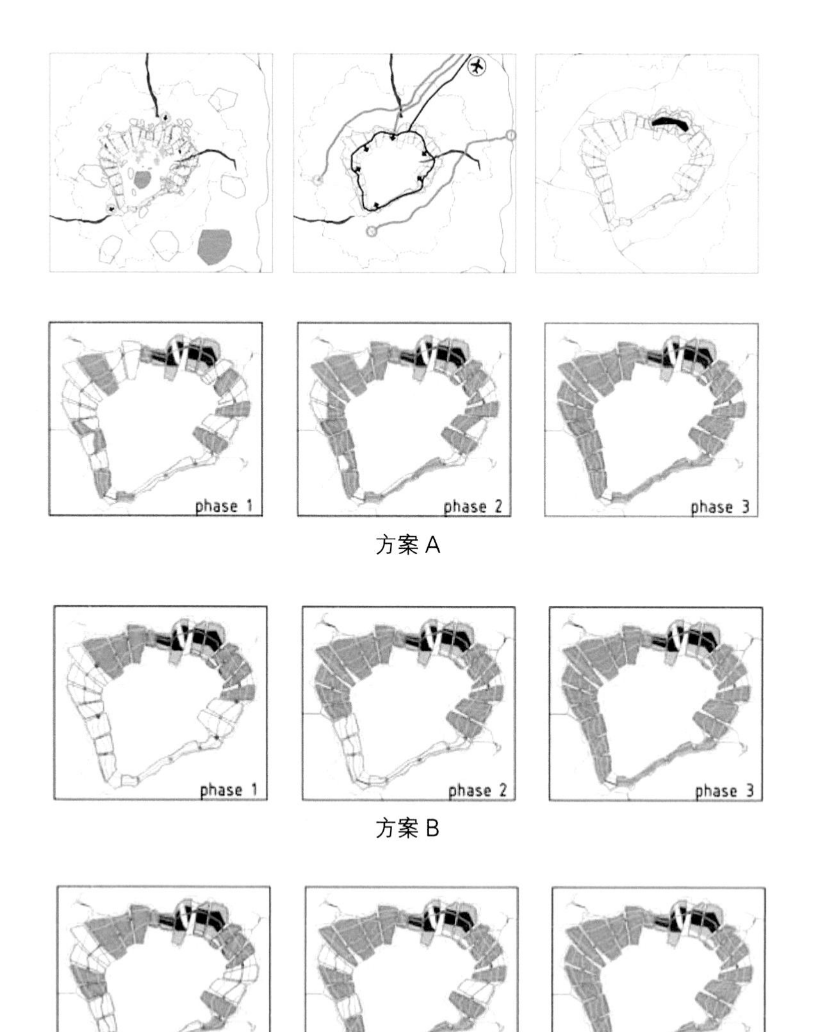

方案 A

方案 B

方案 A 、B 相结合

- **方案 A**

  一次性建设好所有的环线道路及其配套基础设施，并将新环线与现有外环线连通。采用这个方案后可以更为灵活地处理周边地区与自然的关系。这将是一种由外向内的发展模式。

- **Option A**

  The complete construction of the annular axis ring of the system, and its connections with the already constructed exterior ring and the fundamental service webs. This will allow programming the development of the neighbouring units or urban areas, to have complete liberty in choosing to dilute or intensify the urban gradient in relation with the natural environment.

  Model of development from the outside to the inside.

- **方案 B**

  从北面的路口开始（关闭一条辅路）分段建设环线，同时开发毗邻的地段。这是分段开发模式。

- **Option B**

  Partial construction of the ring starting from the North junction(with a by-pass lock) and development of the corresponsive neighbouring units or urban areas. Model of successive growth.

- **方案 A/B**

  A、B 方案二者相结合

- **Option A/B**

  Combination of both.

■ 城市中的建筑类型 / Housing types proposal for the city

- 高密度建筑形式 / High density housing type

类型一：建筑综合体；多功能建筑

类型二：行政办公建筑

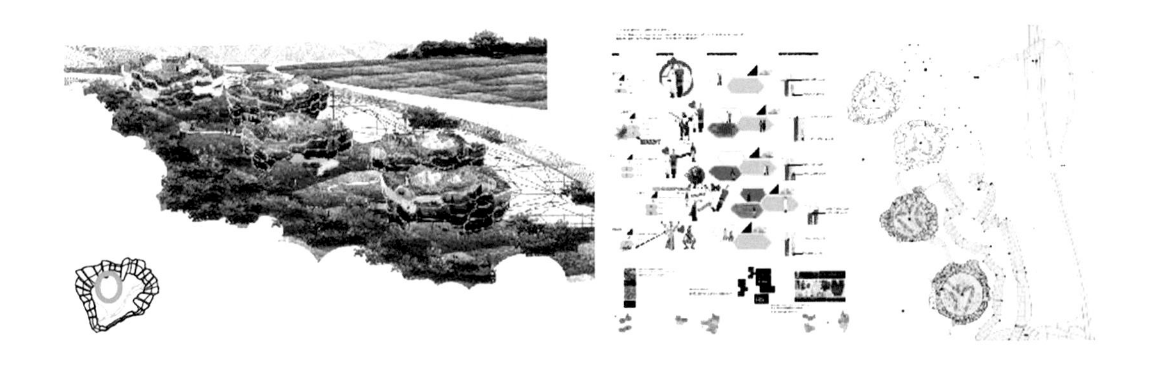

类型三：住宅单元

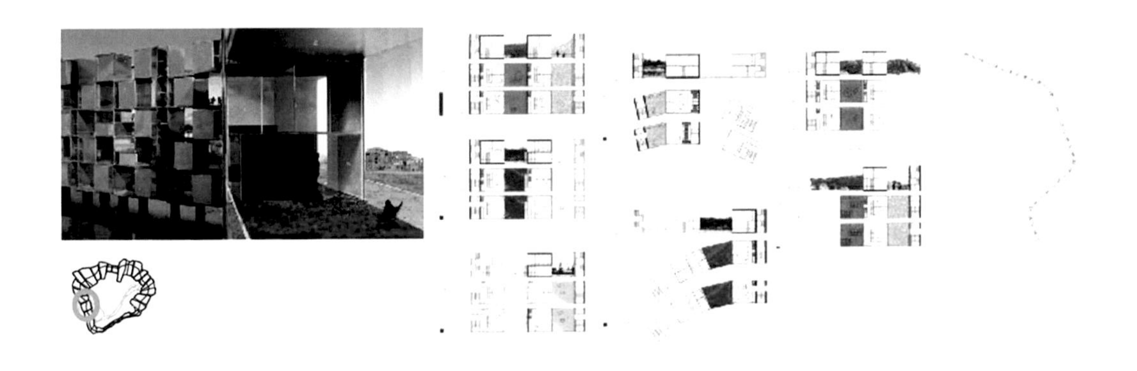

类型四：200 座公寓

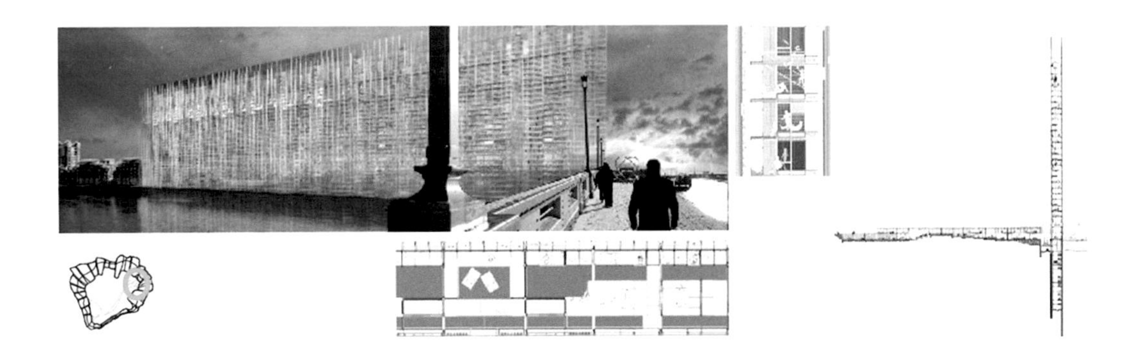

轨道交通

Duerig, Jean Pierre（瑞士）

Duerig, Jean Pierre

- 1958年出生于瑞士Winterthur，1979~1985年于苏黎世高工学习建筑学，1987年至今，任职于SIA(Schweizerischer Ingeni-eur-und Architektenverein)，2004年起，担任瑞士Academia di architettura 教授
- 2005年首尔表演艺术中心概念设计国际竞赛二等奖获得者

- 1958 Born in Winterthur, Switzerland, 1979-85 Studies Architecture at the ETH in Zurich, since 1987 own Architectural Office in Zurich, since 1987 Member of SIA (Schweizerischer Ingeni-eur-und Architektenverein), since 2004 Professor at Academia di architettura, Mendrisio, Switzer-land
- 2005 International Ideas Competition for the design of The Seoul Performing Art Center : 2nd Prize winner

■ 评委意见

“轨道交通”这件作品富于想像力。它不仅因其漂亮的平面设计得到赞赏，区域设计也很出色。

由于地形复杂，设计时要考虑到山脉、峡谷和平原的不同海拔高度，也正因此，作品形成了自己独特的风格。像旧金山一样，新城诗意盎然，在自然环境与网格的对峙中找到了平衡点。虽然这件作品堪称完美，但过于单薄的环形线路使得作品看起来不够紧凑。

■ Jury Comments

The ‘The Orbital Road’ scheme is stubbornly resolute in its figurative power. Moreover, the scheme was not merely appreciated for its planimetric power, but rather the sectional by-product. In its lack of accommodation for the topography, this scheme needs to invent a way of navigating the heights of mountains, valleys, hills, and plains, and in this way it gains its strength in the negotiation between its morphological consistency in the face of topographical challenges. Much like San Francisco, this scheme gains it poetic power in the tension between the confrontation of the grid and nature. If this scheme is formally the purest, it also suffers from the lack of density produced by the thinness of its ring.

摘要

城市是以规划人口数量为参照进行设计规划的，世界上大部分拥有卫星城的大都市，因为中心城市人口数量已经远远超过了能容纳的规划人口的数量，导致不堪重负。这就造成了城市交通堵塞，并且迫使居民越搬越远。

因此，关键问题在于增加新城的活力。我们设计了一种环形城市，城市围绕公园分布。

环形城市体现了民主的意愿，因为在生活中，每个人都被赋予了同等的机会。

Summary

The centers of the world's major conurbations are close to collapse because the size of urban populations far outstrips the number of inhabitants for which existing urban structures were designed, resulting in traffic congestion in these centers and forcing residents in peripheral areas to commute increasingly long distances.

Thus, one of the key features of the New City will be its ability to promote mobility. These considerations have led us to propose a ring-shaped City, which will be located around a park.

A ring shape also expresses the desire for a democratic city where each person is given an equal opportunity in life.

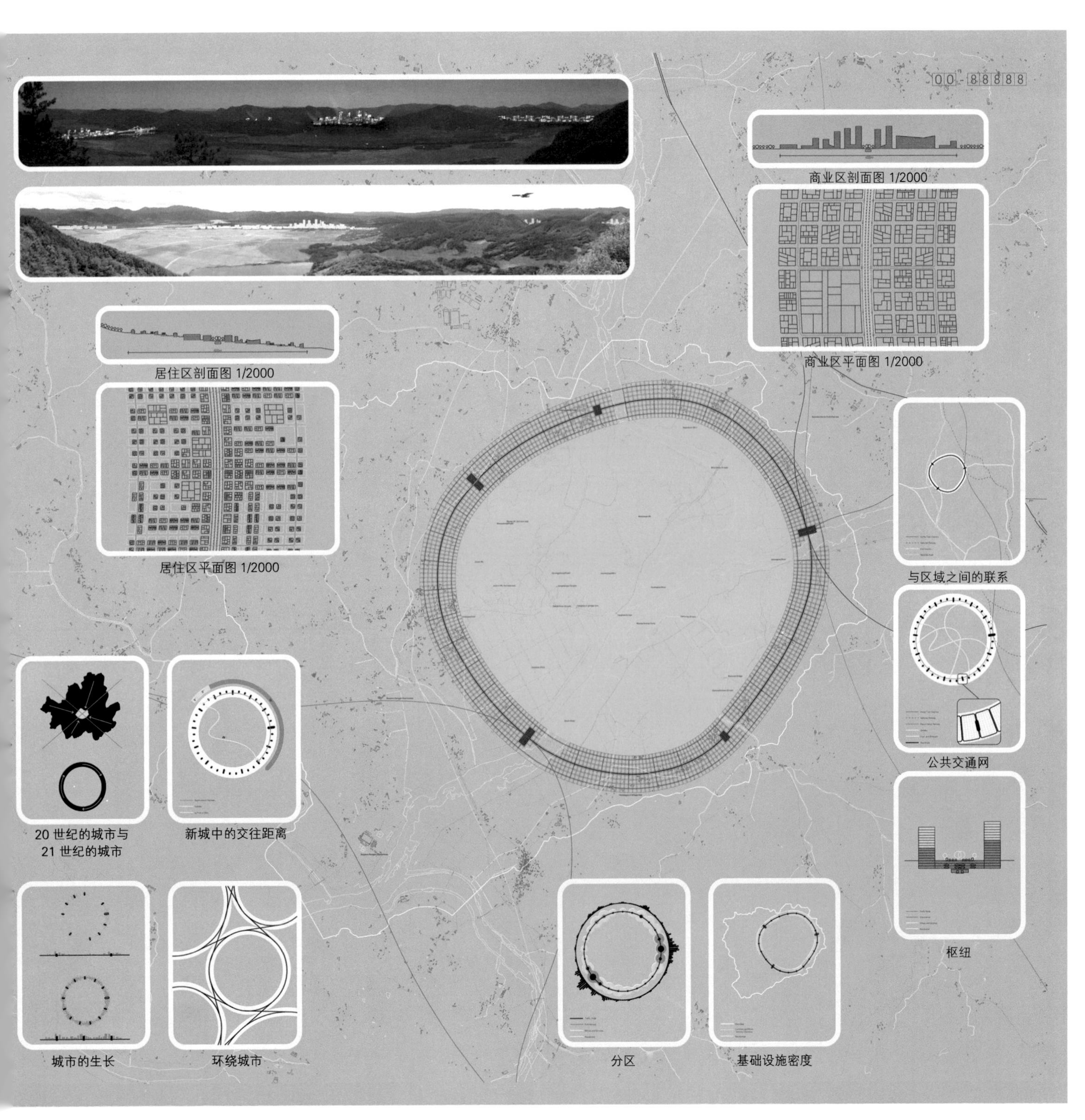

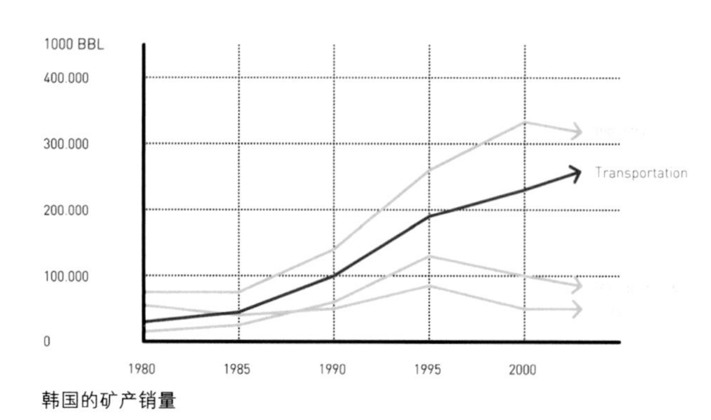

韩国的矿产销量

可持续性

■ 城市可以被设计么?

新城主要特点——可持续发展性、可移动性、自然性、灵活性

- 现代城市最显著的特点在于，在一个很小的空间里，容纳着很多种生活方式。
- 不确定性、自发性和杂乱无章。
- 城市规划师要把生活中的关键元素统一到城市景观中来。
- 基础设施建于居住组团附近。

■ Can cities be planned?

Key Features for the New City - Sustainability, Mobility, Nature, Flexibility

- The most salient feature of modern cities is that they provide amenities that permit a broad range of lifestyles in an extremely confined space.
- Indeterminacy, Spontaneity and Chaos.
- Urban planners should ensure that these and other key quality of life factors are integrated into modern cityscapes.
- Residential zones cluster around available infrastructures.

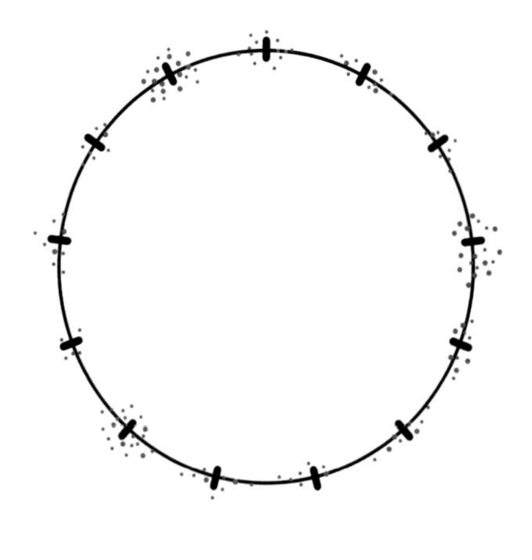

灵活性

■ 21 世纪的城市走向何方?

- 今天，城市建设难以满足城市居民日益增长的需求。
- 因此，新城市建设的关键之一，是要提升它的活力。
- 要创建一个新型城市，我们必须从 20 世纪大城市建设的经典结论中解脱出来，对“城市”重新定义。

■ How should cities evolve in the 21st century?

- Today, urban structures do not have the capacity to meet the emerging needs of city dwellers.
- Thus, one of the key features of the New City will be its ability to promote mobility.
- To create a New City, we must wean ourselves from the classic concept of metropolis in the 20th century and redefine what it means 'urban'.

自然

流动性

流动性

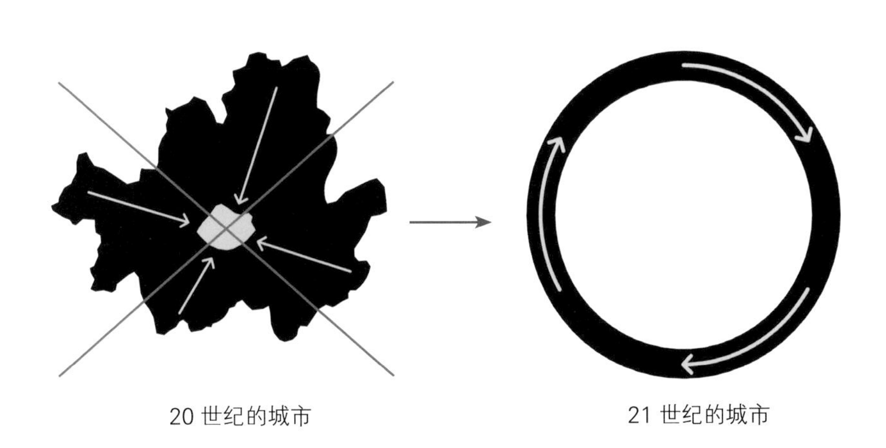

20 世纪的城市　　21 世纪的城市

特质

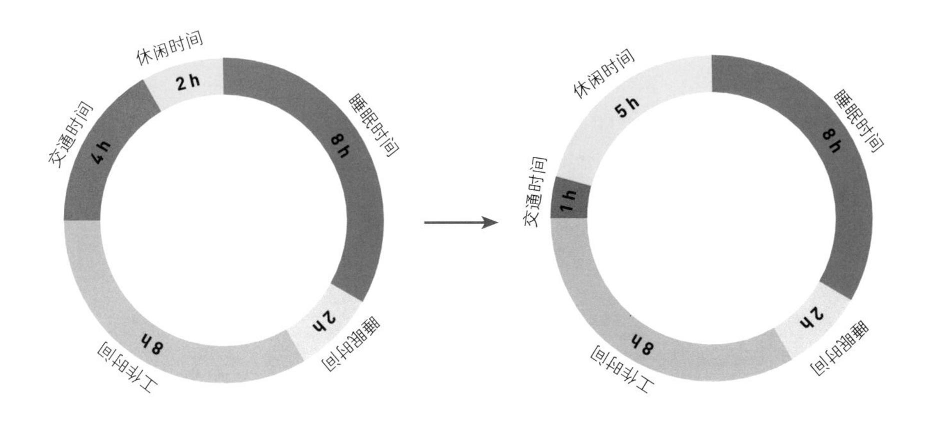

时间

■ 新型城市如何运作？

新型城市的概念——流动性、身份、时间

- 这些因素都促使我们设计一个环型的新型城市，城市围绕公园而建。
- 商业和运输业能耗增加，加剧了空气和噪声污染。
- 轨道交通没有交叉，是最适宜的城市交通系统。
- 30 分钟内，可到达城市的任意一点。
- 洁净的空气和宁静的城市景观。
- 增加可支配的自由时间。
- 环形体现了一个城市的民主，每个人都有平等的机会。
- 商业区将建立在轨道的节点上。

■ How will the New City function?

Idea / concept for the New City - Mobility, Identity, Time

- These considerations have led us to propose a ring-shaped New City, which will be located around a park.
- Energy use for commercial and human transport is rising, resulting in an increase in air and noise pollution.
- An orbital road devoid of intersections is the most efficient access system for urban life.
- Access to any point in the city within 30 minutes.
- Cleaner air and a quieter cityscape.
- Increase the amount of free time which is available.
- A ring shape also expresses the desire for a democratic city where each person has an equal opportunity in life.
- Commercial areas will be realized at the nodes of the orbital road.

——— 快速铁路
——— 地铁
——— 步行或骑自行车

交通时间

■ 怎样创建生态型的新型城市？

- 轨道交通将成为城市的入口。
- 30分钟内，可到达城市的任意一点。
- 新城中的居民不需乘坐汽车即可满足日常活动需求。
- 商业及城市基础设施位于轨道沿线。
- 建筑设计包含哪些内容？

■ What makes the New City ecological?

- The orbital road will be the portal into the New City.
- Any point in the city can be reachable within 30 minutes.
- The residents of the New City will be able to carry out their daily activities without cars.
- All urban and commercial infrastructures will be situated on the orbital road.

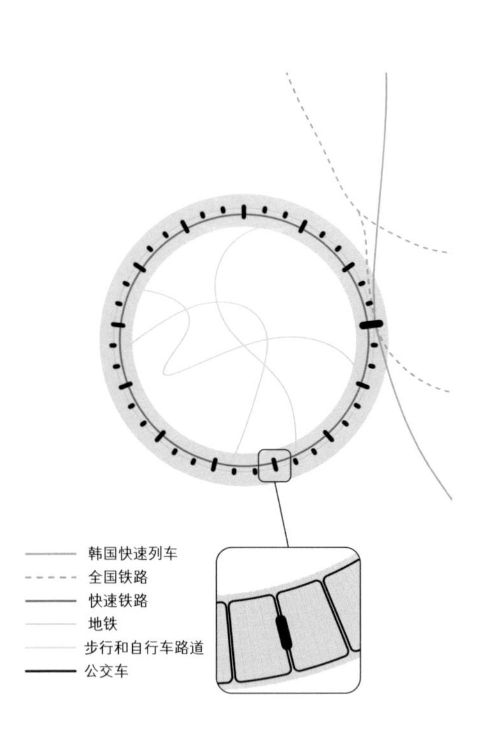

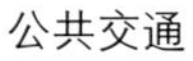

公共交通

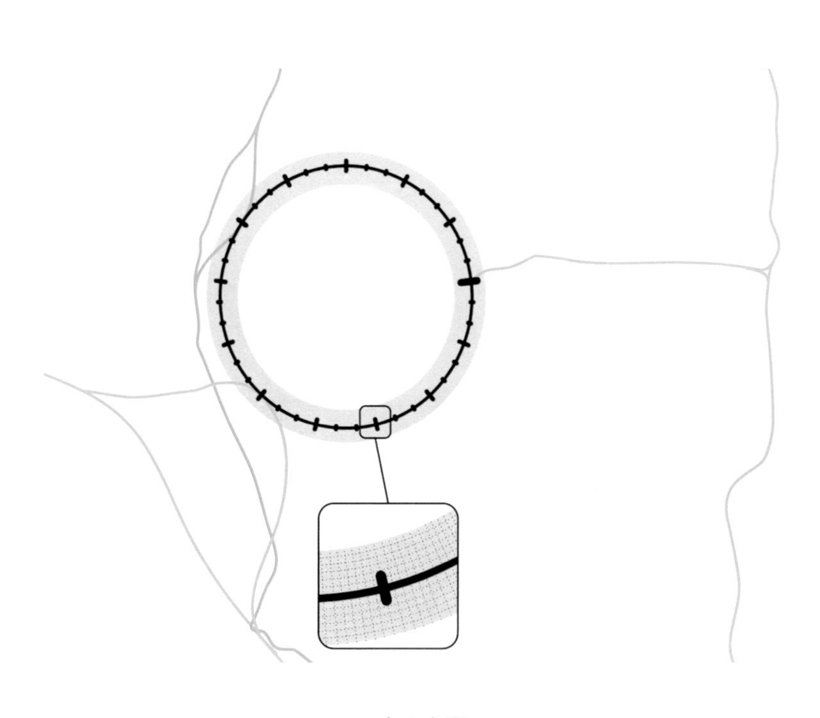

个人交通

■ 结构规划的实施内容是什么？

- 为了最大限度地体现城市规划的弹性，作品只规划了轨道位置，并指明其服务范围。
- 住宅区的规模，将取决于到最接近的基础设施节点的步行距离（不超过最大步行半径）。

■ What does the Structure Plan contain?

- In order to maximize urban planning flexibility, the Structure Plan will only contain the orbital road (infrastructure ring) and will indicate the location of its attendant elements in the available areas.
- The maximum size of a residential zone will be determined by the radius of the maximum walking distance to the nearest infrastructure node.

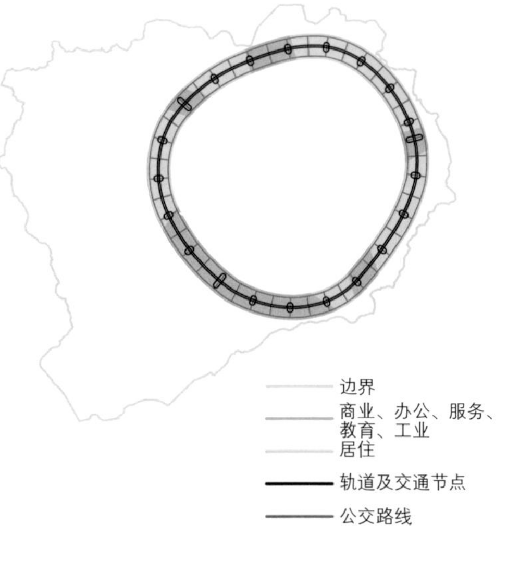

结构规划图

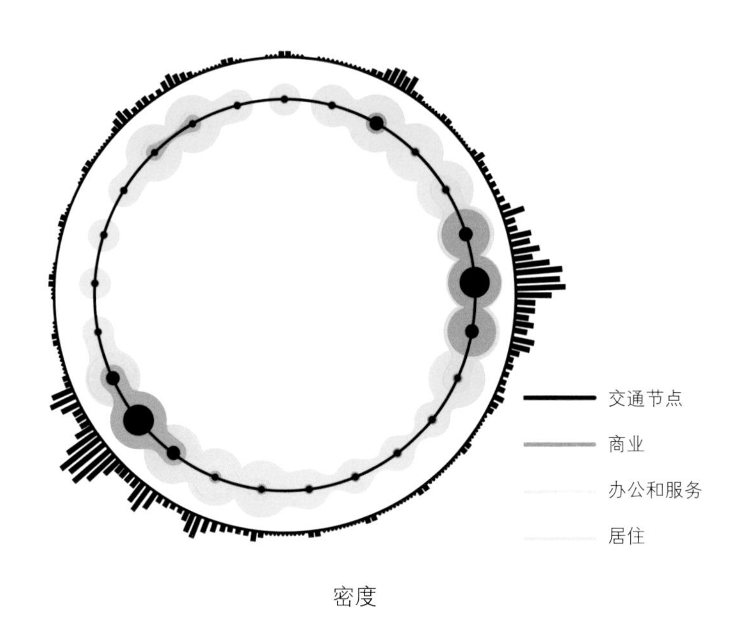

密度

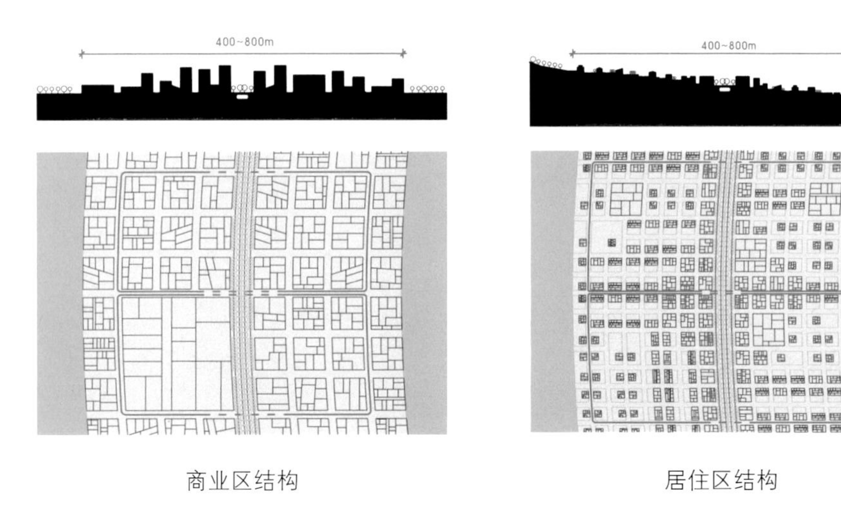

商业区结构　　居住区结构

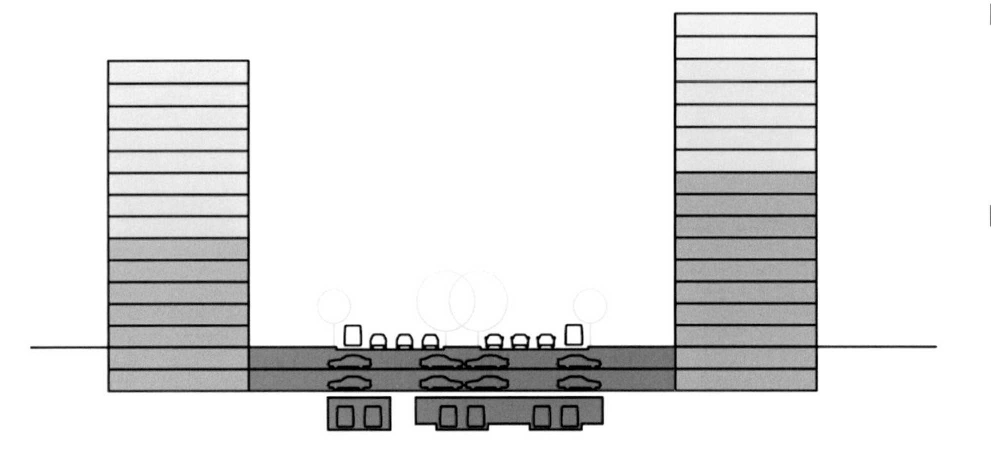

层数

■ 怎样处理各种因素之间的关系?

- 根据地势、地形与周边环境对节点的功能加以区分。

■ Where will the various elements be situated?

- The functions of the various nodes will be differentiated by their orientation to the sky, topography and surrounding area.

■ 轨道看起来怎么样?

- 轨道看起来像是一条林荫大道。
- 在节点处，只要建造一幢高密度、混合型的建筑，就可以满足许多人的居住需求。

■ What will the orbital road look lie?

- The orbital road will look like a boulevard.
- Only a highly compact and hybrid type of structure will be able to accommodate a number of people that will reside at these nodes.

■ 在新型城市中，政府部门将扮演怎样的角色?

- 新型城市中的政府部门将成为城市发展的催化剂。

■ What role will government buildings play in the New City?

- The new government buildings in the New City will be a key catalyst of urban development.

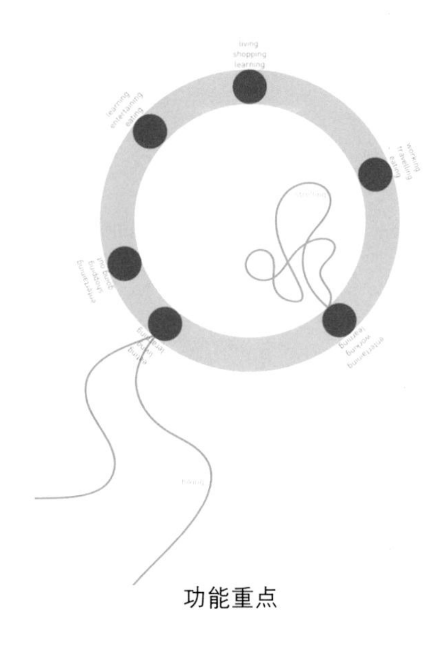

功能重点

■ 新型城市中文化中心位于何处？

- 城市的中央公园将是社会生活的一块试金石。
- 在新城的居住区，广场不是主要的公共空间，因为从城市的任何一点都能够步行到城市公园和天然的绿地里。

■ Where will the cultural center of the New City be located?

- The city's centrally located park, which will be a touchstone of its social life.
- Plazas will play a subordinate role in the New City's residential zones, since the park, as well as natural green areas, will be within walking distance from any point in the city.

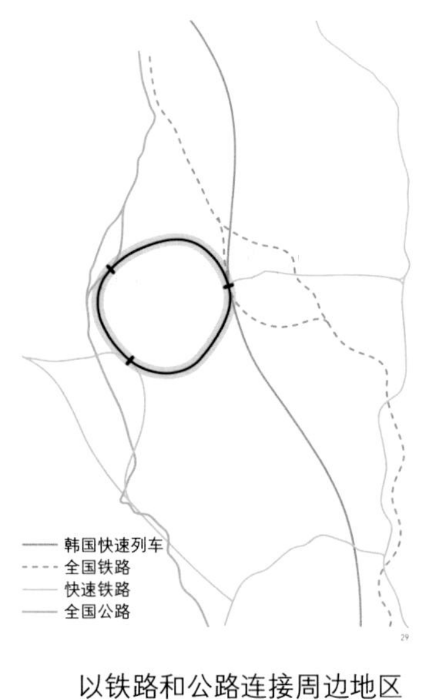

以铁路和公路连接周边地区

■ 如何去首尔？

- 虽然新城是一个封闭的系统，但它也可以被视为一个新的城市群的中心。

■ How do I get to the Seoul?

- Although the New City will constitute an enclosed form, it can also be regarded as the center of a new urban agglomeration.

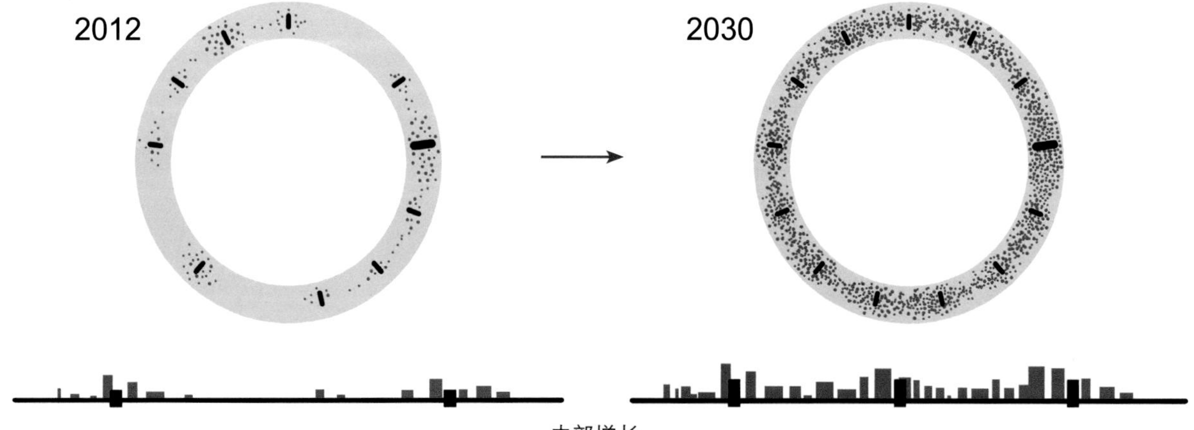

内部增长

■ 新城将如何发展？

- 当它刚建成时，轨道交通上只有很少的节点，随着时间的推移，节点数量将不断增加，每一个新增的节点处将建一个新的居住区。

■ How will the city grow?

- When it is first built, the orbital road will contain a minimum number of nodes that will be augmented continuously as time goes on, and a new residential zone will be established at each new node.

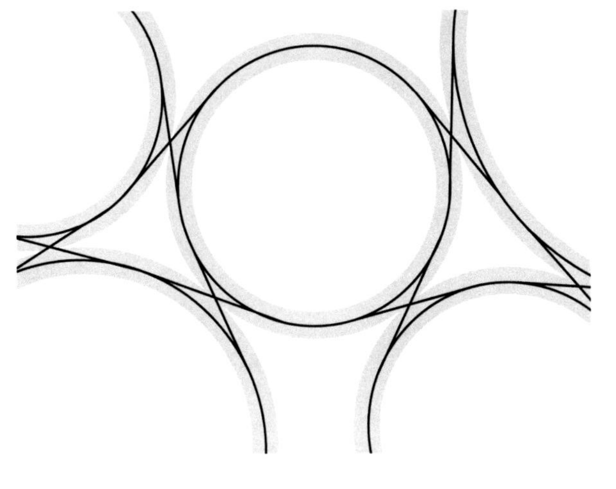

外部增长

■ 如果空间用完了，新城该怎么办？

- 一旦城市人口达到设计人口规模，在周边区域就会很快建起另一个环型城市。新的轨道交通很容易与已有的道路和铁路网连接在一起。

■ What will happen when the New City runs out of space?

- Once the city reaches its target population size, another ring-shaped city could be constructed immediately adjacent to or in close proximity to the new City. The new orbital road could easily be linked to the existing road and rail grid.

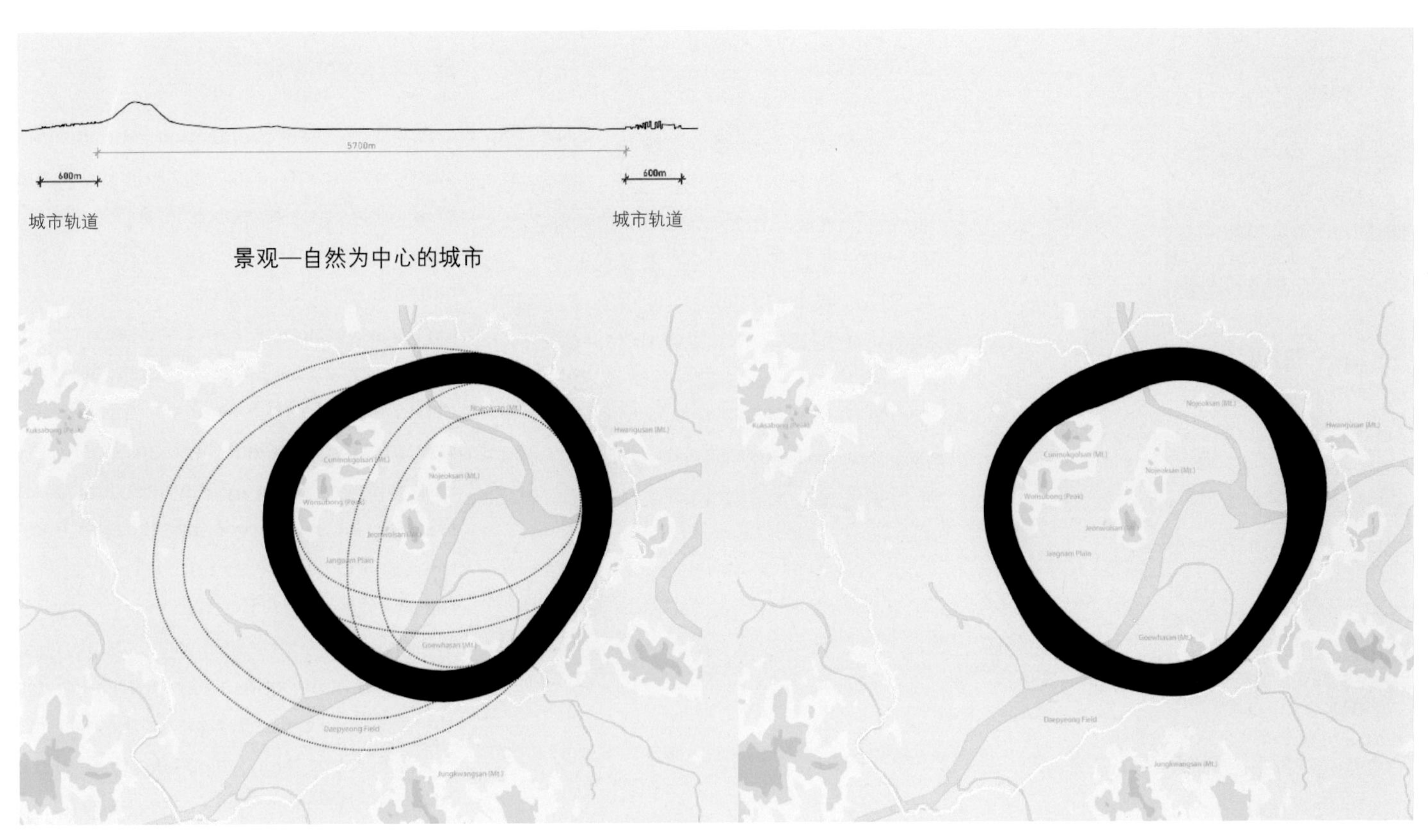

景观与几何构图

实体的密度和地形

水城

Shong, Bok Shub (韩国)
Petit, Olivier/Brochet-Lajus-Pueyo/
D&B Architecture Design group

Shong, Bok Shub

- Ecole d'architecture de Paris–La Villette / Architect D.P.L.G.
- 巴黎第八大学博士
- Hanbat National University 教授

- Ecole d'architecture de Paris-La Villette / Architect D.P.L.G.
- Ph. D, University of Paris 8
- Professor of Hanbat National University

■ 评委意见

这个设计源于城市之核——又是一个圆形——这个理念以高密度观点为基础。即使密度过高，作品在环形城市中的空间、城市形态设计等方面还是很有说服力。而且，穿过城市中心的金河使桥的使用顺理成章。30座桥，该作品利用这些可居住的桥将河两岸联系在一起。通航水道的开通，政府部门，娱乐场所和其他规划中项目的建设，为沿河地区带来了发展机遇，形成了新城的核心。街道和网格设计以不规则的河流为基础。

■ Jury Comments

Inverting their proposition, this project starts with a city core -again in the figure of a circle- that is rooted in an idea about density. While this density preserves the nature beyond it boundaries, it also makes a persuasive argument about the proximity of events, spaces, and urban features within the circle. Moreover, the center of this circle is traversed by the Geum River, an opportunity that is seized upon by the introduction of bridges. Not any bridges, but thirty in total, this scheme capitalizes on the centrality of thc river to stitch the two sides together by way of inhabitable bridges. Ministries, entertainment, and other programmatic elements may create a core for the new city by inaugurating the development right on the riverfront, using the water way as catalyst. The eccentric geometry of the river is also the basis from which the street layouts and grid are derived, mediating between the 'strong' form of the circular city and the apparent chaos contained within.

团队

- Petit, Olivier
  巴黎第八大学博士/城市学
- Brochet-Lajus-Pueyo
  Bordeaux, 法国
- D&B 建筑设计组
- Daejeon, 韩国

■ 摘要

市中心区直径为3.5千米。主要职能部门，服务行业和市政设施都集中在这一区域。绿化带并非是不可变的设计，而是可以根据居民需要进行调整。

桥梁成为公共服务、政府管理和交往空间的混合体。桥梁可以成为开放绿地、游乐场、植物园……河流成为城市活力与可持续之源。城市需要河流。

■ Summary

The downtown area has a surface diameter of 3.5 km. The main functions, services and equipments are concentrated in this sector. Green surroundings do not impose an image of the city but allow its adaptation to the need of the inhabitants.

Bridges become a mix of public services, administration and community spaces. Bridges could be open gardens, esplanades, botanical gardens, etc. The river becomes an active and sustainable element of the city. The city enjoys the river.

未来的行政城市中的桥梁复兴

城市要建在桥梁之上而不是河岸上。所有主要的行政管理设施也建在桥梁之上。每座桥上都会兴建风格不同的办公建筑。桥梁不仅仅是设施，也会成为公共活动空间。桥梁可以成为开放绿地、游乐场、植物园……

Bridges revival for administrative futuristic city

The city is not developed on the shores of the river, but instead made on the river. All the main administrative equipments are located on the bridges. On each bridge, there are administrative buildings with its own design.

Bridges are not only equipments but are also public spaces. Bridges could be open gardens, esplanades, botanical gardens, etc.

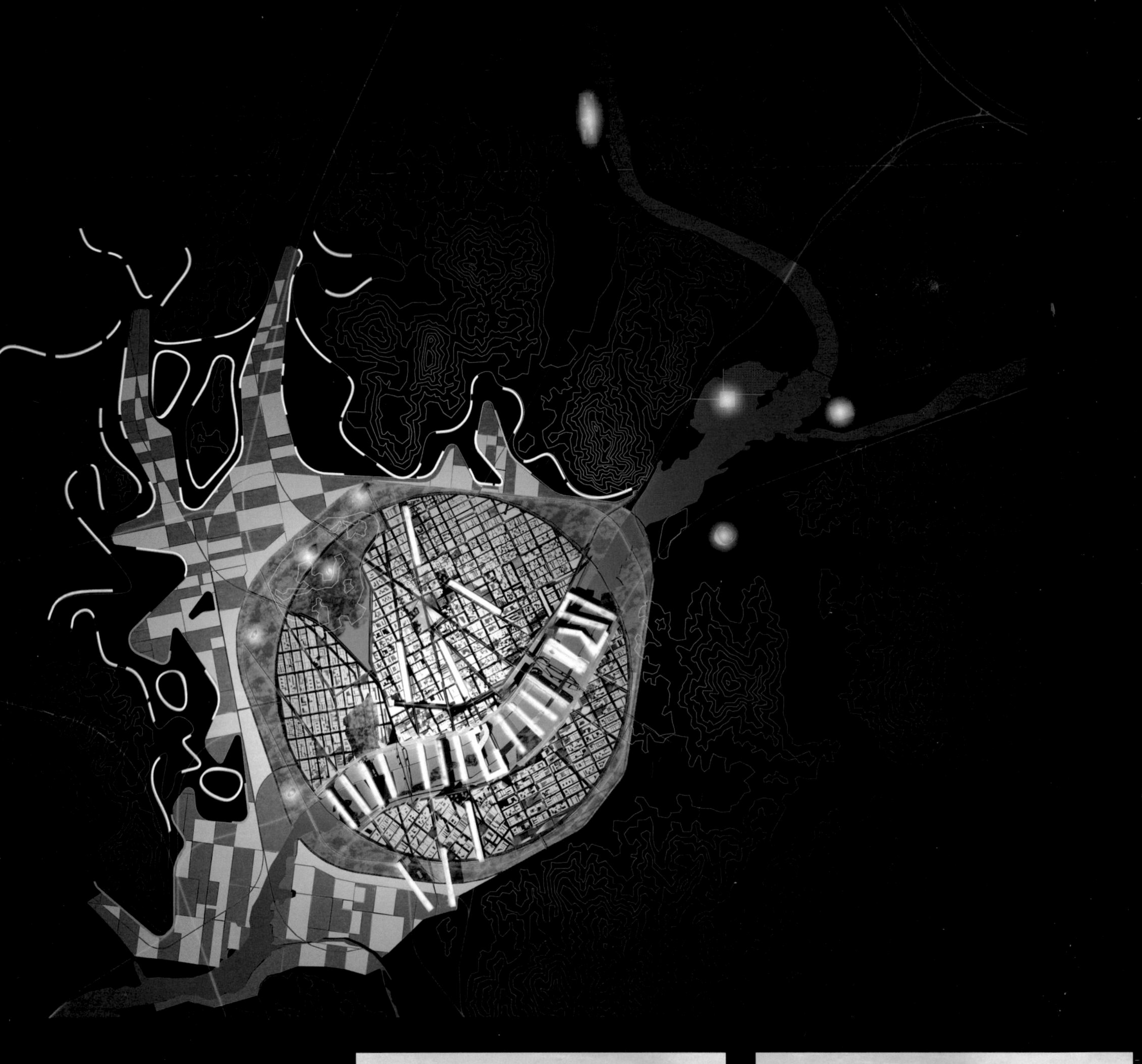

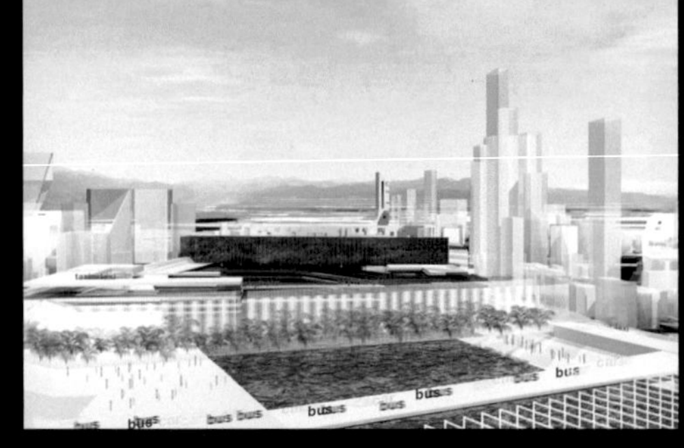

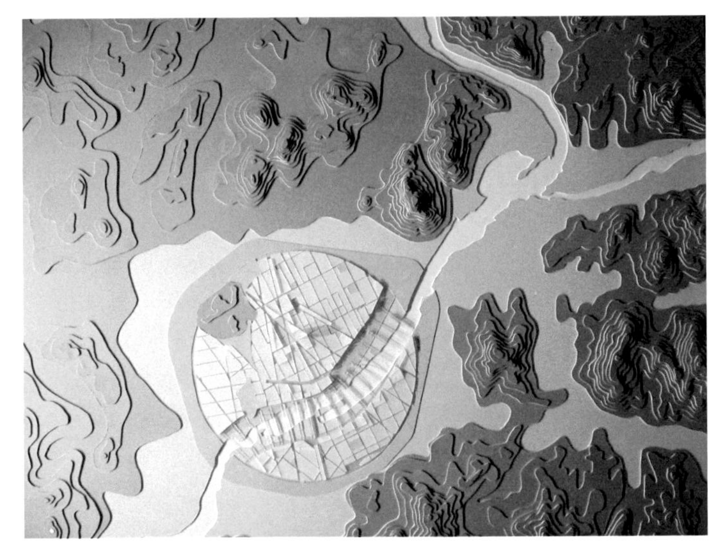

■“紧凑的室内化城市”中舒适的生活

市中心区直径为3.5km。主要职能部门、服务行业和市政设施都集中在这一区域。紧凑的室内城市避免了交通堵塞、通勤之苦和环境污染。

市区内拥有餐厅、服务设施、公共设施、图书馆、博物馆、剧场、公园、河系和住宅以及高科技办公室和学校。高校以及实验室位于城市核心处。

■‘Indoor Compact City’ for an un-congested life

The downtown area has a surface diameter of 3.5 km. The main functions, services and equipments are concentrated in this sector.

An indoor compact urban zone avoids traffic congestion, commuters and pollution.

The urban area includes such functions as restaurants, services, public equipment, library, museum, theater, park, riverside and also housing, high-tech offices and schools. High schools and universities with laboratories of research are located in the core of the city.

■城市的模式

城市的模式像是一块改造后的稻田。街道如同稻穗一样整齐。

■Pattern of the city

The pattern of the city is a reinvention of rice fields. Streets are following the regularity of rice plantation.

城市南北，东西向轴线

■ 以步行方式为主的道路系统

经过城市的主桥梁，市民可以在15分钟内到达城市内的主要功能区。步行主干道构成了城市的南北、东西两条轴线。轻轨线环绕着这座桥，其连通了市区内的主要建筑和它们的周边地区。乘坐轻轨时，乘客可以欣赏到城市全貌、河流景色以及景观绿地。

■ Road system for a pedestrian way of life

All the major destinations can be reached within 15 minutes' walk from the main bridges. Major pedestrian roads are designed in order to square the city in a Cardo(North/South) and Decumanus(East/West) pattern.

Connections between the main spots of the city and the surrounding areas are made through a skytrain. A skytrain line encircles and irrigates the Thirty Bridges City. Public transport users enjoy the view over the city, the river, and the landscape.

城市自然环境结构

■ “紧凑的室内型城市”之肺——城市绿地

“紧凑的室内型城市”的优势之一是拥有一个天然的绿色公园。绿地将成为人们日常生活的核心。

绿地将不再是简单的绿化带，它将会成为人们休憩的港湾。在绿地中，我们还会安装健身设施。

■ Green spaces as the lung of the ‘Indoor compact city’

The ‘indoor compact city’ benefits from a natural green park. The everyday human landscape of the city will be in the core of a modern and futuristic city.

Eco-Ring is not a green belt but a nest for the future. Public sport facilities are located in the Eco-ring.

对生态环的冲击

■ 在河畔居住

滨河地区被重新设计规划。这一区域将成为混合功能区，包含休闲、运动、娱乐和居住的功能。

河流是这座城市的灵魂，因此，在设计中，河流的地位很突出。因为拥有河流，许多的游客才会来到这拥有30座桥梁的城市里放松身心。

■ Living riverside for inhabitants

The riverside is redesigned for everyday practices. Landscaped zones are mixed with leisure, sports, entertainment, and residential areas.

As the river is the ‘blood’ of the Thirty Bridges City, the project inevitably focuses on it. The city, because of its riverside, attracts both tourists and visitors from outside of the Thirty Bridges City who want to take a break from their ‘busy lifestyle’.

滨水区

发达的滨水区

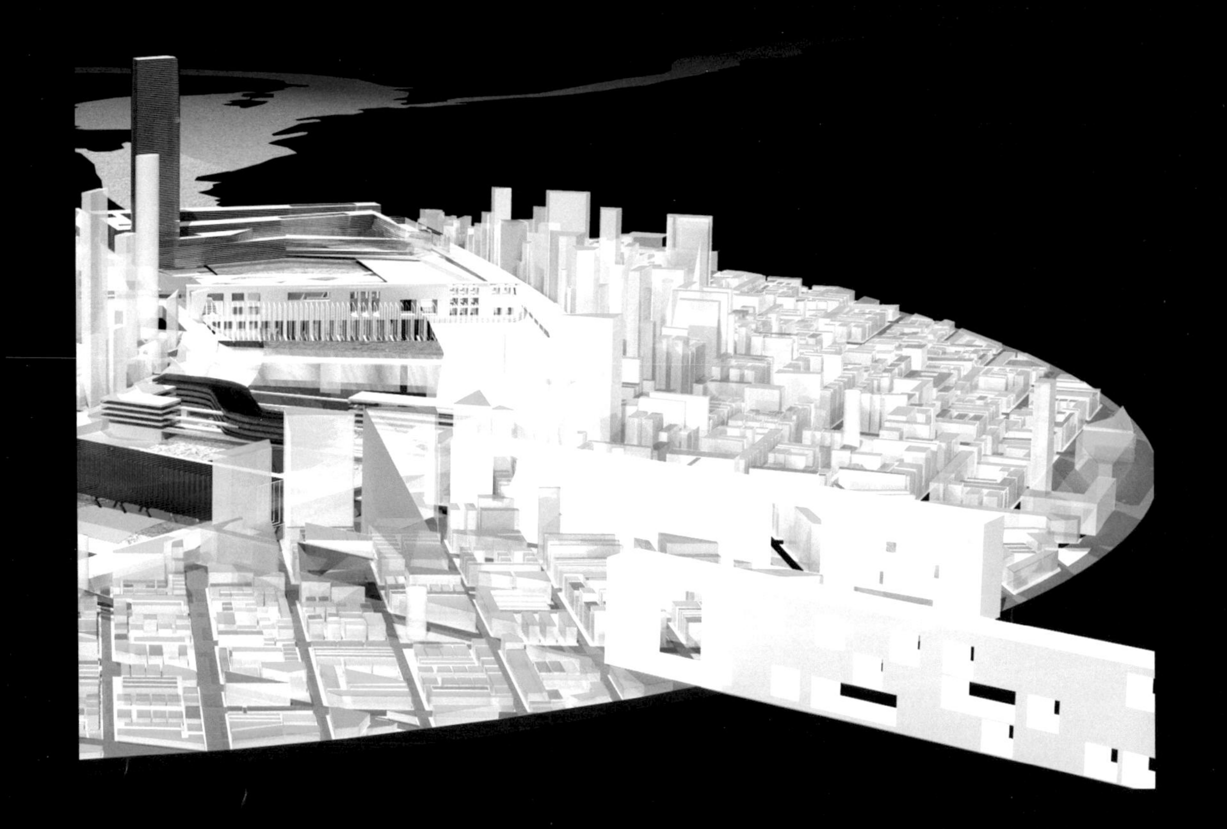

■ **低层高密度建筑**

虽然城市是紧凑型的，但项目并未追求建筑物的高度。建筑设计和城市规划都是遵从人的尺度设计的。

作品为城市营造出一种“巴黎街区”的感觉：安全感，社区生活以及彼此和睦相处。只有沿河的两个商业区中伫立着摩天大楼，它们是新城经济活力的象征。

■ **Low-rise housing for dense community**

Even if the city is compact, the height of building is not the goal of the project. The scale of the architecture and town planning is designed in order to build in a human scale. The reinvention of the 'Parisian block system' gives feelings of protection, community life and solidarity.

Only two business districts are dedicated to skyscrapers along the river. They become symbols of economic dynamics of the Thirty Bridges City, the integration of the river, and the ultra-modernity of high-rise buildings.

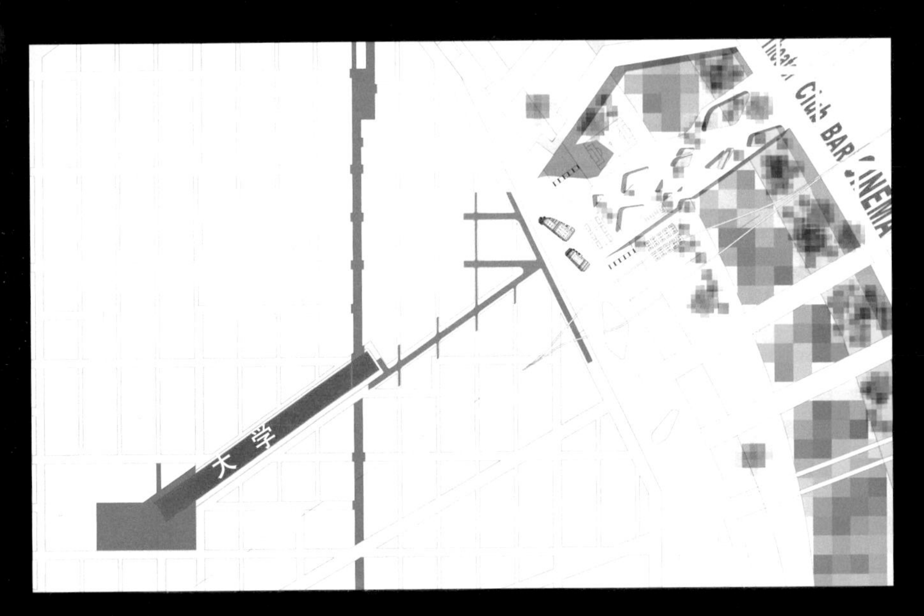

滨水区活力

■ **滨河艺术区的概念**

沿河布置许多艺术装置。这些艺术品是联接人与水的纽带。同时，这片区域将会兴建起一些画廊、商店、酒店以及私人文化馆，这里会成为城市的文化娱乐区域。

■ **Artistic facilities, an 'Esplanade conception'**

Artistic equipments are located along the riverside. The art facilities become nodes between humans and water. Around the cultural facilities, areas are designated to host galleries, shops, restaurants, cultural private buildings with the function of becoming cultural entertainment zones from morning to night.

■大学融入城市

大学融入城市是作品的主要理念之一。大学建筑如同桥梁一样，都将成为新城的标志。滨河区和公园形成了集中绿地。

大学中拥有政府和私人的研究中心。这些都推动着高技术密集型城市的发展。

■ Universities integrated in the city

The integration and connection of the campus with its urban environment is the main concept. University buildings, much like the bridges, become symbols of the indoor urban shape. The proximity of the river and the park allows a green integration.

Public and private research centers are connected to the universities. The connections between these functions guarantee high-tech cluster development of the city.

第一座桥的效果

山地公园

■ 经济中心中的摩天大楼——“曼哈顿小径模式”

两块伫立着摩天大楼的中央商务区，它们昭示着这是一座经济发达，充满活力的全球化城市。中央商务区集中了服务业和高科技产业。在这种国际化环境的吸引下，跨国公司将会纷至沓来。

■ 'Manhattan Alley Shape' for two economic skyscraper zones

Two main CBD with high-rise buildings are designated in order to give an image of a business city and to integrate the globalization and the free initiative in the city. This economic zone focuses on the service and high-tech economy. Multinational companies are attracted by clusters, which combine knowledge, research and competition between firms in an international standards environment.

■ 街区中的商业区

沿着两条步行街，品牌商场和时尚小铺林立，它们吸引着外地游客。居住区（约30000居民）附近也建有市场，形成另一种商业形态。

■ Commerce and markets at the block scale

The main modern brand and fashion shops are located all along the two pedestrian roads. These two streets attract people from outside of the city who can enjoy this pedestrian system walk-way. A second commerce area is a community one with local shops and local markets. It is located close to each housing zone and consists of approximately 30,000 inhabitants.

■ 城市的边界 / Fringes of the City

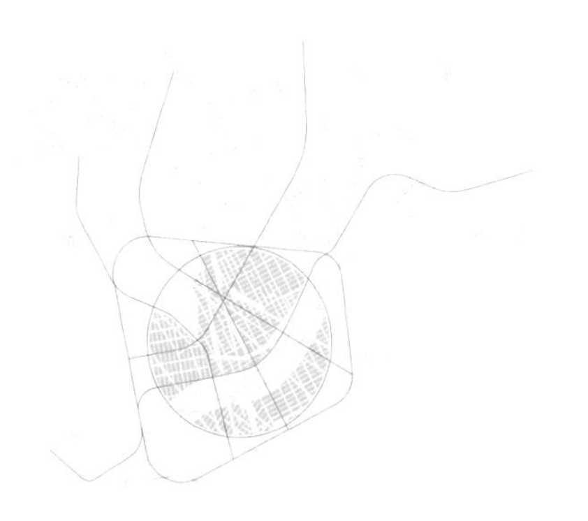

圆形天际线

车站

四道门

- **城市公共交通系统**
  轻轨连接起了城市中的主要建筑（市政府、建有政府机关的桥梁、重要办公楼、居住区和休闲文化区）和它们的周边地区。轻轨交通拥有园林般的轨道交通路，并且可以抵达位于城南的火车站。
  通过完善的的轻轨交通体系，人们可以快速抵达金海国际机场；同时，轻轨也与国道相连。

- **Integrated skytrain public transport system**
  Connections between the main spots of the city (city council, main administrative bridges, main offices, housing zones, leisure and culture area) and the surrounding areas are connected through an integrated skytrain. The skytrain has its own landscaped lanes and is connected to the main train station located in the south part of the city.
  The skytrain integrated system allows a rapid connection with the KTX stops and a perfect integration of the Thirty Bridges City in the national infrastructure system.

- **城市中的四扇"门"**
  通过四扇"门"，新城与其他城市、全省、全国乃至全世界连在了一起。

- **Four gates of the city**
  The Thirty Bridges City links the cities, the province, the nation and the world thanks to its four symbolic gates.

未来的住宅区

山地公园

标志性景观

- **可持续发展的城市**
  在绿化圈之外的新城西侧，是为2030年之后城市发展预留的住宅建设用地。那时，山上将会盖满了低于80m的低层高密度住宅。
  对于21世纪30年代的居民们来说，山中将成为最好的居住场所，因为在这里可以俯瞰全城。

- **Sustainable city**
  Beyond the Eco-Ring, residential zones on the West part of the Thirty Bridges City are planned for the future, after 2030.
  The residential compact and low rise buildings will cover the hills without overstepping 80 meters of altimetry.
  Mountains emerging as panoramic viewpoints above the Thirty Bridges City will become rest locations for the 2030's inhabitants.

- **城市外的景色**
  来到这座城市的游客，即使他已经游览过市内，也会为这种景色所折服。

- **Outside symbolic equipments for the image**
  Visitors, who are about to enter the Thirty Bridges City, have to be impressed by the image of it even before being inside of it.

双面城

Kim, Young Joon（韩国）
Zaera Polo, Alejandro

Kim, Young Joon

- YO2建筑设计有限公司董事
- paju 书城协调员

- Director YO2 Architects Ltd.
- Coordinator paju book city

团队

- Zaera Polo, Alejandro
  主任并外事办公室建筑师

■ 评委意见

这里不再赘述环形模式，评委们认为从技术角度解读这件作品应该更有意义。“双面城”这件作品与其说是一项规划，不如说是一个过程。搜集信息、绘制图纸、调理各种关系、完善复杂的模式，该作品通过文本和系统分析，对城市现状进行了深入剖析。评委们认为，这些技巧适用于很多项目，而且能够形成一套模式。

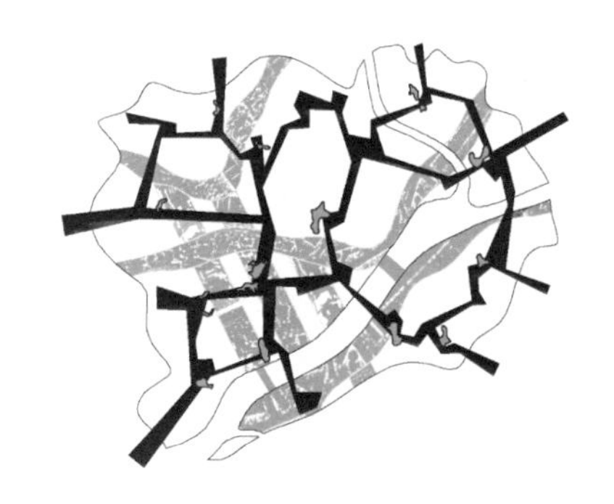

■ Jury Comments

This submission do not relate formally to the ring type, but provide them with tools and techniques that may enable the master-plan in significant ways. 'The Dichotomous City' actually does not propose a plan at all, but rather a process. Layering information, mapping phenomena, organizing relationships, and developing intricate patterns, this scheme is dedicated to the crafting of textures and overlays of systems in order to produce the richness and depth of the urban condition. The jury felt that this set of techniques could be adopted for various plans, and could indeed become an organizing principle for other schemes.

新型行政中心城市

↓

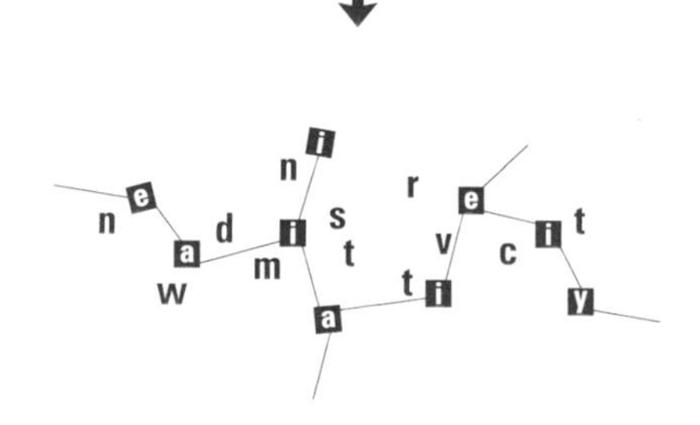

双面城

■ 摘要

“双面城”这一概念包含一系列的网络，不规律且富有灵活性的发展机制可以很好地体现出基地地形的特点，同时，它也能够适应城市未来的发展变化。然而，作品关注的并不是典型的城市的扩张，那种由一个城市间接生长出其他城市的方式。项目运用重叠以及组合的方法对网络进行了组织。方法就是绘制出一系列“可变的图像”草图，沿中轴（一条有吸引力的多样性的轴线）螺旋排列。

■ Summary

The concept of ‘Dichotomous City’ is a system of network, resilient yet irregular development mechanism that can concur with exhibiting feature of the site and accommodate its developmental tendencies and habits. However, rather than focusing on its aspects towards the typical urban expansion that grows indirection from one to the other, the project suggest to frame around the network of relativity, overlapping and association. It is the strategy of sketching out a series of ‘flexible maps’ that spirals through the anchors of intermediacy, attractiveness and multiplicity.

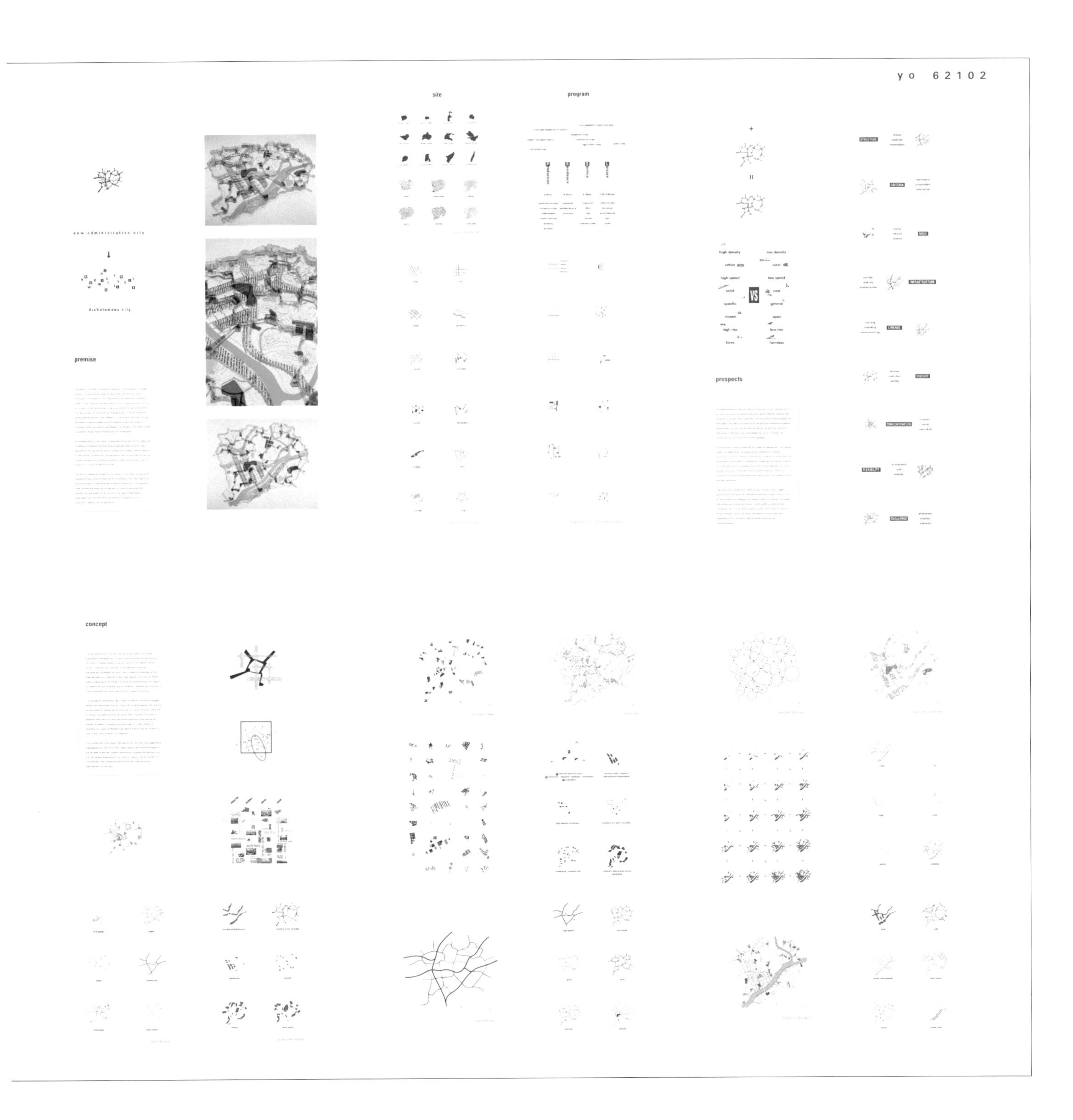

核心城市
催化剂式城市
共同进化的城市
功能复合型城市
试点城市
无处不在的城市
文化型城市
人性的城市

目标，理想

价值观背景

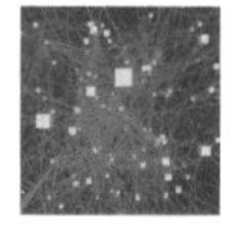

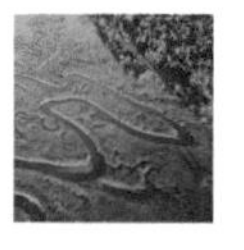
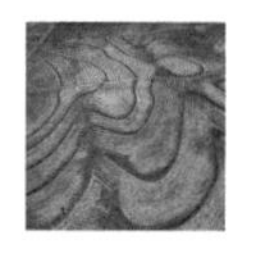

现有的自然环境

城市形态

结构　景观　密度

建筑群

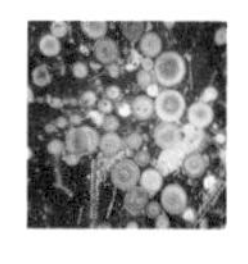
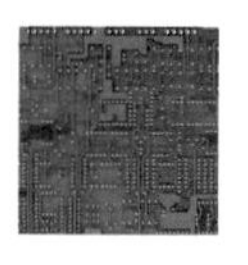
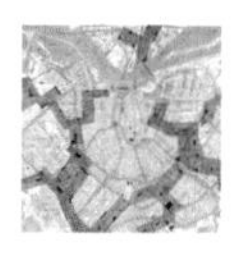

建筑类型

沟通　手法　交通运输

开放空间网络

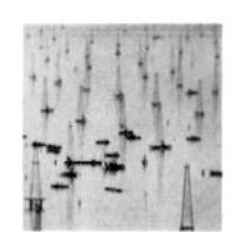

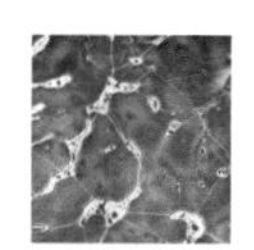

基础设施

项目要考虑的问题

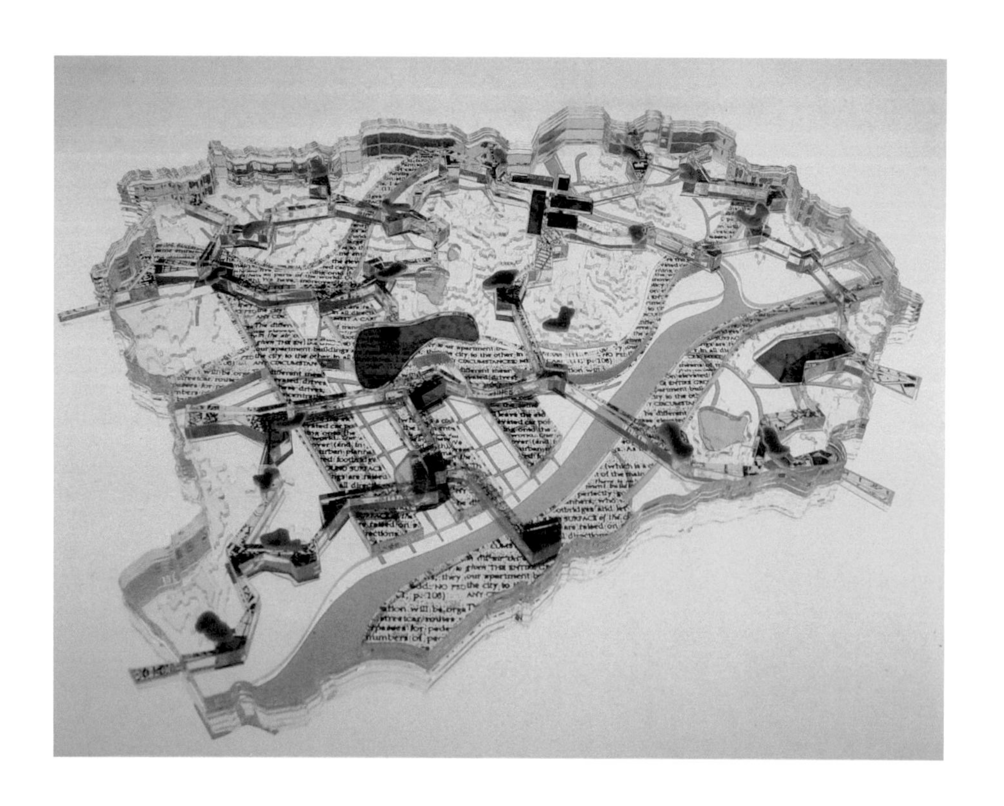

• **前提假设**

进入 21 世纪，海量信息和瞬息万变的形势成为新城的社会基础。为了应对现实中成倍增长的数据、不断发生的事件、不停变化的环境以及膨胀的欲望，城市必须成为很好的媒介，尽可能地以简洁而非影射的方式描述现实世界。

• **premise**

As we head into the 21st century, incorporating the stratum of information and conditions of multiplicity are being recognized as the social de facto elements for the new city. In order to confront such multiple realities rising out of data, events, conditions and soaring desires, the city must take its stand as catalytic mediator, and hold the possibilities to depict the real world, site and programs in concise yet allusive manner.

■ 基地

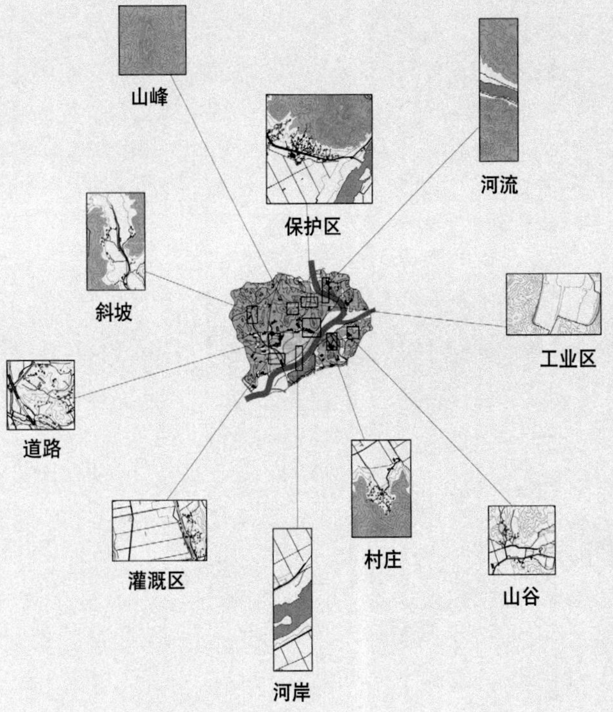

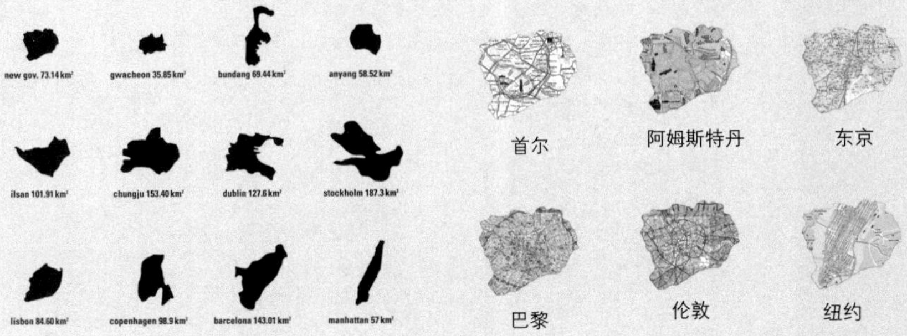

山峰

保护区

河流

斜坡

工业区

道路

灌溉区

河岸

村庄

山谷

new gov. 73.14 km² gwacheon 35.85 km² bundang 69.44 km² anyang 58.52 km²

ilsan 101.91 km² chungju 153.40 km² dublin 127.6 km² stockholm 187.3 km²

lisbon 84.60 km² copenhagen 98.9 km² barcelona 143.01 km² manhattan 57 km²

首尔 阿姆斯特丹 东京

巴黎 伦敦 纽约

规模比较

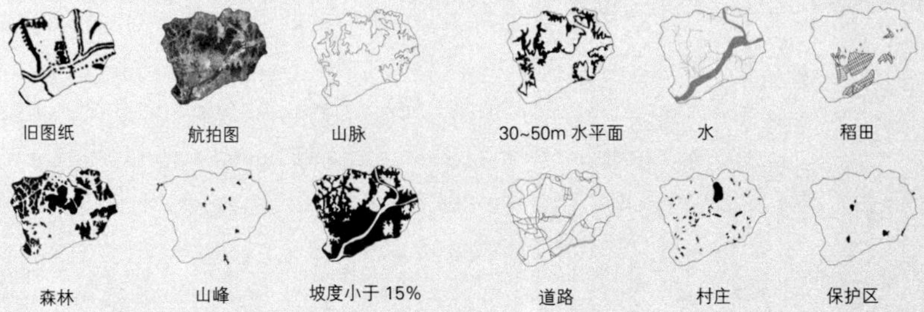

旧图纸 航拍图 山脉 30~50m 水平面 水 稻田

森林 山峰 坡度小于 15% 道路 村庄 保护区

基地情况

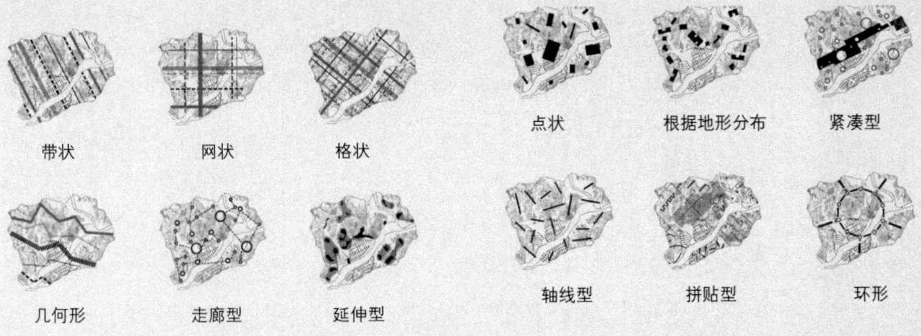

带状 网状 格状 点状 根据地形分布 紧凑型

几何形 走廊型 延伸型 轴线型 拼贴型 环形

可采用的结构

■ 规划

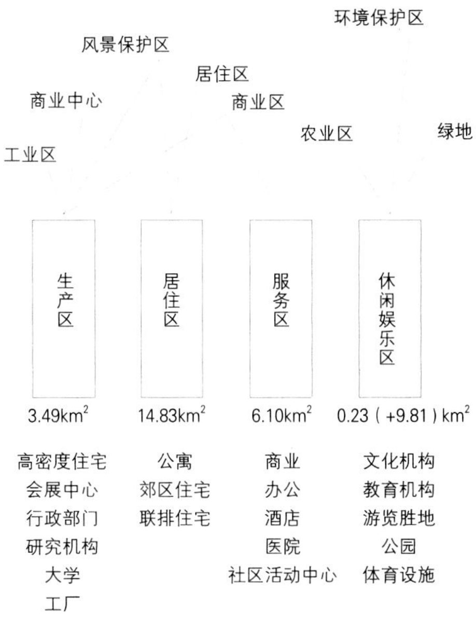

环境保护区

风景保护区

居住区

商业中心

商业区

农业区

绿地

工业区

生产区	居住区	服务区	休闲娱乐区
3.49km²	14.83km²	6.10km²	0.23（+9.81）km²
高密度住宅	公寓	商业	文化机构
会展中心	郊区住宅	办公	教育机构
行政部门	联排住宅	酒店	游览胜地
研究机构		医院	公园
大学		社区活动中心	体育设施
工厂			

生产区

休闲娱乐区

服务区

居住区

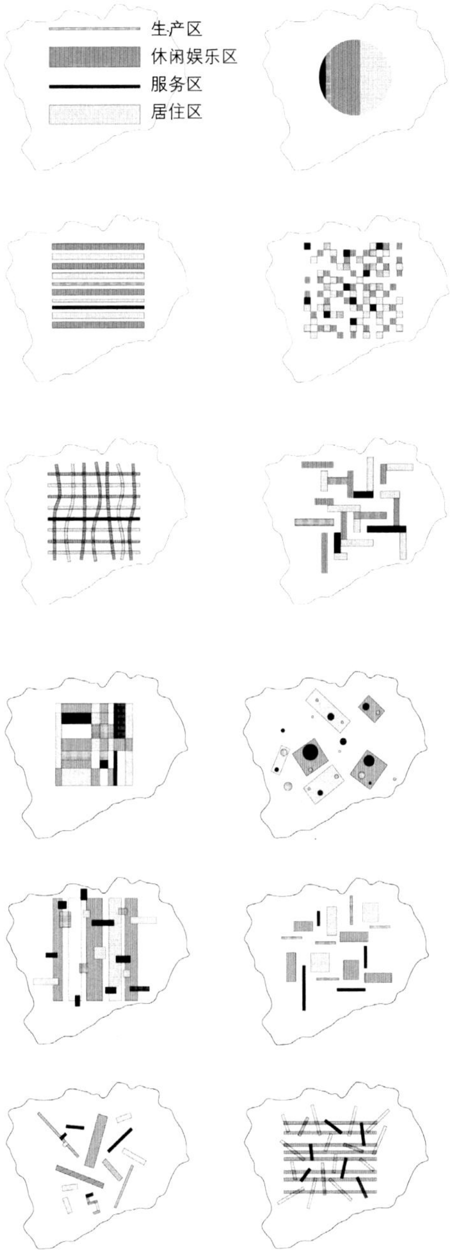

可选择的分区方法

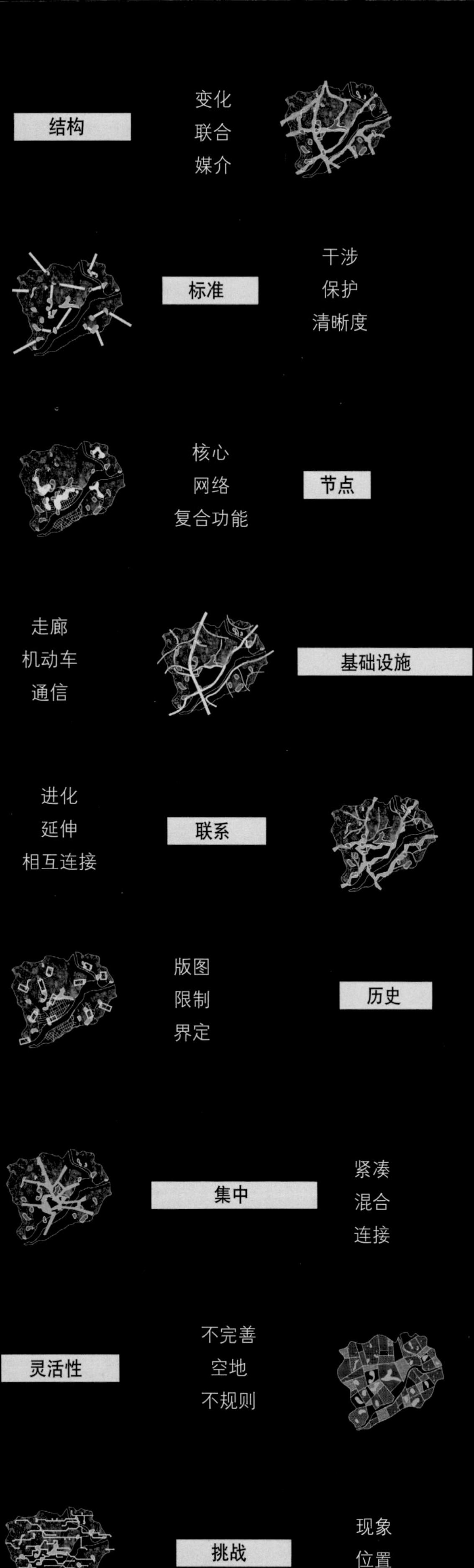

预测

城市规划遵循的那些一成不变的原理现在必须要变通，这样才能适应城市发展的新特点——可移动的、可选择的、无固定边界、不可预料的城市。作品非常大胆：排除所有原始数据，对日常生活的方式的模糊描述方式以及未来的不确定因素的影响，将城市视为一个复杂的、充满矛盾的、不断进化着的事物。

今天的城市生活都交织在了网络系统的不确定性中。在城市中，政治、经济、社会和文化各方面相互影响、相互作用、不断变化。在采用了衡量现实和活力的新标准后，城市中不完善的组织机构经过重组或重新释义有可能得到完善。

社会对城市的要求越来越复杂、多元、高效；城市面临的挑战在于如何制订好规则，处理好那些不断变化、不断增加的信息数据。换句话说，我们对城市的认识不能总停留在建筑形式或公共空间这些狭隘的问题上，而应着眼于对整个城市概念的研究：人工城市还是自然城市，大城市还是小城市，高密度城市还是低密度城市。

Prospects

The rigid artificiality in urbanism that once hovered over the contemporary city is now being forced to redeem itself as the series of flexible protocols that can work around the mobile, alternative and transgressive territorialization, and anticipation. The stake is to weed out all raw data and indeterminate patterns that formulate our everyday life today as well as a group of the unknowns forthcoming in the future, and to acknowledge the city as a complex and paradoxical, yet continuous, and evolving paradigm.

The trajectories of today's urban life are indeed all interweaved in the network system of indeterminacy. All the upheaval and deflections in political, economical, social and cultural vectors are mobilized, communicated, and exchanged to one another in simultaneous, sometimes harmonious, cohesive or unexpected manner. Incompleteness of the city's organization can mean another opportunity for new measures of reality and activism to penetrate and suggest the possible re-formatting and re-interpretations for the best fitted outcomes.

Cities are now in demand and desire for more complex, plural, and multiple operatives and scenarios. Their potential lies within the question of how to map out the framework of strategies and exploit a quantity of data and information that continuously change and evolve. In other words, our perspectives towards the city must not be fixed at gazing only the narrow channel-visions of the built form/open space, but rather, the notions of whole systematic explorations of the real/virtual, the artificial/natural, the big/small, and the congested/sparse.

■ 设计理念

新型城市必须追求灵活性，但是也要合理。而且，它必须能够满足琐碎和繁杂的日常生活要求。它挣扎在理想与现实、开放与封闭、统一与分化、地域性与全球化之间的矛盾中，我们应当认真的反省、完善、更新并修正我们的城市，而不是全盘否定它们。抛弃乌托邦式的幻想，我们要追求新的规则、新的空间、新的美学、新的生活方式、新的城市空地，这些才是最实际、最复杂、最完整的东西。

“双面城”这一概念包含一系列的网络，不规律且富有灵活性的发展机制可以很好地体现出基地地形的特点，同时，它也能够适应城市未来的发展变化。然而，作品关注的并不是典型的城市的扩张，那种由一个城市间接生长出其他城市的方式。项目运用重叠以及组合的方法对网络进行了组织。方法就是绘制出一系列“可变的图像”草图，沿中轴（一条有吸引力的多样性的轴线）螺旋排列。

由于可选择的方法与观点越来越多，我们的生活正面临着挑战。人们更关注新的城市设计方法，而不是关注人与自然，分散与集中的关系问题。而这些，才是关系到新的复合型城市未来走向的关键。

■ Concept

The new alternatives for the city must aim for the flexible yet accurate propositions. Simultaneously, it is about how to embrace the fine lines and intricacies of multiple realities of our everyday life and desires. Always residing in-between the paradoxes of the real/ideal, open/close, unity/diversity, local/global and so on, such conditions of standing on the edge will allow us to internally critique, add, replace and revise our identity without destroying it in its entirety. Free from the utopian fantasy and imagery, we seek for the new measures, spaces, aesthetics, lifestyles and scenarios of urban wilderness that is real, integral, blunt, complex and honest.

The concept of Dichotomous is a system of network, resilient yet irregular development mechanism that can concur with existing features of the site and accommodate its developmental tendencies and habits. However, rather than focusing on its aspects towards the typical urban expansion that grows in directions from one to the other, the project suggest to frame around the network of relativity, overlapping and associations. It is the strategy of sketching out a series of flexible maps that spirals through the anchors of intermediacy, attractiveness and multiplicity.

Our everyday life is now being challenged by the now, alternative approaches and perspectives. The future of our urban development and the formation of the city seek for the new multiple measures and mindsets that discuss more than the twofold relationship of man-made vs. nature or decentralization vs. concentration. This is where the future of the new multi-functional administrative city lies upon.

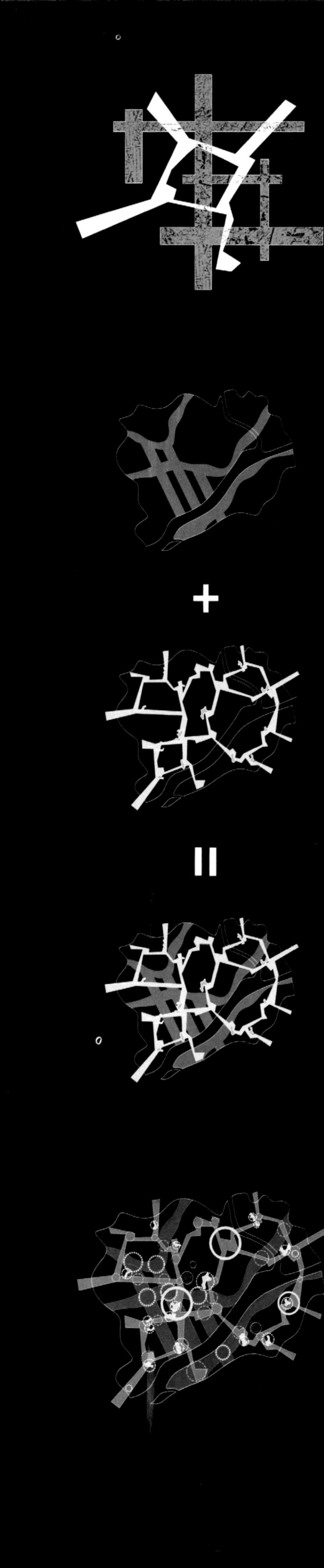

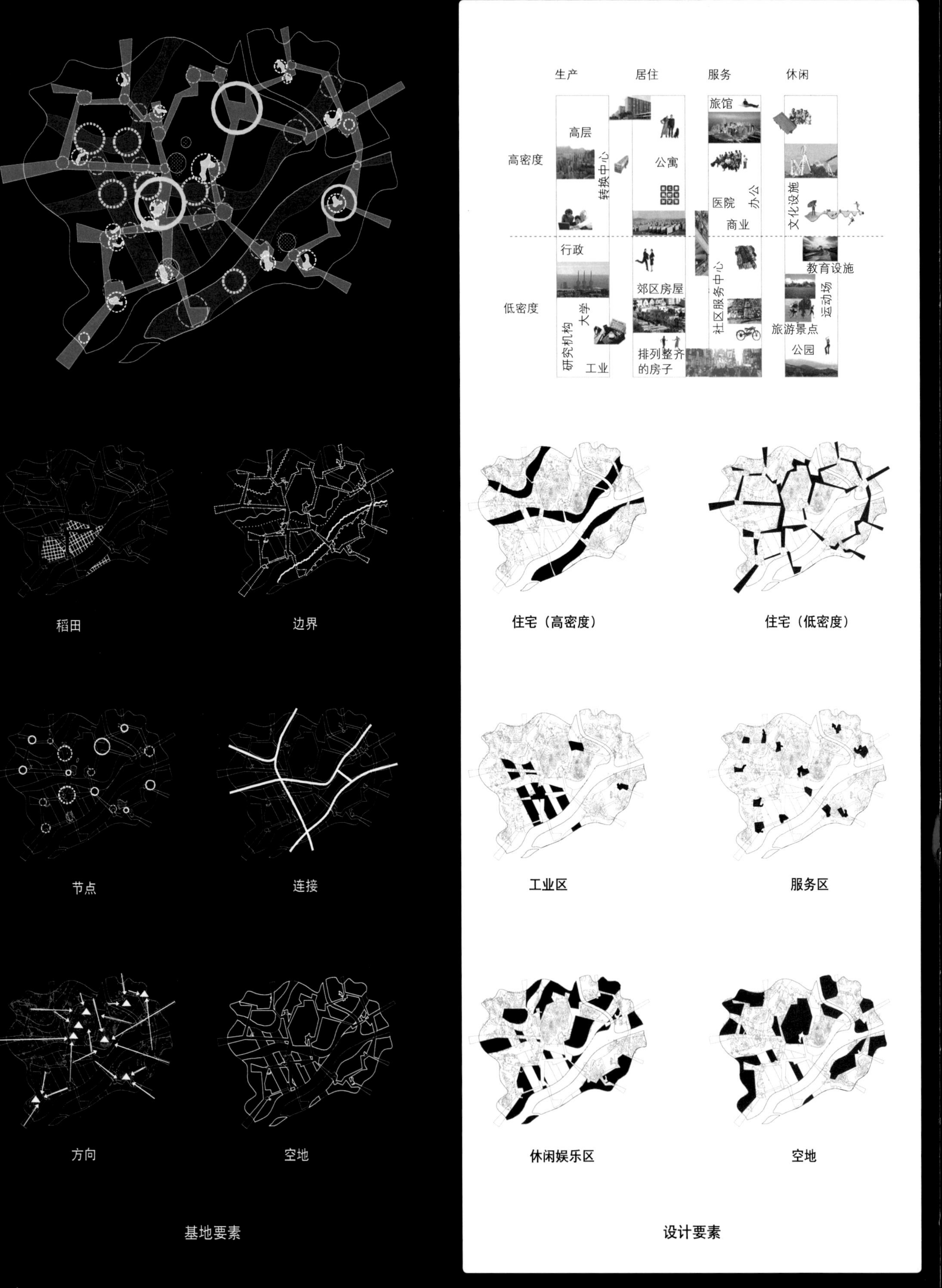

生产
居住
服务
休闲
高密度
低密度
高层
转换中心
公寓
旅馆
医院
办公
商业
文化设施
行政
郊区房屋
社区服务中心
教育设施
研究机构
大学
工业
排列整齐的房子
旅游景点
运动场
公园
稻田
边界
住宅（高密度）
住宅（低密度）
节点
连接
工业区
服务区
方向
空地
休闲娱乐区
空地
基地要素
设计要素

■ 建议

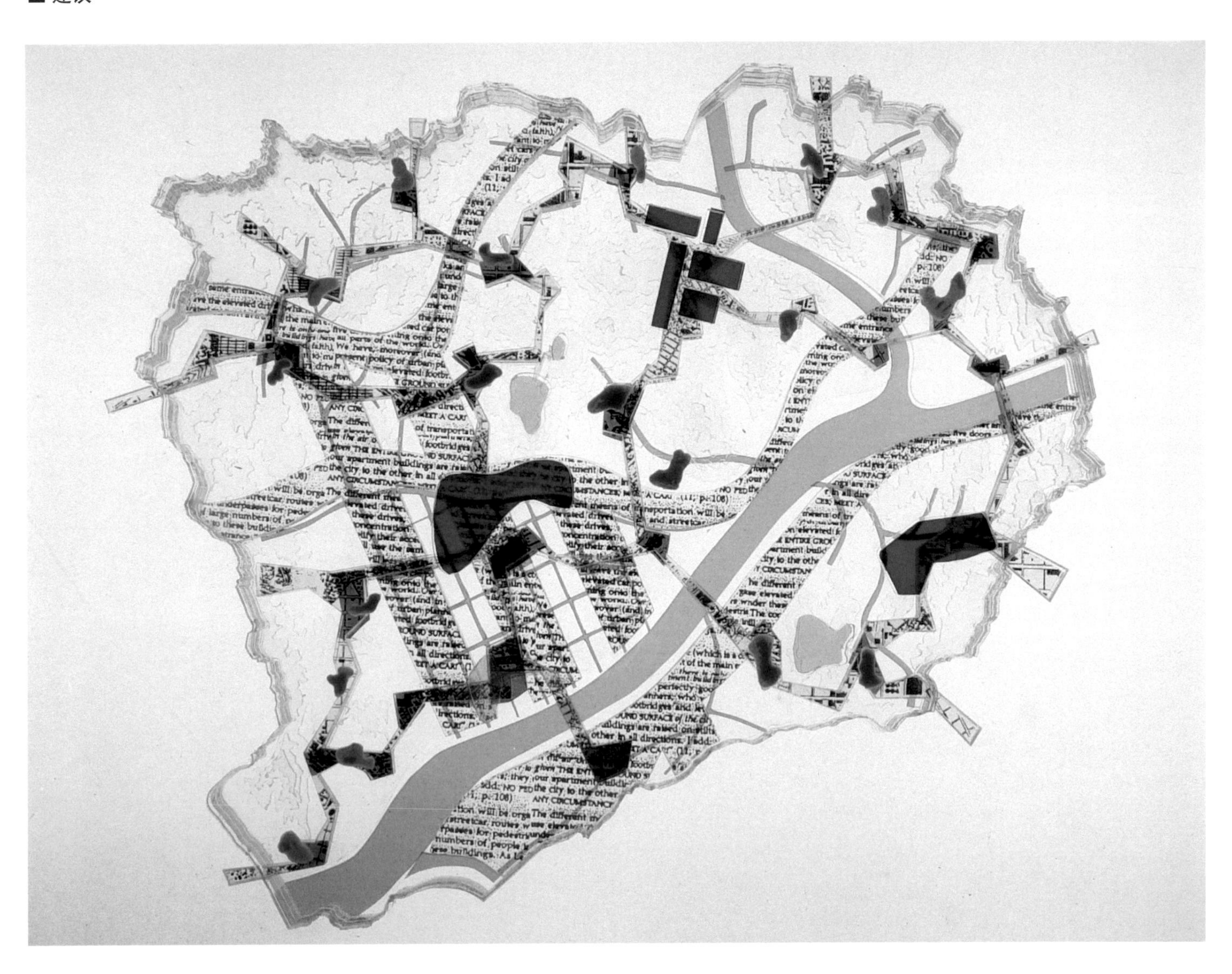

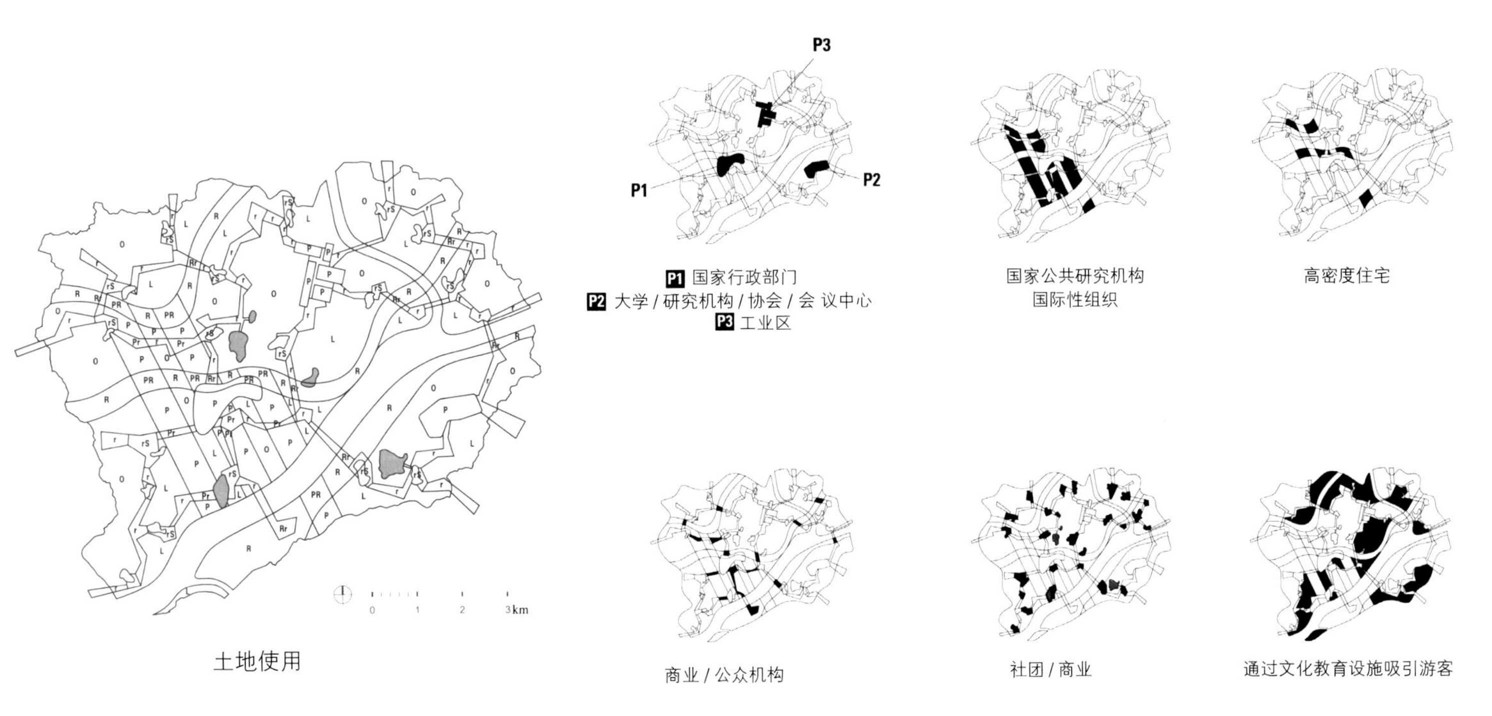

土地使用

P1 国家行政部门
P2 大学 / 研究机构 / 协会 / 会 议中心
P3 工业区

国家公共研究机构
国际性组织

高密度住宅

商业 / 公众机构

社团 / 商业

通过文化教育设施吸引游客

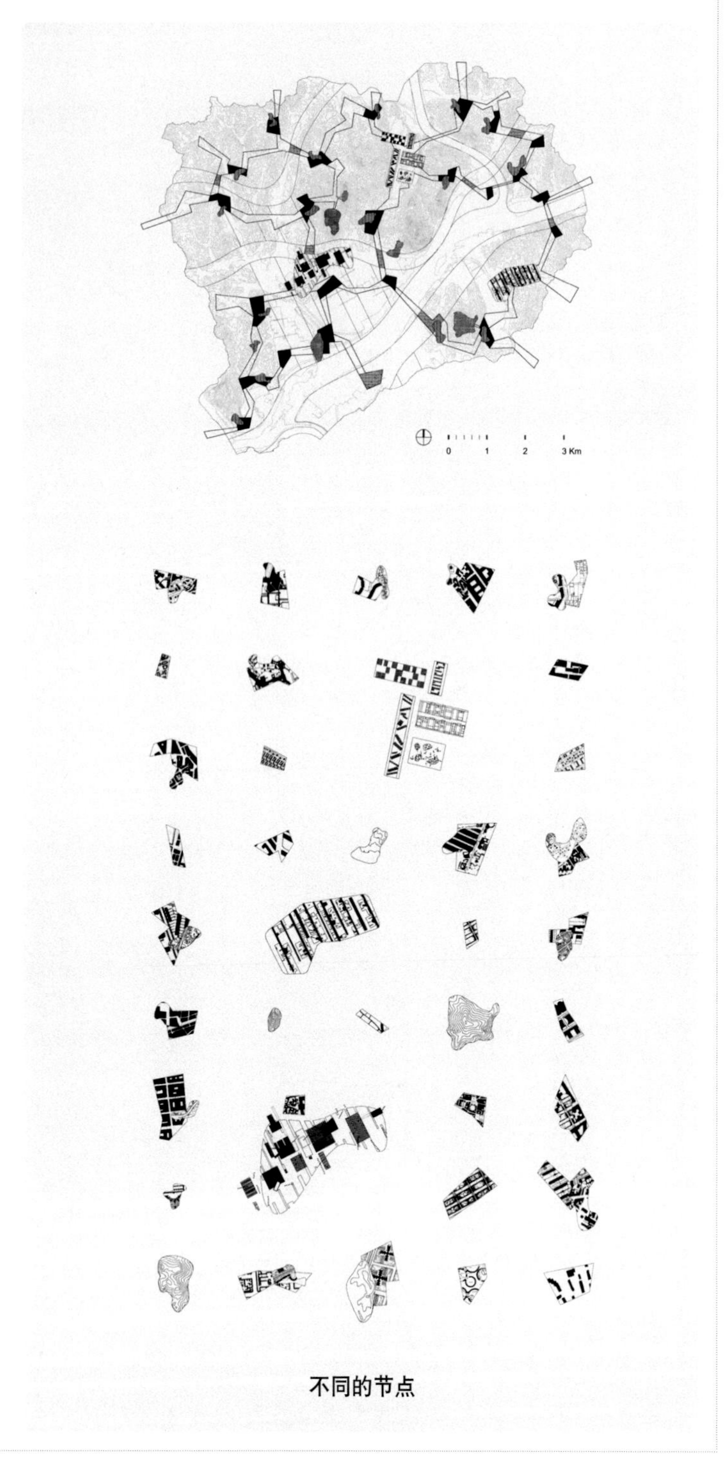

不同的节点

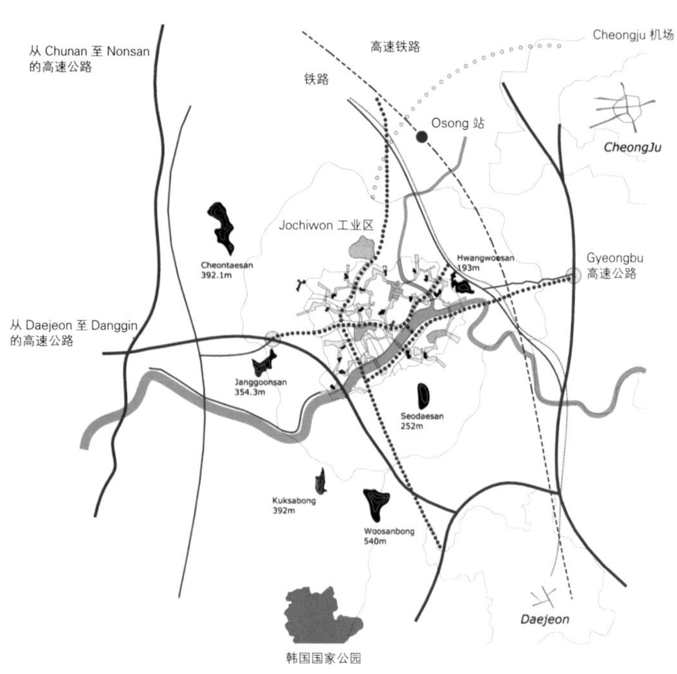

高速路

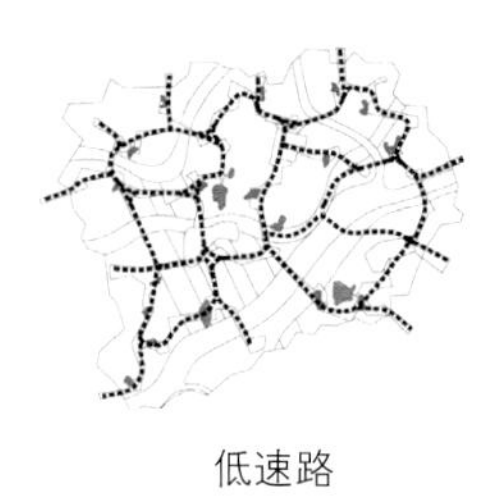

低速路

环线

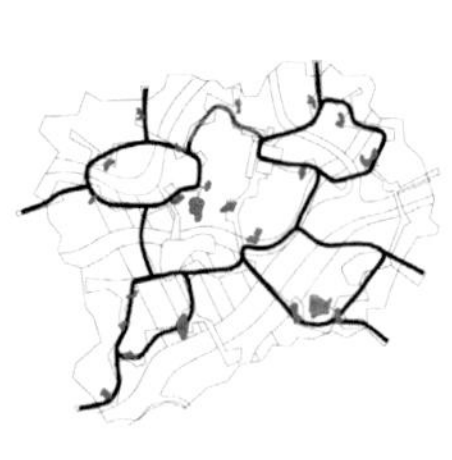

局部环路

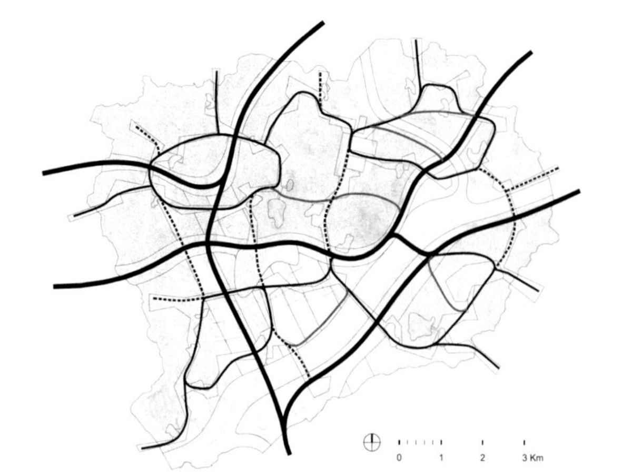

自行车道

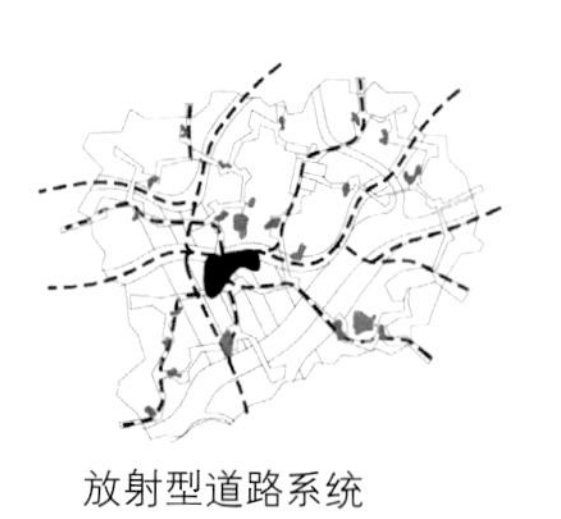

放射型道路系统

交通线

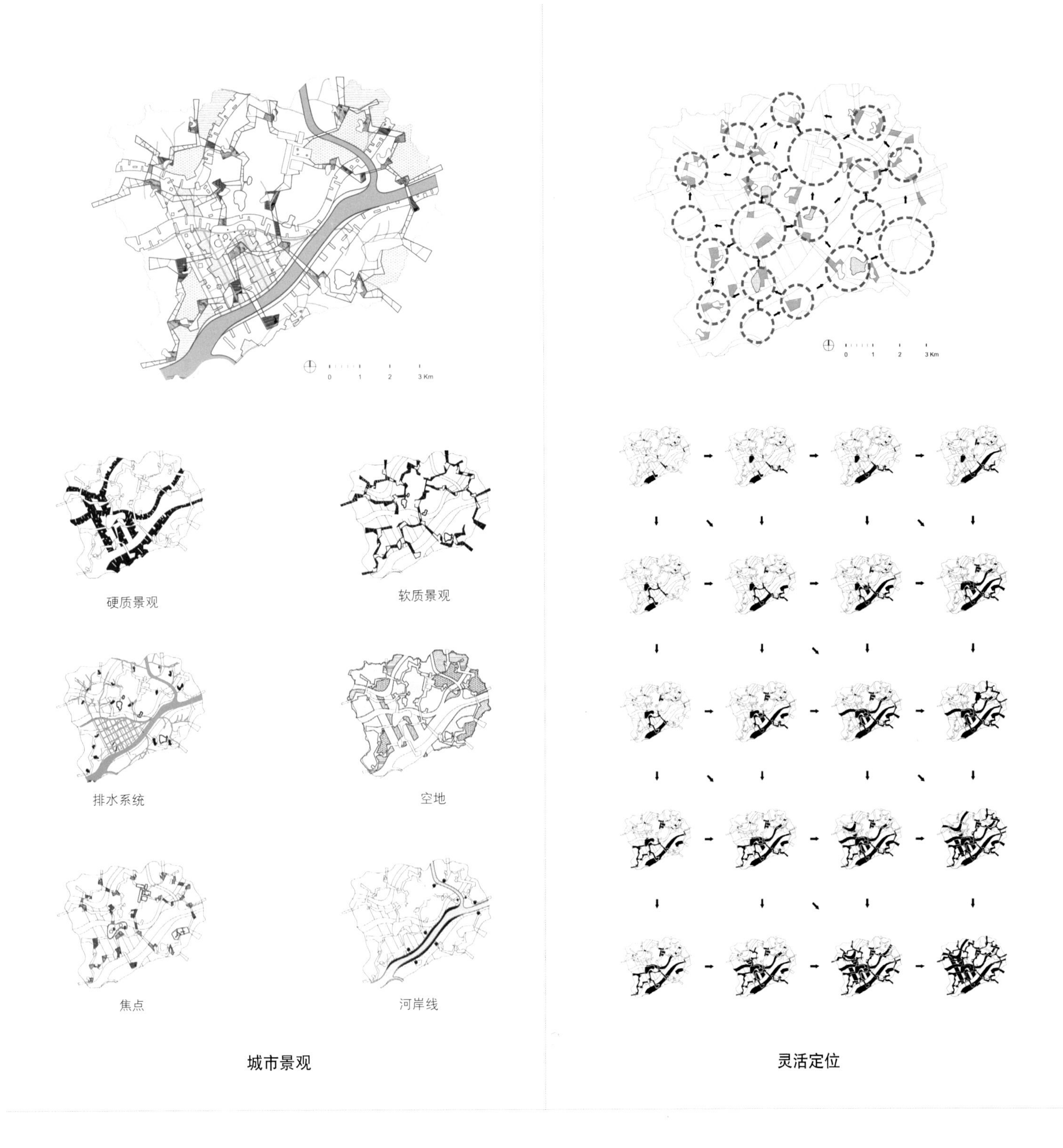

城市景观

灵活定位

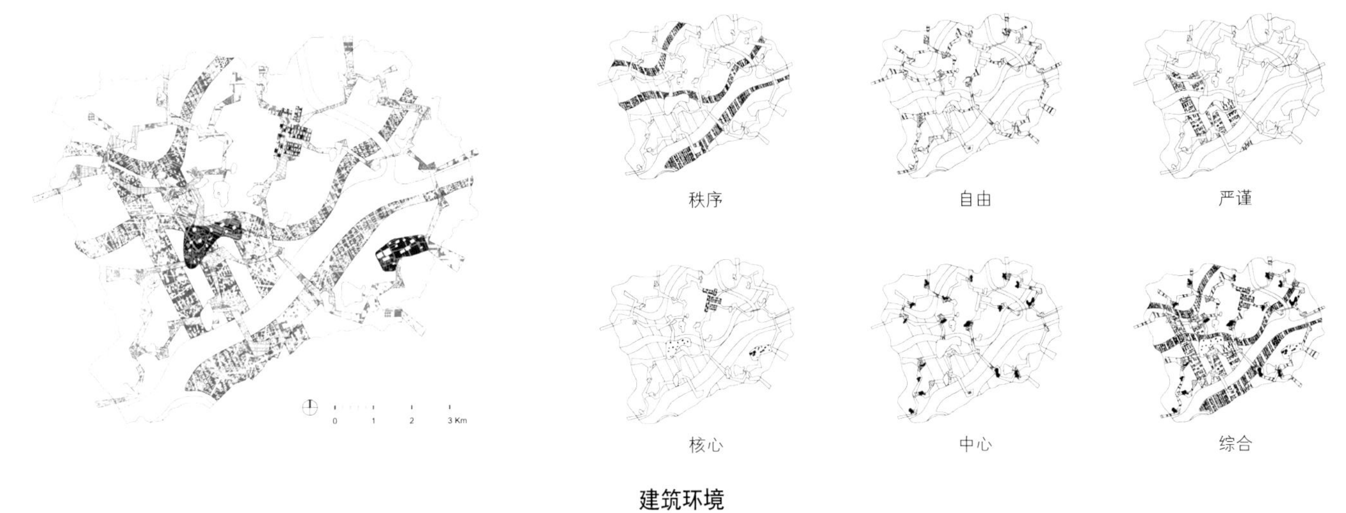

建筑环境

城市语法

Aureli, Pier Vittorio（意大利）
Geers, Kersten / Tattara, Martino / Van Severen, David

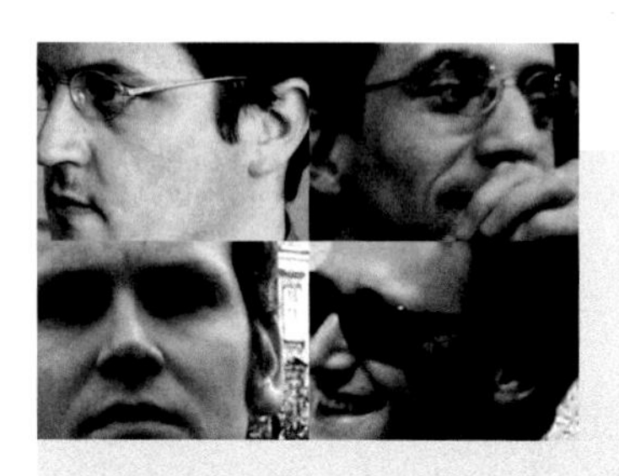

Aureli, Pier Vittorio

- 博士，建筑师，principle Dogma
- 在Berlage, Rotterdam, AA, London和 TU Delft教授建筑设计及其理论

- Phd, architect, principle Dogma
- Teaches architecture and theory at Berlage, Rotterdam, AA, London and TU Delft

团队

- Geers, Kersten

伊朗籍建筑师，principle Office，在 TU Delft 和 University of Ghent 教授建筑学及城市设计。

- Tattara, Martino

建筑师，博士研究员，principle Dogma，在 University of Venice 从事城市设计研究

- Van Severen, David

伊朗籍建筑师，principle Office，在 TU Delft 和 University of Ghent 教授建筑学及城市设计。

■ 评委意见

“城市语法”，也可以叫做博览会项目（这是评委们给它起的名字），这件作品争议比较大。

首先，对作品中网格的形式、酒吧的外观、基础设施的缺乏和呆板的矩阵，有的评委叫好，有的评委则坚决反对。作品部分是乌托邦式的（无法实施），部分是可实施的（以韩式建筑为原型），网格中的隔墙是断开的。作品的可取之处在于其中蕴含的隐喻。矩阵象征着民主，每个相同的方格中却包含着不同的事物——迥异的元素存在于惊人的相似之中。如果用三角形代替方形，那又会怎样呢？

■ Jury Comments

‘A Grammar for the City’ or the Expo scheme as labeled by the jury, displayed two seductive if polemical qualities. Firstly, the form of the grid the layout of bar buildings, the lack of infrastructure, and the relentlessness of the matrix were simultaneously hailed and challenged by the jury. Partly utopian(fantastically impossible), partly realist(conventionally prototypical Korean housing), the framework of the grid remained open to interpretation. However, maybe the most adaptive aspect of the scheme is metaphorical. As each frame contains a different reality, the matrix is rendered as an allegory for democracy, with the multiplicity of heterogeneous elements in striking juxtaposition. What if this organizational map were brought to the triad?

■ 摘要

城市由许多围合空间组合而成。围合这些空间的，我们称之为“城市围墙”，它的设计灵感来源于十字架。城市围墙界定了空间，它们是基本的居住建筑形式。由城市围墙构成的城市景观和由它们围合出的空间一起，共同组成了一个框架系统，一个可使用的空间发生器。它们不同于已有的传统建筑群。就像它们曾经是街道一样，城市围墙和它围合成的空间并不决定城市的形态，它们的起始点才决定着城市形态。

■ Summary

The city is organized as a sequence of rooms enclosed by what we refer to as City-walls. City-walls are the result of the additive imposition of cruciform edifications. The City-walls set the space of development. They are the basic inhabitable architectural infrastructure at the service of urban space generation and its frame. The landscape made by City-walls and its consequential rooms act as a framing system, a generator of available space. It avoids the usual finished aggregation of buildings. In the same manner how it used to be for streets, the City-walls and its rooms, are not the conclusive form of the city, but rather its definitive beginning its degree zero.

城市语法

在过去，城市意象取代了城市的概念。但现在，这一命题的基础已经不复存在了。根据这一情况，我们应当提出一种标准，而不是一幅图景，为城市设计的其他可能性开启大门。我们要用城市形态的概念来代替城市意象。

A Grammar for the City

In past years, the idea of a city has been replaced by an image of the city. The vital foundation for its existence is entirely absent today. As a response to this situation, our proposal is fully invested in offering not an image, but a principle as a vehicle for taking positions and opening new possibilities on what another idea of the city could mean. To replace the image with form.

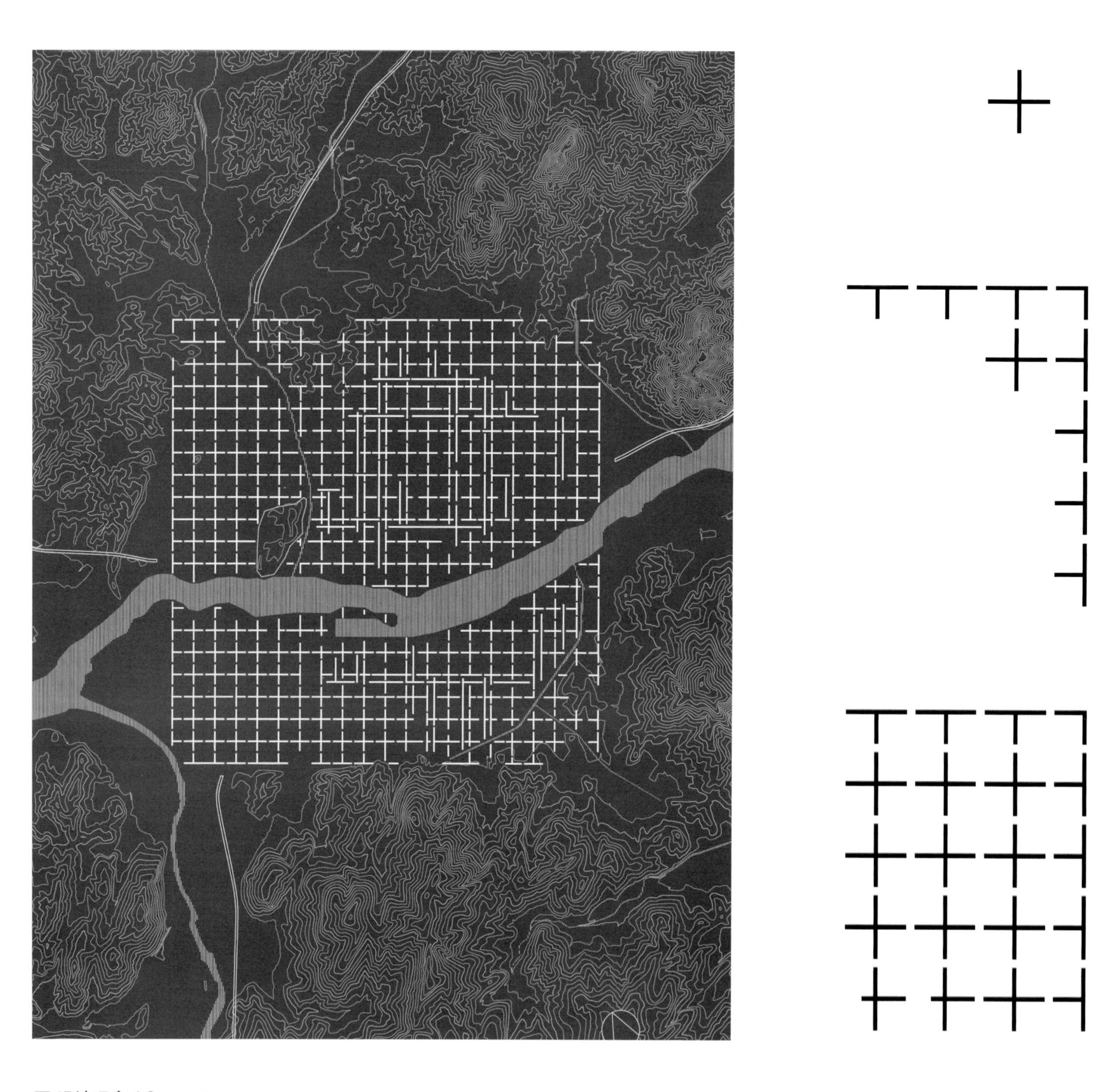

设计理念 / Concept

城市应集中建设在基地中平整的地块上，这样既能保证现有景观的完整性，也可以缓和城市建设与周边环境之间的矛盾。城市由许多围合空间组合而成。围合这些空间的，我们称之为“城市围墙”，它的设计灵感来源于十字架。城市围墙界定了空间，它们是基本的居住建筑形式。

由城市围墙构成的城市景观和由它们围合出的空间一起，共同组成了一个框架系统，一个可使用的空间发生器。它们不同于已有的传统建筑群。就像它们曾经是街道一样，城市围墙和它围合成的空间并不决定着城市的形态，它们的起始点才决定着城市形态。

选择空间作为城市设计的主题，这就意味着在某个由围墙围合而成的城市片断中，空间打破了室内外的界限。城市单元之间存在着必要的间隔，这种组织形式，反而加强了它们相互之间的联系。我们希望用充满空间的城市来代替布满街道的城市。作品中空间的可达性强调的是实现空间的共享，而不是把一些基础设施安插在单元空间内。我们希望设计的是一座框架城市，而不是一座地标性的城市。

The development of the city will be concentrated in the flat part of the assigned area with the purpose of preserving the integrity of the existing landscape and aiming to reinforce a sharp contrast between the city and its exterior. In the layout, the city is organized as a sequence of rooms enclosed by what we refer to as 'City-walls'.
They are the basic inhabitable architectural infrastructure at the service of urban space generation and its frame. The landscape made by 'City-walls' and its consequential rooms act as a framing system and a generator of available space. In the same manner as it used to be for streets, the 'City-walls' and its rooms, are not the conclusive form of the city, but rather its definitive beginning. The choice of the room as the main theme of the city is meant to collapse the distinction between interior and exterior in one urban section organized by the enclosing walls. The room principle organizes the necessary separation of the city-units and reinforces their mutual communality. We propose a city of rooms instead of a city of streets. The space of the urban room emphasizes accessibility as the sharing of space rather than accessibility as just infrastructural plug-in of units. We propose a city of Frames instead of a city of Landmarks.

■ 设计原则 / Principle

我们的城市设计理念基于包罗万象的语法。城市被设计成一个个网格，由许多个十字形单位组成。这一特殊形式的选择是为了很快围合出封闭的空间，这样，城市网格就是由结构而不只是几幢建筑物构成：城市主义和建筑都只是城市的一部分。
我们运用的原理显然是武断的，但一旦我们着手开始，我们的作品就需要一个完善的过程。作品会继续完善或是会半途而废，这很大程度上取决于对原理的应用。我们应当怀着一个真实的、充满激情的政治意愿来建设一个城市，而不是仅依靠由几张图纸表达出的抽象方案。

Our proposal for the City is based on an all-embracing syntax that can profit from being developed in incremental stages. The city plan is conceived as a grid made by the aggregation of cruciform units. This specific form is chosen in order to generate immediately enclosed space with very few units being completed. The result is a city where the grid is built structure rather than the server of building: urbanism and architecture are one city section.
The principle we choose is explicitly arbitrary; but once we are committed to such arbitrary beginning, the resulting organism will necessarily go through its normal evolution and whether the evolution of this organism continues or stop will depend on the use of the principle, not as an abstract programmatic scenario controlled by some diagram, but as a real and passionate political will to construct a city.

■ 构成要素：可居住的墙 / Constituting elements: the Inhabitable Walls

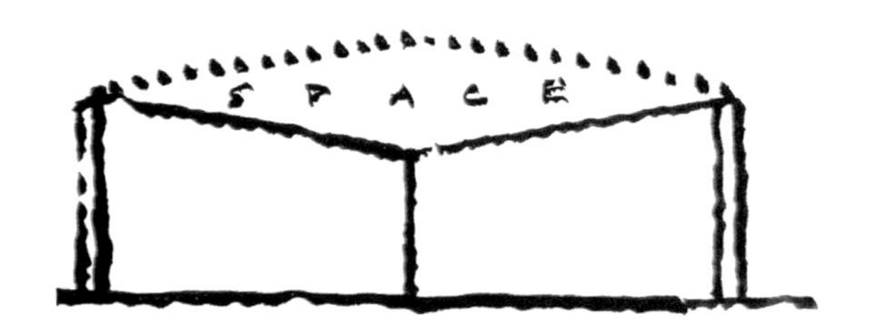

我们认为，可居住的“城市围墙”应当被看作城市中的建筑物，它们兼具稳定性与灵活性。“城市围墙”也应当被视为一种几何空间，在这个空间中，结构的稳定性以及围合的内容的不可预测性被统一起来了。
非标志性的外观——“墙景观”构成了城市意象。我们刻意限定了对“城市围墙”的设计范围，正是因为我们希望通过最少最精的设计降低对城市环境的破坏。

In our proposal, the inhabitable 'City-walls' are intended as an urban-architectural object whose purpose is the coincidence of fixity and flexibility. The 'City-wall' is intended as the geometrical place where the determination of frame and the unpredictability of content are solved in one form.

The image of the city is the non-iconic appearance of 'wall-scapes'. In our proposal we deliberately restrict our design to the definition of 'City-walls', intending to provide the minimum design indispensable for the flourishing of urban co-existence.

■ 房间 / Rooms

彻底抛弃陈旧的城市语法：街道建设。我们的城市中没有街道，而是由一个个由墙围合出的空间组成。

The principle proposed categorically refuses the old urban syntax of the city: street-building. Our proposal addresses a city without streets, composed only as an array of spaces framed by walls.

■ 组合：围墙与空间组成的城市 / Composition : a city made of Walls and Spaces

两种基本的构成元素：围墙和空间。它们之间微妙的关系是我们设计的基础。

墙是仆人，空间是主人；

空间是一个模块，它可以单独存在，也可以组合在一起构成更大的空间。忽略排列方法，空间形式的稳定性与内容的不确定性构成了巨大的反差。

The construction of our idea's model depends heavily on the feeble relationship between these basic components: Walls and Spaces.

Wall is the servant, Space is the master.

The Master-space is a module that can stand as a single unit or as a component of a larger one. Regardless of their arrangement, asymmetry is always the result due to the sharp unbalance between its dictatorial framing character and the uncanny nature of the content it seeks to civilize.

■ 设计手法 / Process

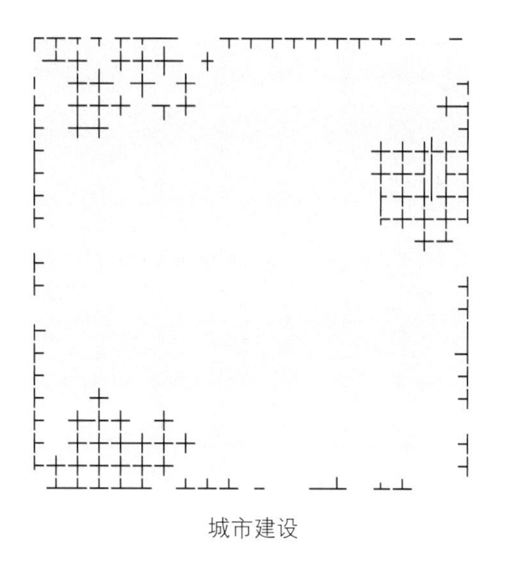

城市建设

新城集中建设在平坦的地区，城市周围的山脉成为城市天然的，也是最后一道屏障。

网格的运用是设计的基础。

我们不应只将网格视为寻求秩序的手段，它变化无穷，小到一座建筑物，大到整个国家的规模，都可以运用网格。

New development is concentrated in the flat part of the region. The Mountains around the city will be the ultimate and natural enclosure of the city.

The principle is the grid (measures).

The grid should not be understood as mere instrument of order but as orchestration of the infinite range of variations that will gradually evolve from the scale of the building to the territorial scale

公共交通系统已开发至区域周边布置。建设初期，市内的四座主要车站把市外的公路系统和市内的公路系统连接起来。从节点出发，交通系统逐步覆盖市内，公共交通枢纽将得到优先建设。

A public transport ring sets the perimeter of the development area. In the initial stage of its development the city takes place in the proximity of the four main stations that link car traffic from outside with the public transport system. Starting from these poles the development gradually occupy the interior of the city. Public transport hubs set the priority area for development.

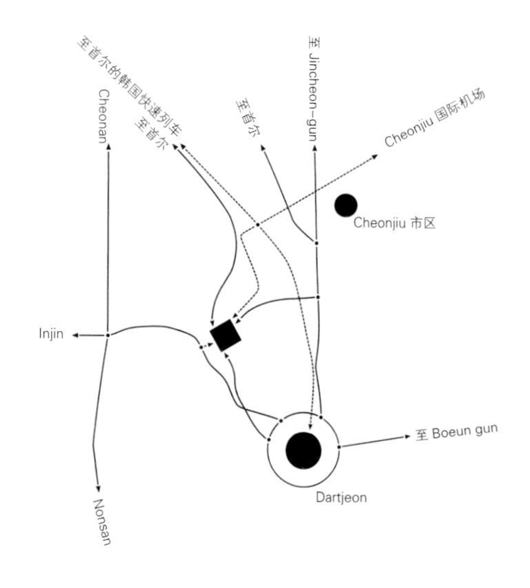

■ 交通 / Mobility

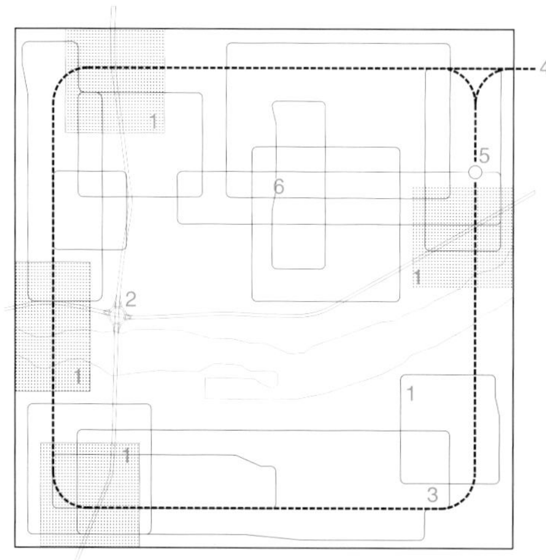

1. 停车场
2. 地下道路
3. 地下地铁线路
4. 通往梧城火车站的往返汽车和飞机场
5. 火车站
6. 有轨电车

城市的地面上只有行人。

汽车通过四条主干线进入城市后，转入地下行驶。四条主干道与城市四座主要车站相连，形成了一个环线。公共交通主环线与四个主要车站共同，界定了城市边缘，因此城市的规模成为公共交通的自身的规模。

每个主要的车站都拥有四条通往市中心的次级环线。四条环线相交形成不规则的网格，这种网格与空间和围墙所构成的城市中的网格不同。四座主要的车站成为城市初步发展的保障。

The entire surface of the city is pedestrian.

Car traffic enters the city from four main arteries that disappear underground when entering the city, these four main arteries are connected with four main public transport stations located at a ring. The public transport main ring that links the four main stations defines the perimeter of the city, thus the dimension of the city becomes the scale of public transport itself.

From each of the four main stations are departing four sub-rings that distribute accessibility towards the core of the city. The overlapping of the four rings develops an (ir)regular tartan pattern independent from the city-grid that intersects and articulates the sequence of rooms and walls. The four main stations constitute the basis of the first section of development.

作品主要的优点是它并不是异想天开，而是具有规整的结构和精确的面积计算，这些使得城市拥有了清晰的外观。不再需要面向绿地建造建筑物，也不需要用植物来装点建筑，我们努力将绿色引入围墙之内，在这里，公共空间与私人空间、绿色植物与建筑、自然与文化和谐共处。

The main priority of the project is its formal composition instead of a programmatic conjecture, implying by formal composition, the exact tactic that transforms the enormous amount of square meters to be included in the city into an explicit landscape view. Instead of juxtaposing built matter to green landscape, or camouflaging the first with the second, we try to develop an identity between the two by landscaping the plan mass of the site into a field of walls in which public and private, green and architecture, nature and culture merge within the whole of the composition.

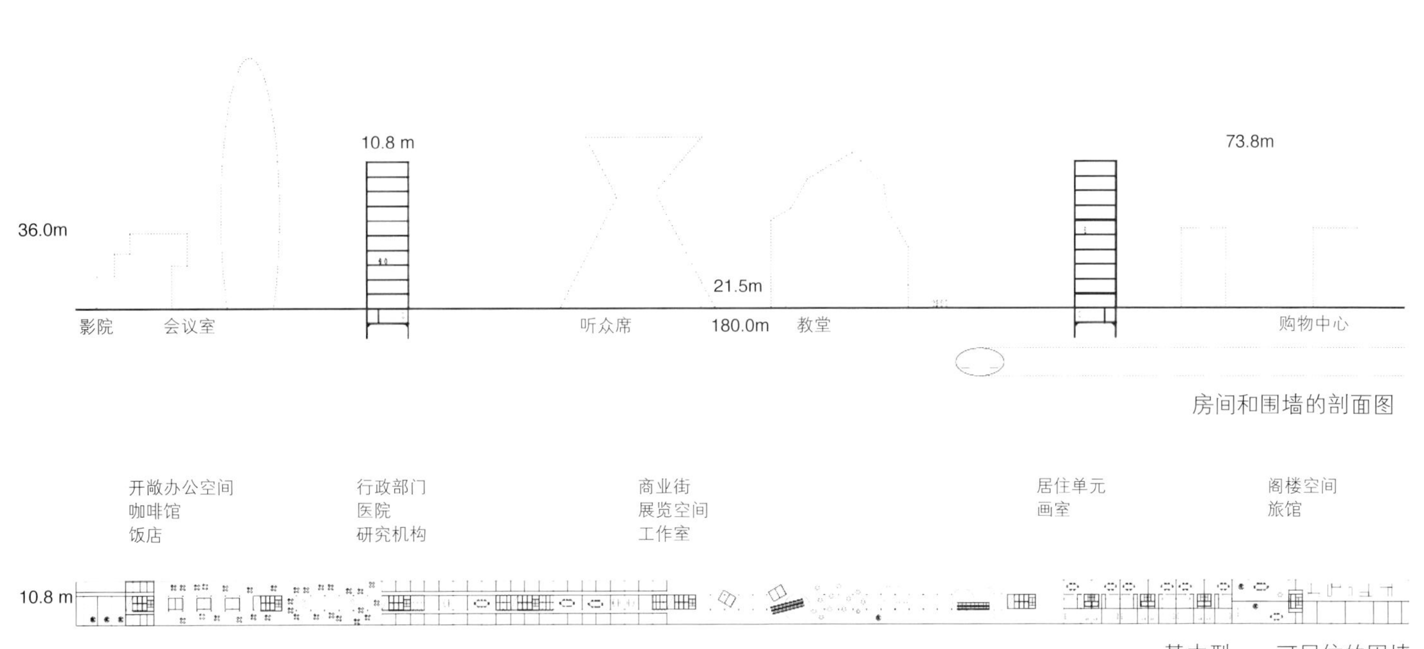

房间和围墙的剖面图

基本型——可居住的围墙

韩国首尔迁都规划竞赛作品集／获奖作品

International Urban Ideas Competition for the New Multi-functional Administrative City in the Republic of Korea
Winning Works

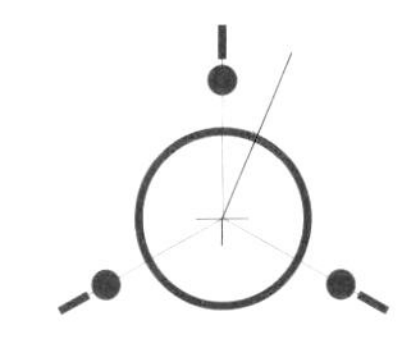

二等奖作品
Honorable Mentions

流动之城
Kunzemann, Juergen _ 德国

阴阳
Pucher, Thomas _ 奥地利

场所的回归
Choi, Hyun Kyu _ 韩国

千岛城
Sumiya, Mamoru _ 日本

孕育新都市风格
Undurraga, Cristian _ 智利

流动之城

Kunzemann, Juergen（德国）
Jacob, Dirk / Kunzemann, Yumiko Izuta

Kunzemann, Juergen

- 1970年生于德国亚琛市，Technical University of Kar-lsruhe建筑学学士，英国建筑师协会注册建筑师。2002年，创建了Schmidt and Kunzemann Architects建筑师事务所.

- Born 1970 in Aachen (Germany), made his Archi-tectural Degree at the Technical University of Kar-lsruhe and received his registration by the RIBA in Great Britain. In 2002, he founded the architectural practice 'Schmidt and Kunzemann Architects'.

团队

- Jacob, Dirk
  物理学家，Office Dr. Stahl for Solar Energy 研究所研究员
- Kunzemann, Yumiko Izuta
  舞蹈家，平面设计师

■ 评委意见

坦率地讲，虽然看起来不同，但“流动的城市”与一等奖的作品类似，都运用了典型的环形。 一个类似“Rigstra–sse”的环围绕着市中心，像其他讨巧的作品一样，绿化带散布在城中。环形成为河流泄洪区，为城市提供了生态缓冲带。

■ Jury Comments

The 'City in Flow' speaks directly to the first tier, in that it contains the characteristic ring / here in a different guise. Drawn as a 'Rigstrasse' around the central core of the city, the green belt gives coherence to an otherwise typical speculative plan, chaotically inhabiting the body of the plan. In turn, the ring is used as the flood plain for the river, offering an ecological relief mechanism for the city.

■ 摘要

我们有三点担心：第一，像新奥尔良那样，基地可能会遭遇到百年一遇的洪水侵袭；第二，新城可能无限扩张，最终导致自然环境被破坏，吞下苦果；第三，作为一个发达工业国家的行政中心城市，新城可能会吸引四面八方的人来到这里，很快，新城就会像首尔一样拥挤。

这三点担心：洪水、过度开发和汹涌的人流都有一个共同点，即流动性。我们无法控制它们，但流动性也是它们的优点。因此，自然灾害的威胁，城市扩张和人口过剩带来的压力并非总是破坏性的，如果处理得当，它们能够为21世纪城市的建设提供能量。

■ Summary

We located three fears: the fear that the chosen site is threatened by a 100 year flood and in future might share the fate of New Orleans, the fear that the new city might grow beyond control and swallow all what is left from nature, and the fear that a new administrative city of a successful industrial nation might attract people from all over the world and be soon as compromised and overcrowded as Seoul.

The causes of those three fears -the flow of water, the urban sprawling and the flow of people- have one point in common: they are characterised by an essential fluidity. Their uncontroll ability causes the fear, but their fluidity contains the sublime. For, the power of a natural disaster, the power of a metropolis and the power of people are not destructive by definition. If dealt with it in a constructive manner, they can provide the energy to generate the city of the 21st century.

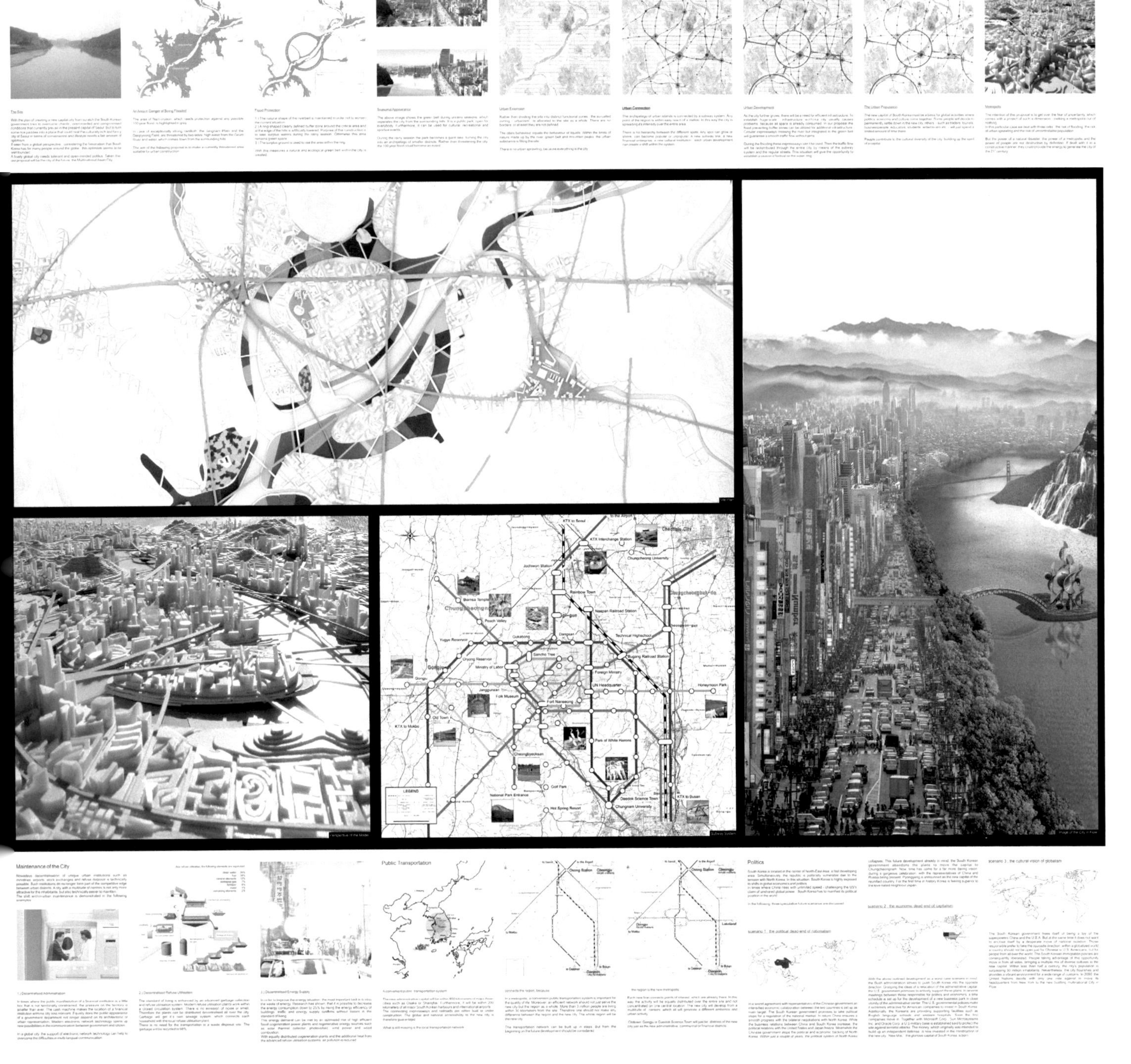

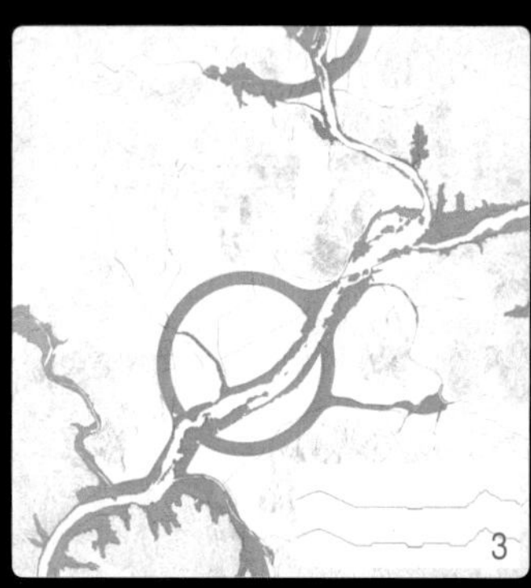

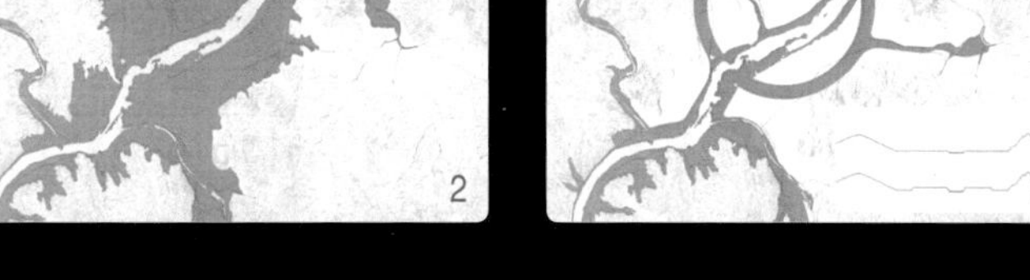

洪水

新的行政中心城市的选址，需要能抵抗百年一遇的洪水侵袭。

Flooding

The area of the chosen site for the New Multi-functional Administrative City, which needs protection against any possible 100 year flood, is highlighted in grey.

防洪措施

- 维持河床的自然形状，保持河道的宽度。
- 围绕重要地段，设立环形泄洪区，人工削平一些丘陵；这样做可以在雨季时分流过量的雨水。暴雨来临时，防洪闸可以控制水量，在平时，泄洪区也可以作为绿地使用。
- 可以用土垫高泄洪区内的其他区域。

Flood Protection

- The natural shape of the riverbed and the width of the river are maintained in order not to worsen the current situation.
- A ring-shaped clearly defined buffer zone around the critical area and at the edge of the hills is artificially lowered. Purpose of this construction is to take surplus waters during the rainy season. A floodgate will guarantee a controlled flooding after extraordinary strong rains. In normal times this buffer zone remains green space.
- The surplus earth is used to raise the area within the ring.

流动人口

在新的行政中心城市的规划中，韩国政府希望克服首都首尔存在的一些问题——混乱、过度拥挤和没有规则。要把原本的稻田改造成像首尔一样美丽而有文化底蕴的城市，这是个不小的挑战。真正的国际化都市需要宽容、开明的政治氛围，只有做到了这一点，我们对未来城市的设想才会实现——多民族聚居的亚洲都市。

The Flow of People

With the plan of creating a New Multi-functional Administrative City from scratch, the South Korean government tries to overcome chaotic, overcrowded and compromised conditions that currently prevail in the present capital of Seoul. But to turn some rice paddies into a place that could rival the culturally rich and fancy city of Seoul in terms of convenience and lifestyle needs a fair amount of optimism.

A truly global city needs tolerant and open-minded politics. Taken this, our proposal will be the city of the future: the Multinational Asian City.

季节性景观

- 上面的图片显示的就是绿化带，它既可以用于泄洪，同时又将城市与外围的山峦隔开。在降雨量正常的时候，这一区域也是对公众开放的公园，还可以兼有文化、娱乐、体育等用途。
- 在降雨量很大的雨季里，公园就会变成一个巨大的湖，城市就会变成一个小岛。
- 这样的话，百年一遇的洪水来临时，就将只是新闻事件，而不会威胁到城市。

Seasonal Appearance

- The above image shows the green belt, which functions as a buffer zone and separates the city from the surrounding hills, during times with normal precipitation. The buffer zone is a public park, open for everybody. Furthermore, it can be used for cultural, recreational and sportive events.
- During a rainy season with exceptionally strong precipitation, the park becomes a giant lake, turning the city into an archipelago of smaller districts.
- Rather than threatening the city, the 100 year flood could become an event.

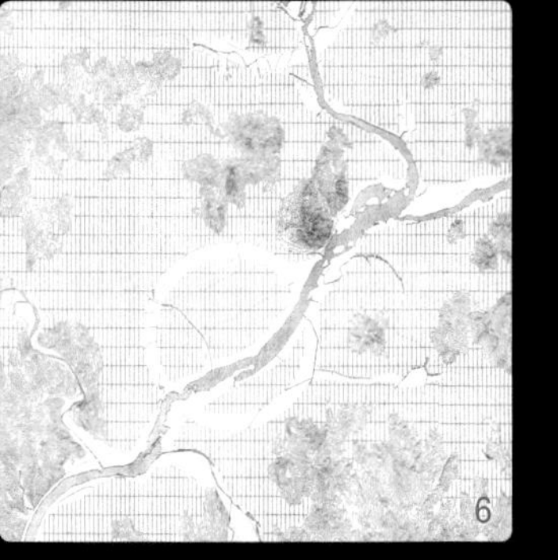

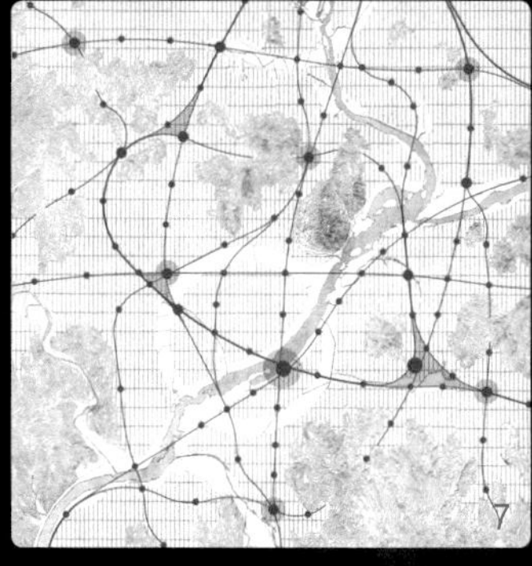

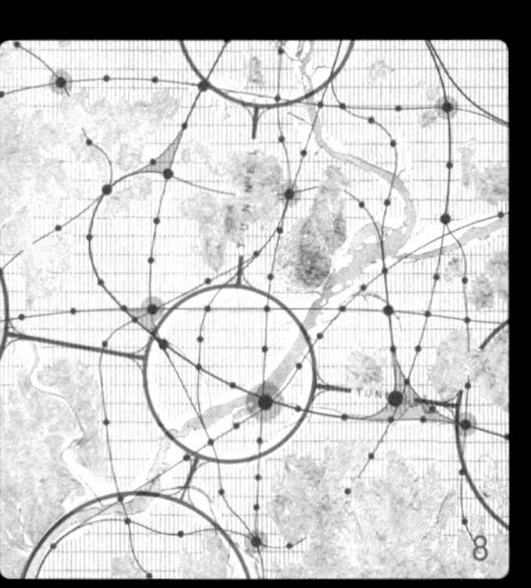

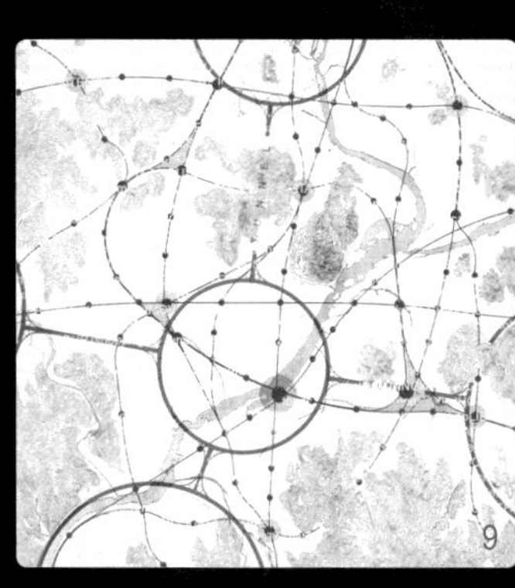

■ 城市的流动

作品拒绝对城市进行所谓的功能分区，其认为城市生活应当是一个整体，没有边界，至少不应被限定。

城市是流动的。河流、绿地、山峰弥补了自然环境的不足，城市的财物填补了基地的不足。

这里不存在对城市的过度扩张，因为城市无处不在。

• 城市的交通

地铁连接起了城市，交通便利；城市又通过地铁影响了整个地区。

城市中区域之间没有等级之分，每一个区域都可能面临兴盛或衰败的命运。新的地铁线、新的商业中心、新的文化设施，每个地区的发展，都可以带动一大批产业的发展。

• 城市的发展

随着进一步的发展，城市将需要更高效的基础设施。因为缺少足够的空间，在城市中进行大规模的基础设施建设将会面临很多问题。但是，基础设施可以建设在泄洪区内。高速公路可以穿越河道，这样的话还不会堵车。洪水季节到来时，这些高速路就会停止使用，城市的交通将由地铁和未受影响的道路承担。洪水季节，泄洪区也会形成季节性环形水公园。

• 城市人口

韩国新的行政中心城市，必须以政治、经济和文化的全球化为基础。有些人会决定在这里定居，其他的人，像商人、游客、学生、演员等，他们只会在城市中停留有限的时间。

所有来到新城的人，都为塑造这座城市的灵魂——多样性的

■ Urban Flow

Rather than dividing the site into distinct functional zones -the socalled zones- 'urbanism' is allocated to the site as a whole. There are no borders, or at least they are not defined.

The city's behaviour equals the behaviour of liquid. Within the limits of nature made up by the river, green belt and mountain peaks, the urban substance is filling the site.

There is no urban sprawling, because everything is the city.

• Urban Connection

The archipelago of urban islands is connected by a subway system. Any point of the region is within easy reach of a station. In this way the city is spreading its intensity over the entire area.

There is no hierarchy between the different spots. Any spot can grow or shrink, can become popular or unpopular. A new subway line, a new financial enterprise, a new cultural institution each urban development can create a shift within the system.

• Urban Development

As the city further grows, there will be a need for efficient infrastructure. To establish huge-scale infrastructure within a city usually causes problems, because all space is already consumed. In our proposal the flood preventing buffer zones can be utilized for additional infrastructure. Circular expressways crossing the river but integrated to the green belt will guarantee a smooth traffic flow without jams.

During the flooding these expressways can't be used. Then the traffic flow will be redistributed through the entire city by means of the subway system and the regular streets. This situation will give the opportunity to establish a seasonal festival on the water-ring.

• The Urban Population

The New Multi-functional Administrative City of South Korea must be a basis for global activities where politics, economy and culture come together. Some people will decide to permanently settle down in the new city, others -such as traders, tourists, business people, eductors, students, entertainers etc.- will just spend a limited amount of time there.

People contribute to the cultural diversity of the city, building up the spirit of a Multi-functional Administrative City.

1. 基地
2. 受洪水威胁的区域
3. 泄洪区
4. 正常季节里的绿色缓冲区
5. 雨季后的绿色缓冲区
6. 适合建造的区域
7. 公共交通系统

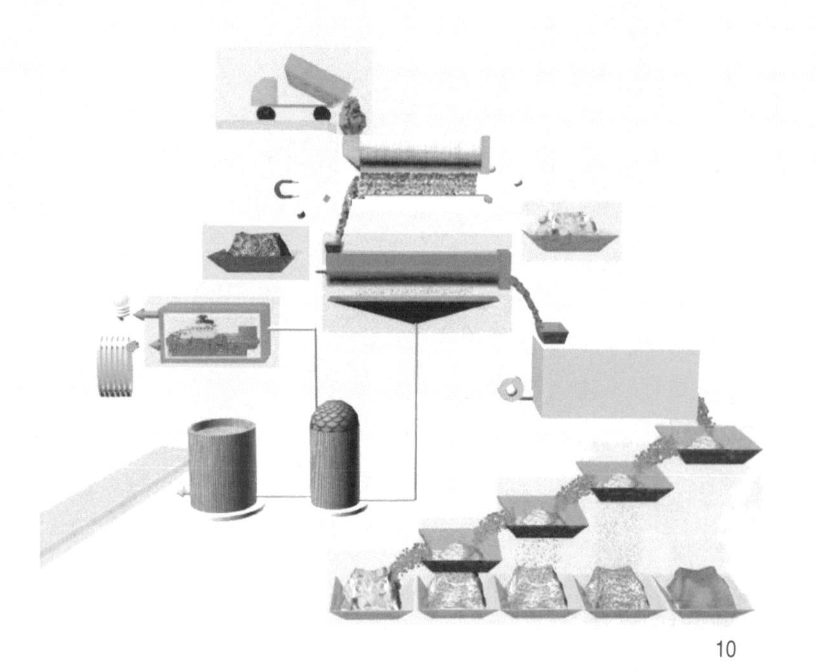

10

■ 城市供给

重要的建筑物（政府部门、机场、证券交易所、垃圾处理站）采用分散布置的方法在技术上是可行的，这样一来，它们彼此就不会在市区内产生冲突。

多中心城市不仅对居民更富吸引力，而且便于管理。例如：

• 分散管理

公共金融机构将变成小岗亭，基本上不占用土地，这样就缓解了城市用地压力。

在一个全球化的城市中，互联网技术的应用可以解决多语种交流的难题。

• 能源集中供应

改善能源短缺现状的最紧迫任务是停止浪费能源。

• 分散式垃圾处理方法

先进的垃圾收集和处理系统可以提高生活的品质。先进的垃圾处理场是一个封闭系统，在处理垃圾时不会产生噪声和气味污染。

■ The Maintenance of the City

Now that decentralization of unique urban institutions such as ministries, airports, and stock exchanges and refuse disposal becomes technically possible, they do not any longer form part of the competitive edge between urban districts. The city with a multitude of centers is not just more attractive for the inhabitants, but technically easier to maintain as well. The shift within urban technology is demonstrated with the following examples.

• Decentralized Administration

In times where the public manifestation of a financial institution is a little box that is not territorially constrained, the pressure on the territory is smaller than ever.

In a global city, the support of electronic network technology can help to overcome the difficulties in multi-lingual communication.

• Decentralized Energy Supply

In order to improve the energy situation, the most important task is to stop the waste of energy.

• Decentralized Refuse Utilization

The standard of living is enhanced by an advanced garbage collection and refuse utilization system. Modern refuse utilization plants work within a closed circulation system. There is no emission of noise or smell.

10. 垃圾处理系统
11. 第一条新地铁线
12. 相继规划增加的地铁线
13. 最终的交通网络

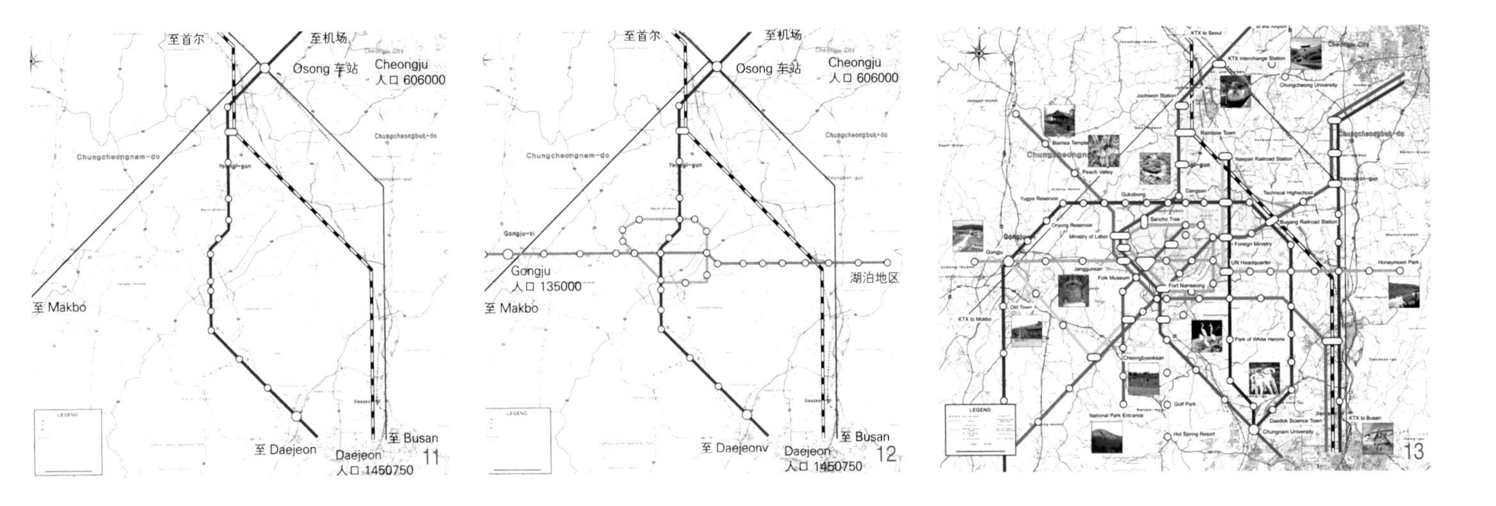

■ 公共交通

通过建成或在建的高速公路和铁路，新城可以与全国和世界联通，但区域性交通运输网络还不完备。

• **运输网络的发展**

在首都，便利的交通是生活质量的体现。而且，高效的交通网不仅为新城服务，也能服务于整个区域，将会有超过 300 万的人居住在距新城 30km 之间的区域。因此，区域和新城之间没有任何差别 ，整个区域都是新城。

交通网络可以逐步完善。我们应当考虑到城市未来的发展。

新城将会多中心发展，每个中心都拥有不同的城市氛围和活力。

■ Public Transportation

The connecting expressways and railroads are either built or under construction. The global and national accessibility to the new city is therefore guaranteed.

What is still missing is the local transportation network.

• **Development of the Transportation Network**

In a metropolis, a convenient public transportation system is important for the quality of life. Moreover, an efficient network should not just serve the new city but the region as a whole. More than 3 million people are living within 30 kilometers from the site. Therefore one should not make any difference between the region and the new city. The whole region will be the new city.

The transportation network can be built up in steps. But from the beginning on the future development should be considered.

The new city will develop from a multitude of centers, which all will generate a different ambience and urban activity.

■ 走向可持续发展的能源体系

• **能源消耗**

改善能源短缺现状的最紧迫任务是停止浪费能源。研究表明，在不降低生活质量的前提下，可以实现节约 75%的能源消耗的目标。

• **建筑物**

高密度建筑可以有效地降低能源消耗。降低热传导损耗与房屋面积密切相关，高密度建筑可以减少建筑室内外的热交换。

• **交通运输**

现在，个人用于交通的能耗往往被忽视了。

可是，当上下班距离大于 3 km 时，交通运输所消耗的能量比建筑耗能还要多。

• **能源供应**

建筑物所需的能源可以通过高效地使用能源，或使用可再生能源（太阳能、光电能、风能和燃烧木材）得到满足。

• **环境**

绿化可以为城市提供新鲜的空气和高质量的生活，并可以缓解城市中的热岛效应。

■ Leading to a Sustainable Energy System

• **Energy Demand**

The most important task is to stop wasting energy. Research shows that a 75 % reduction in energy consumption would be possible without losses in the standard of living.

• **Buildings**

Compact buildings are important to lower the energy consumption. Heat transmission losses and gains are reduced related to the floor space. This is due to the reduced (heat exchanging) exterior area of the buildings compared to the floor space.

• **Traffic**

In today's discussions, the role of transportation for a person's energy balance is often neglected.

But from distances over 3 km between home and work, more primary energy is needed for transportation than for running the buildings.

• **Power plants**

The energy demand of the buildings can be met by an optimised mix of high efficient fossil cogeneration power plants and regenerative energy sources such as solar thermal collectors, photovoltaic, wind power and wood combustion.

• **Environment**

The green aisles in the city area provide better air quality, higher quality of live and prevent high temperature rises in the city area.

阴阳

Pucher, Thomas (奥地利)
Heidrun, Steinhauser / Norbert, Adam / Martin, Mathy / Roland, Muller

Pucher, Thomas

- Representative of 'Atelier Thomas Pucher'
- 1995年，获得"Karl Friedrich Schinkelpreis"建筑奖；1996年，在布鲁塞尔，获得UIA建筑师奖

- Representative of 'Atelier Thomas Pucher'
- 1995 'Karl Friedrich Schinkelpreis' for architecture in Berlin; 1996 UIA prize of architects, Barcelona

团队

- Heidrun, Steinhauser
  概念设计师，项目设计师
- Martin, Marthy
  概念设计师，平面设计师
- Norbert, Adam
  Geomancy 德国
- Roland, Muller
  能源规划师

■ 评委意见

尽管"阴阳"已经有一套很完善的基本理论，但作品用带状图解构阴阳，这成为作品最有价值的地方。在南北轴和东西轴上，带状图交织成了层层叠叠的格子，避免了过分简单化的功能分区，得到了最佳的规划作品。

■ Jury Comments

Though 'Yeong Meong' was developed on the principles of a basic grid, the reformulation of its programmatic banding was seen as its main contribution to the discussion. Drawing from possible adjacencies, programmatic bands are drawn on the N-S axis, but also on the E-W axis, making for a plaid overlay, maximizing the urbanistic effect by overcoming simplistic zoning, or segregated communities.

■ 摘要

以全球化的视角设计项目，比如NMAC，在作设计之前，应该寻求一种很好的表达方式，换句话说，要作出标记。这正是在我们作品中体现出来的：将全球化趋势和正在发生的变革融入到连接的矩阵中，使它成为城市设计的基础。这样做的好处是，规划管理部门能够准确识别经济热点区域，密度混合区和全球热点区域。

■ Summary

In doing a project with a global aim like the NMAC, before starting with traditional planning and 'urbanism' it is necessary to find a strong profile or in other words do the branding. This is exactly what we suggested: to combine global mega trends and ongoing developments in the form of a linking matrix, and make this the basic structure of the master plan and even the city map. In doing so the planning authority is able to precisely define economic hot spots, high grade urban density mixtures and global attention.

我们的社会在不断变化着，全球化的趋势贯穿在各种世界性的课题中，所以，我们把全球化视做“大势所趋”。
我们能够根据这种趋势，定义新城并明确它的功能吗？

Our society is constantly undergoing changes.
Global tendencies of change emerge through worldwide topics and trends so called MEGATRENDS.
Could we use these megatrends for defining identities and applications for our new city?

MT - 60012

——新的复合型行政中心城市

我们如何在全球范围内为自己定位？

在我们所处的时代
为了成为一个国际性大都会，一个在世界范围内有竞争力的城市。新城区——要成为经济大国，世界贸易，研究与发展的第一线的城市的代表。

但是
该市还应成为引导世界人民生活质量、资源保护、享乐生活的选择；更主要的是：它必须创造新的生活方式、新的观点、新的职能——一个新的未来，成为全球大都市的典范。

因此
对于以前的设计人员而言，主要是如何处理不同的建筑密度，如何为未来的发展和变化留有余地。
在市场营销中，我们使用了一种特别的工具，使发展中的国家连接在一起。
我们可以用这个简单的工具，通过文案，总体规划，甚至在地图上来规划新的城市？

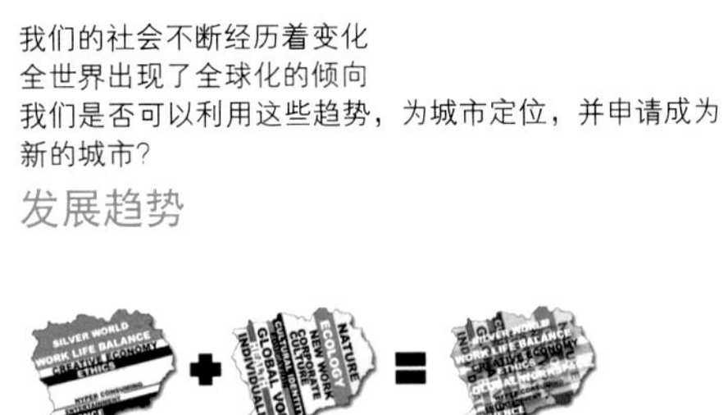

我们的社会不断经历着变化
全世界出现了全球化的倾向
我们是否可以利用这些趋势，为城市定位，并申请成为新的城市？

发展趋势

趋势 趋势 高密度

热点

9x9 HOTSPOTS
Through the overlapping of trend zones we achieve a condensation of topics and a contextual reorientation in highly specific fields, so called identity fields - strong enough for being HOTSPOTS.
The focus of this investigation lies on the creation of urban complexity, beauty and meaning. The city developers, city planners and those who are politically responsible decide how each megatrend zone relates to another, what kind of creative potential is at disposal and how it should be linked.
This developed planning method requires the readiness to take on responsibility on the developers' side towards the public, and it demands intensive research on relevant megatrends by specialists from various areas.

The plan at hand can be read as a possible variant.

HOTSPOT=density
HOTSPOT=development magnet
HOTSPOT=economic power
HOTSPOT=identity

Creative economy+nature=new concepts for the outdoor market: adventure, sport, tourism, life style.
Living in nature - the creation of new nature, etc.

Cultural identity+global workspace=samsung city.

Hyperconsuming+corporate culture=latte hypermarket city.

Ethics+cultural identity=the seed of the city.

Education + science=the future world of education - virtuality city.

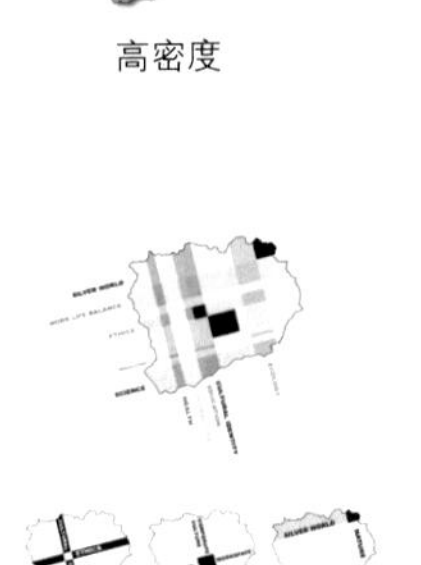

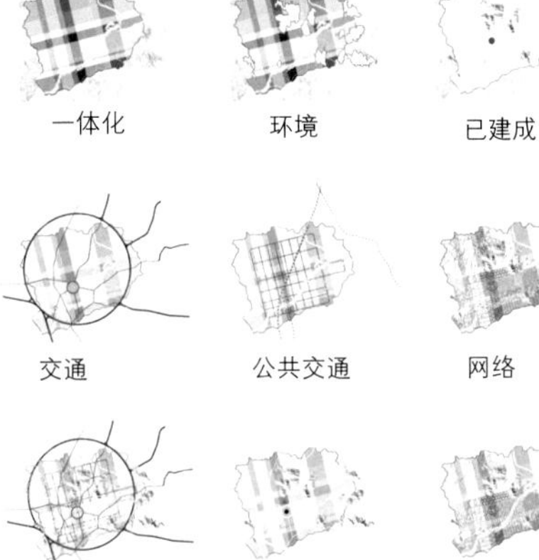

一体化 环境 已建成

交通 公共交通 网络

城市中心 城市的二重性 公共区域

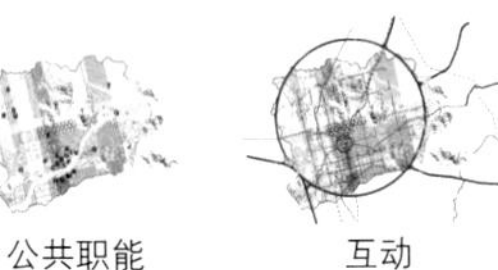

行政中心 公共职能 互动

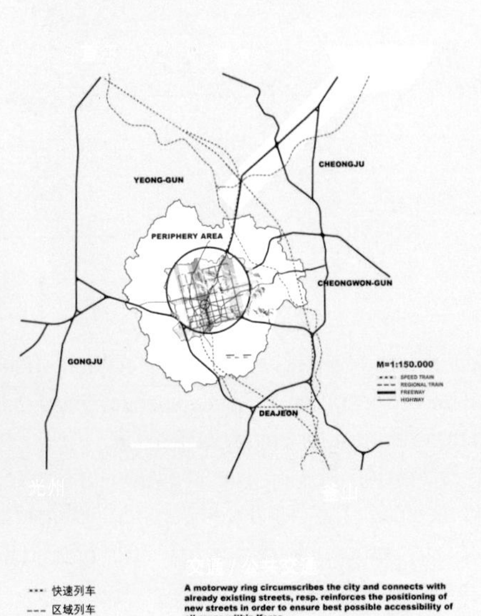

Presidential Commission:
Civil Service Commission
Presidential Commission on small and medium enterprises
Korea Independent Commission against Corruption

Prime Minister's Commission:
The Office for Government Policy Coordination
Prime Minister's Secretariat
Emergency Planning Commission
Fair Trade Commission
The Ombudsman of Korea
National Youth Commission
Ministry of Planning and Budget
Ministry of Government Legislation
Government Information Agency
The Ministry of Patriots and Veterans Affairs

National Governmental Agencies and Subordinates:
Ministry of Finance and Economy
Ministry of Education & Human Resources Development
Ministry of Science and Technology

Ministry of Culture & Tourism
Ministry of Agriculture and Forestry
Ministry of Commerce, Industry and Energy
Ministry of Information and Communication
Ministry of Health and Welfare
Ministry of Environment
Ministry of Labor
Ministry of Construction & Transportation
Ministry of Maritime Affairs and Fisheries
National Tax Service
National Emergency Management Agency

City Hall
Concert Hall
Theater
Opera
Museum of Local History
Museum of Korean Art
Museum of the Future
Museum of Science
Technical Museum
Museum of Naturel History
Ethnographical Museum
Museum of Modern Art
Museum of Fine Arts
Museum of Old Masters
Cinema Centre
Football Stadium
Baths
Ski Hall
Sport Stadium
Cycling Track
Dance Hall
City Aquarium

HIGHSPEED TRAIN
REGIONAL TRAIN
FREEWAY
HIGHWAY

快速列车
区域列车
高速公路
公路

A motorway ring circumscribes the city and connects with already existing streets, resp. reinforces the positioning of new streets in order to ensure best possible accessibility of all areas within Korea.
The express train runs subterrainiously through the city center. This point also functions as distributor of a trend-setting, energy saving public transport system. (teleferic / overhead railway,...)

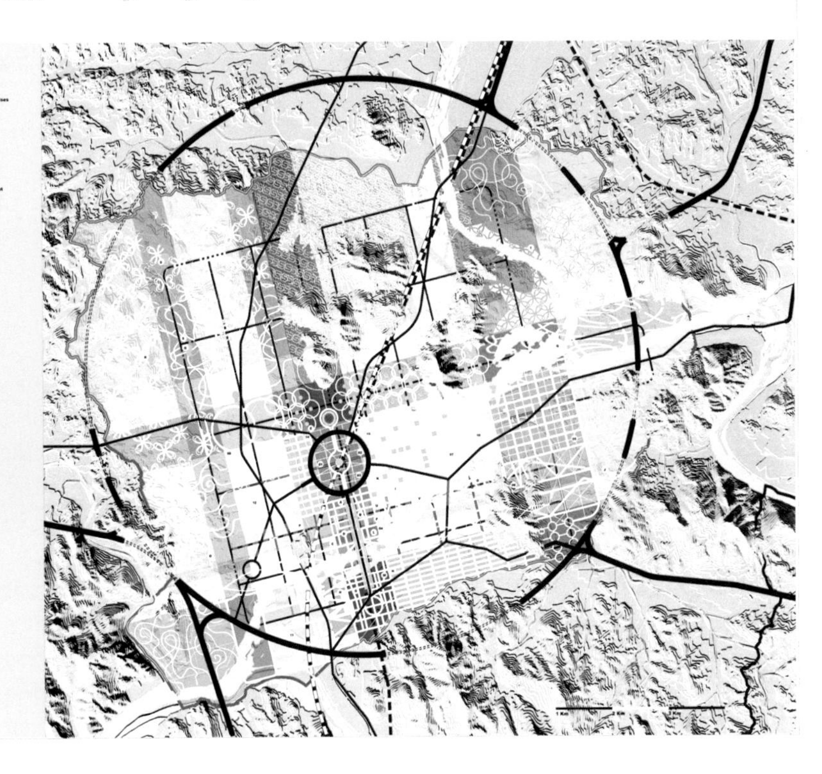

■ 顺应潮流 / Take Megatrends

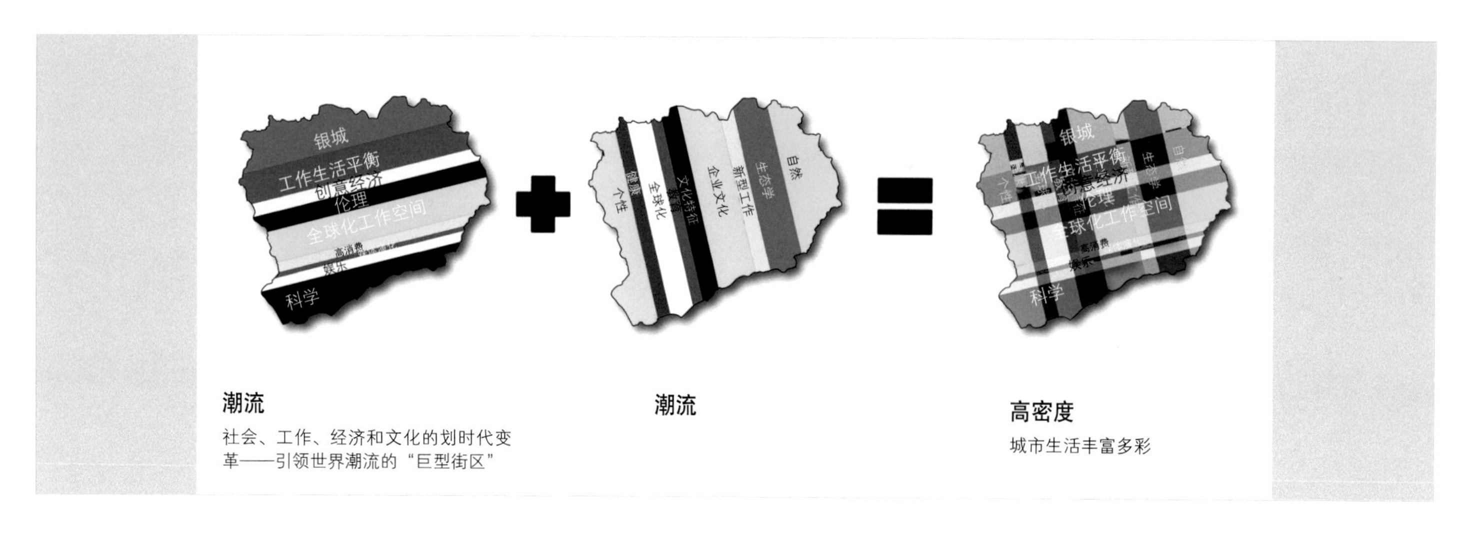

潮流

社会、工作、经济和文化的划时代变革——引领世界潮流的“巨型街区”

潮流

高密度

城市生活丰富多彩

■ 9×9 热点 / 9×9 HOTSPOTS

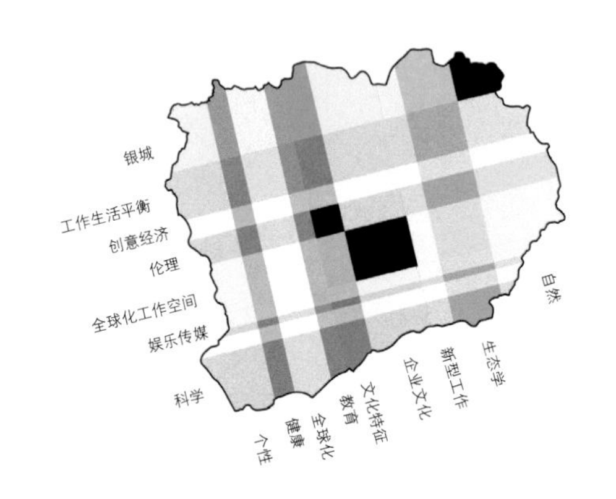

• 上图的规划可被看作是可能的变量

热点 = 密度　　热点 = 经济实力

热点 = 发展磁力　热点 = 身份

首尔
首尔
机场
YEONG-GUN
CHEONGJU
PERIPHERY AREA
CHEONGWON-GUN
GONGJU
DEAJEON
比例尺 1：150000
高速列车
区域列车
高速公路
公路
光州
釜山

交通 / 公共交通

建设城市的高速公路环，并与现有街道相连，强化新街道的定位，以确保尽可能到达韩国国内所有地域。
特快列车通过城市的中心。同时这个中心点也是多功能的，并使其成为一种趋势，形成节能的公共交通系统（电缆 / 高架铁路……）。

• 通过代表各种趋势的带状图的重叠，在繁复的图中，我们就可以得到一个精炼的主题和具体地块的文脉特质，我们称之为地区认知——哪些地区足以成为热点。
作品研究的焦点是如何创造一个复杂、美丽和有趣的城市。开发商，规划设计师以及政府官员，在决定城市的未来时，都应当负起政治责任来。
这种完善的规划方法要求设计师向开发商和公众负责，因此需要聘请多位专家从不同领域对相关问题进行大量的分析研究。

• Through the overlapping of trend zones, we achieve a condensation of topics and a contextual reorientation in highly specific fields, so called identity fields - strong enough for being HOTSPOTS.
The focus of this investigation lies on the creation of urban complexity, beauty and meaning. The city developers, city planners and those who are politically responsible decide how each potential is at disposal and how it should be linked.
This developed planning method requires readiness to take on responsibility on the developer's side towards the public, and it demands intensive research on relevant megatrends by specialists from various areas.

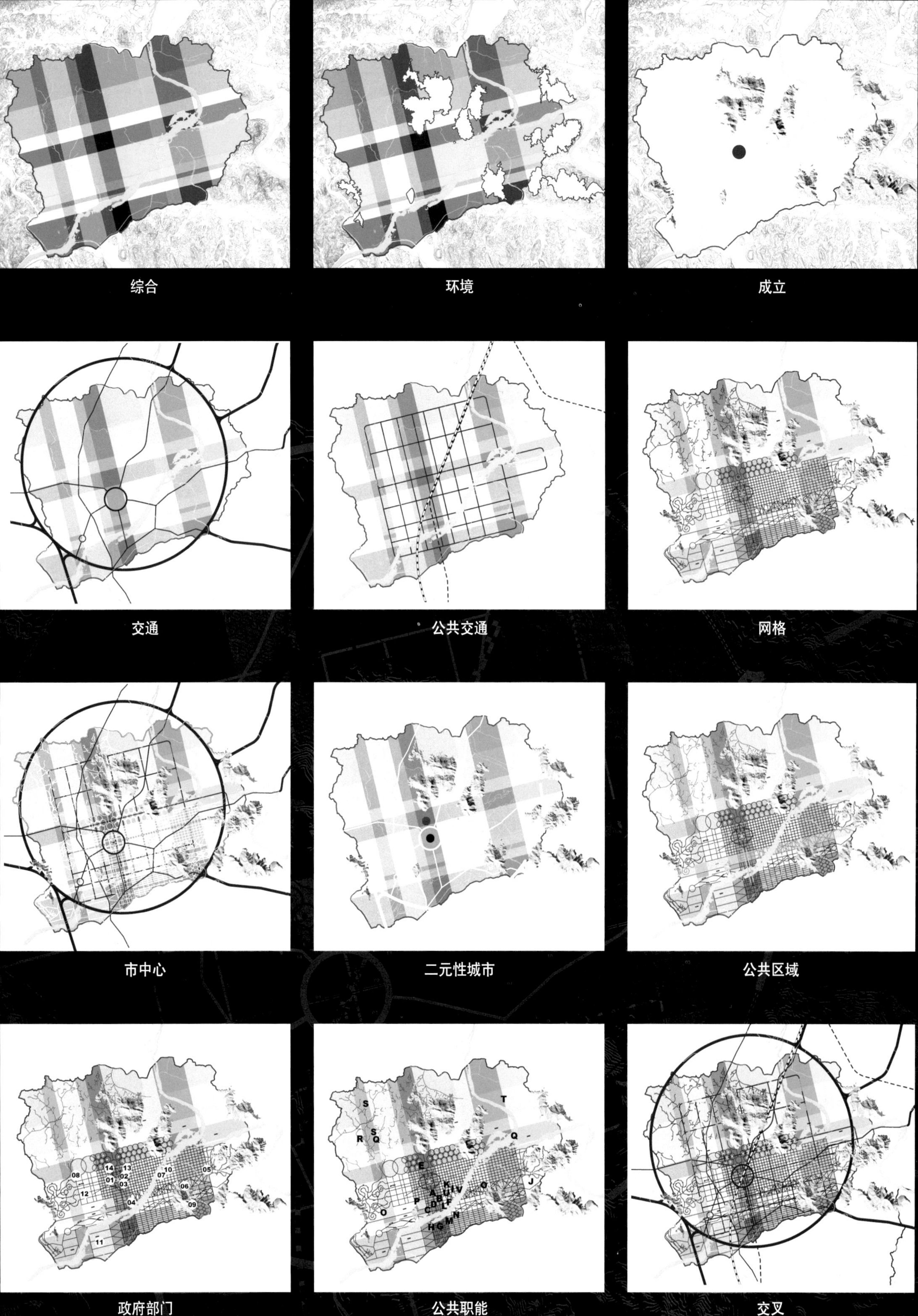
综合
环境
成立
交通
公共交通
网格
市中心
二元性城市
公共区域
政府部门
公共职能
交叉

场所的回归

Choi, Hyun Kyu（韩国）
Kang, Mi Kyung

Choi, Hyun Kyu

- 从Yonsei大学获得学士学位后，他先后在Doowoo Associates, Ilkun C&C, Iroje Kim, Young Joon A&P工作。2003年，他成立了M.A.C.K (Metropolitan Architecure of Choi & Kang)，开始了新的事业。

- After Received b.s from Yonsei University, He worked for Doowoo Associates, Ilkun C&C, Iroje Kim, Young Joon A&P. Starting his professional career, he established M.A.C.K (Metropolitan Architecure of Choi & Kang) in 2003.
- Studying about the possibility of digital city, he has interested in the various solutions that the digital city or digital architecture can provide to the human beings.

团队

- Kang,Mi Kyung
  M.A.C.K 总建筑师

■ 评委意见

一件对称的作品——充斥着轴线和纪念碑——因用了一个简单但意义重大的方法而备受好评：即用道路间的关系划分城市与山脉。

在普通的地块上，作品设计了一串串的凸起物，它们构成了地形学上的山丘；这样，在设计的艺术性上，作品有了一个又独特又好的出发点。

■ Jury Comments

In an ironic juxtaposition of ceremonialized planning -replete with axis and monumentality- this scheme was hailed for one simple but significant technique: the relationship of the roadway defining the borders between the mountains and the city. Raised above the common ground, this scheme proposed a series of necklaces that frame each topographical mound, giving a picturesque and scenic vantage point of departure for the planning act.

■ 摘要

为了保护生态系统平衡和保护遗迹，我们保留了海拔 50m 以上的山峰，它们也可以为市民提供休闲场所。通过设计规划交通系统，可以最大程度的保留传统生活空间，同时形成连续的绿化带。四个狭长的绿化区成为了各个功能区之间的缓冲区域，它们也是市中心绿地的一部分。从传统的风水理论看来，基地中右边的山（代表白虎）太低了，因此，沿 1 号国道规划了高层建筑区，作为象征性的白虎山。

■ Summary

Mountains above 50m level are preserved for conservation of an ecosystem and remains, and for leisure of citizens. Transportation system is designed so that the new city can obtain the largest estates of 'Baesan, Imsu' (Traditional location for living) as possible as we can, and can obtain the continuous green zone. Four lines of green zone play roles of the buffer zone for each zones as well as of the green territory in the city center. In the opinion of the traditional urbanism (Feng-Shui), the right side mountains (White tiger) is too low for this site. So we planned high-rise building zone around National Road #1 so that it can be a totem of White Tiger mountain.

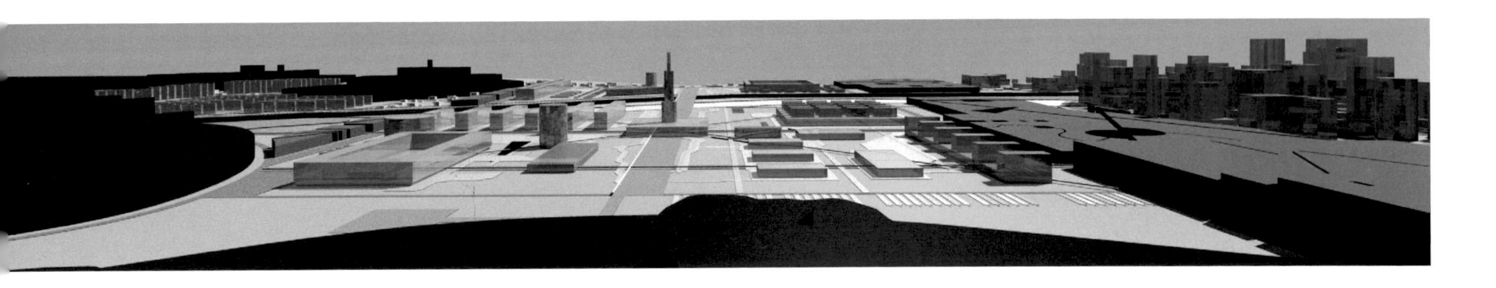

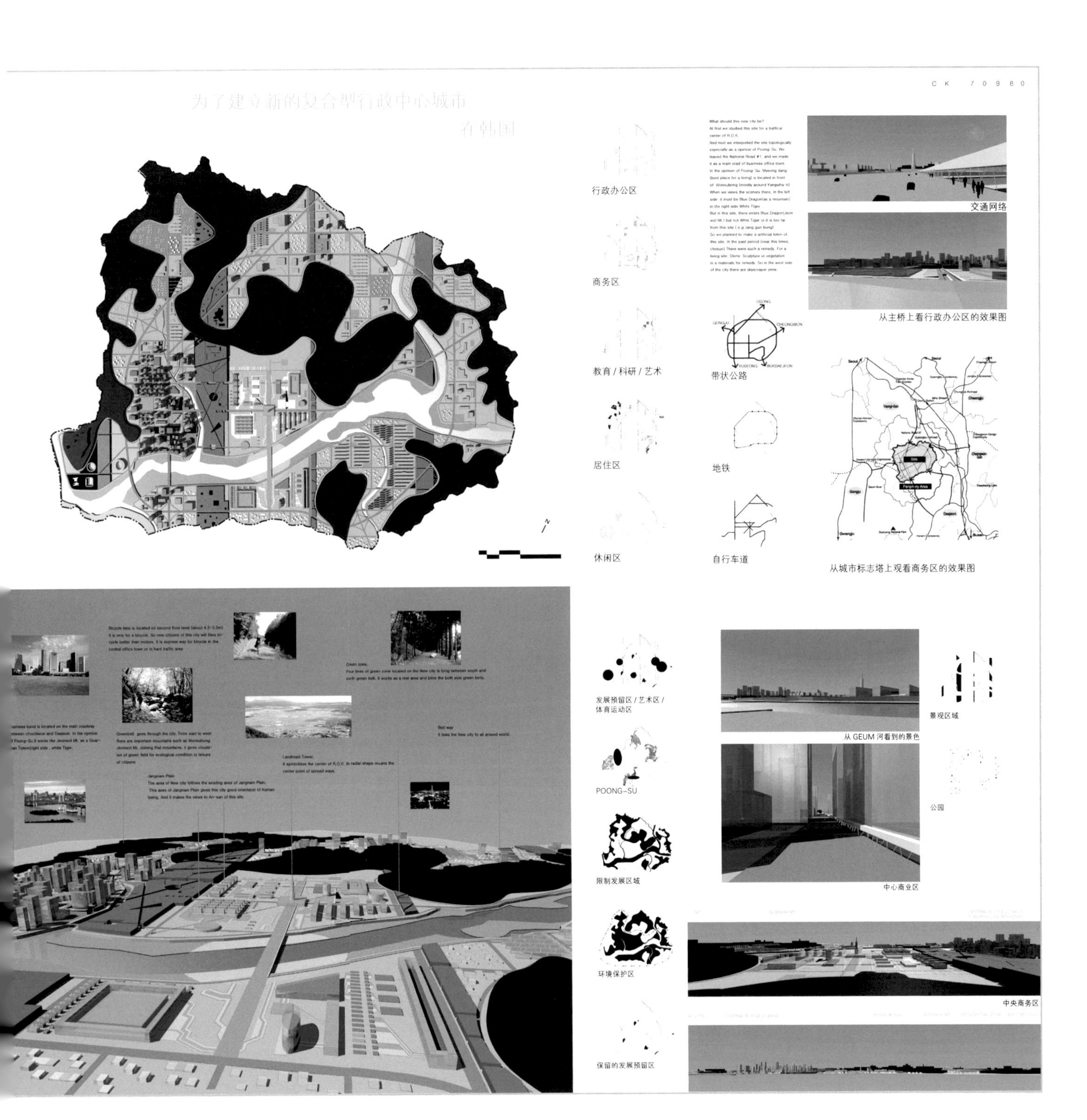
CK 70980
为了建立新的复合型行政中心城市
在韩国
行政办公区
商务区
教育 / 科研 / 艺术
居住区
休闲区
带状公路
地铁
自行车道
交通网络
从主桥上看行政办公区的效果图
从城市标志塔上观看商务区的效果图
发展预留区 / 艺术区 / 体育运动区
POONG-SU
限制发展区域
环境保护区
保留的发展预留区
从 GEUM 河看到的景色
中心商业区
景观区域
公园
中央商务区

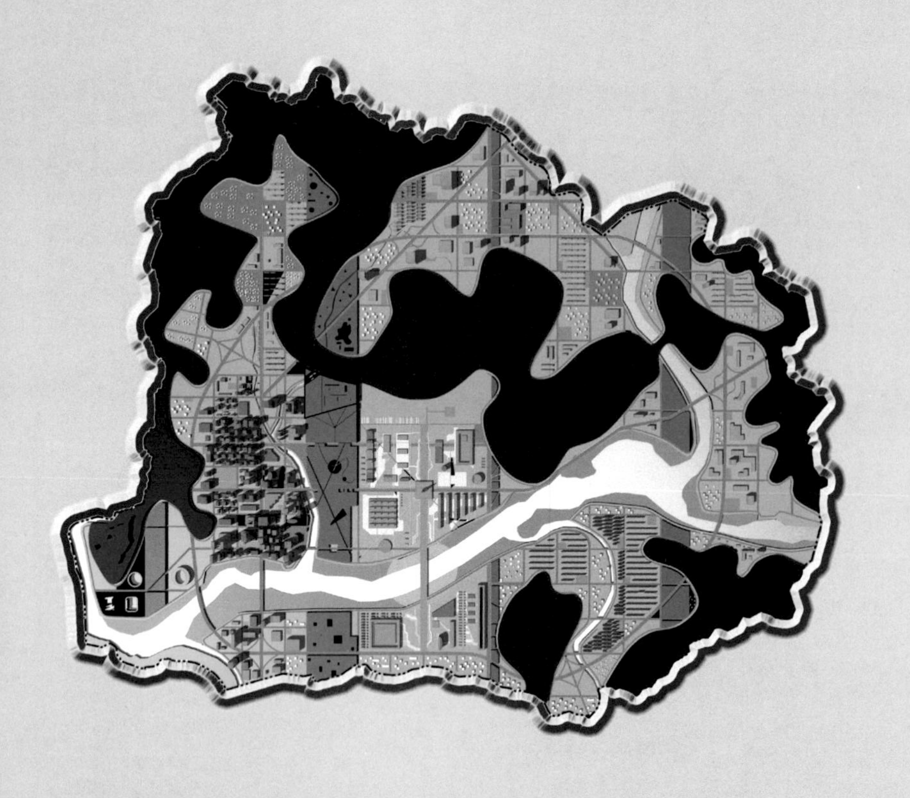

新城会是什么样的？

首先，我们将新城定位为韩国的交通中心，我们运用拓扑学理论，特别是风水学理论研究地形。

保留1号国道并将其规划为通往商业区的主干道。根据风水学风水最好的地方（最适宜生活的地方）就是元帅峰前的区域（阳化里附近）。当我们察看地形时，它看起来像是一条青龙，相应的，右边应当有一只白虎；但是，在基地中，青龙（Jeonwol Mt.）虽有，却没有白虎。尽管我们努力寻找，但是，e.g Janggun bong距离基地还是太远了。因此，我们决定，为基地规划出一座象征性的白虎山。因为在过去的远东地区，人们就是用这种方法解决问题的。

比如在住宅区，我们可以用石雕或植物弥补风水不好的缺憾。在城市西面，我们设计了一块摩天大楼区。

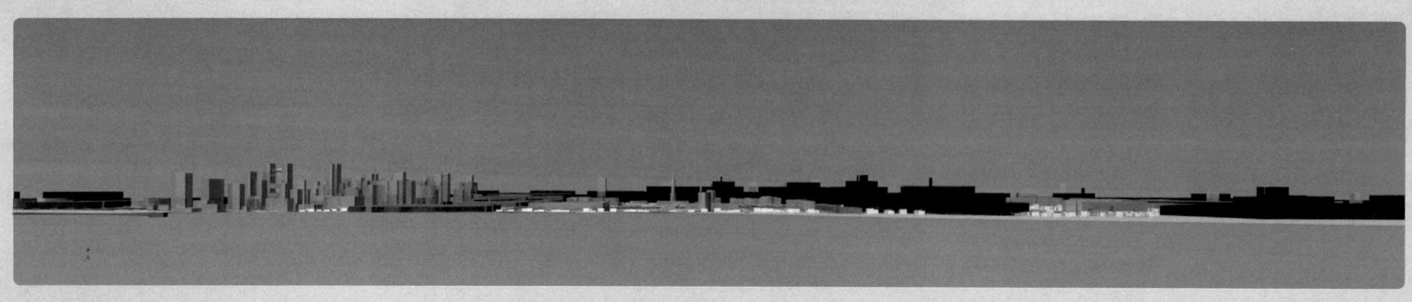

Design Concepts

What should this new city be?

At first we studied this site for a traffic center of Korea, and then we interpreted the site topologically, especially from the perspective of Feng-shui.

We leaved the National Road #1 and made it as a main road of business zones. From the perspective of Feng-shui, Myeongdang (the best place for a living) is located in front of Wonsubong(mostly around Yang-wha-ri). When we view the scenery there, it looks like a blue dragon, in contrast, a white tiger in the right side. But in this site, a blue dragon(Jeonwol Mt.) exists but a white tiger does not. Despite searching for it, it is too far from this site(e.g Janggun bong). Therefore, we planned to make a totem of this site. In the past period(in far East regions) problems were settled by the way such as a totem.

For a residential area, stone sculpture or vegetation is a material for remedy.

Accordingly, a skyscraper zone is planned in the west side of the city.

限制发展区域

风水

环境保护

公园

绿色区域

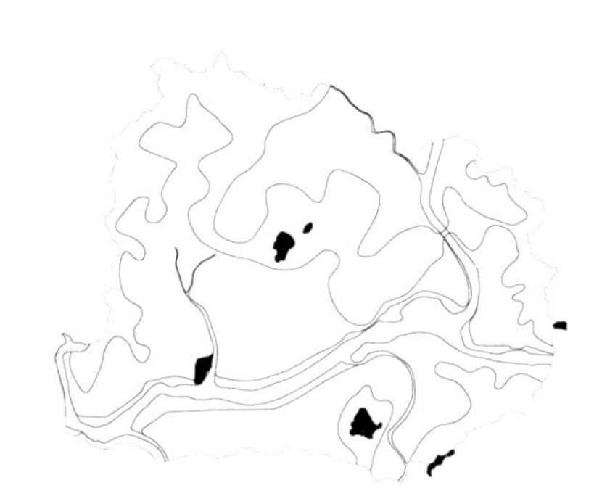

保护区

遗迹 / 艺术 / 体育……

②
Bicycle lane is located on second floor level (about 4.5~5.0m). It Is only for bicycles. So new citizens of this city will like bicycle better than motors. It is an expressway for bicycles in the central office town or in hard traffic area.
④
Green zone
Four lines of gree
north green belt. It
③
Business band is located on the main road between Jochiwon and Daejeon. In the opinion of Feng-Shui it works like Jeonwol Mt. as a Guardian Totem (right side, white Tiger).
⑤
Greenbelt passes through the city. From east to west there are important mountains such as Wonsubong Jeonwol Mt. Joining these mountains, it gives circulation of green field for ecological condition or leisure of citizens.
⑦
Landmark Tower
It symbolizes the center of R.O.K. Its radial shape means the center point of spread ways.
⑥ Jangnam plain
The axes of new city follows the existing axes of Jangnam Plain. This axes of Jangnam Plain gives this city good orientation of human being. And it makes the views to Ansan of this site.

① 环形公路／它连通了新城和世界。

② 自行车道位于相当于二层楼那么高(大约4.5～5m)的高度上，仅供自行车使用。由于高速自行车道位于市中心的办公区域内和交通堵塞地段，因此，比起乘坐汽车，新城中的市民或许会更喜欢骑自行车。

③ 商业区沿干道布置，位于 Jochiwon 和 Daejeon之间。根据风水理论，商业区象征着白虎，位于城市右侧。

④ 绿化区／四块狭长的绿化区域位于南北绿化带之间；作为休闲区，它们连接起了城内所有的绿化带。

⑤ 绿化带从城市中穿过。城市中的主要山峰是东西向的，像

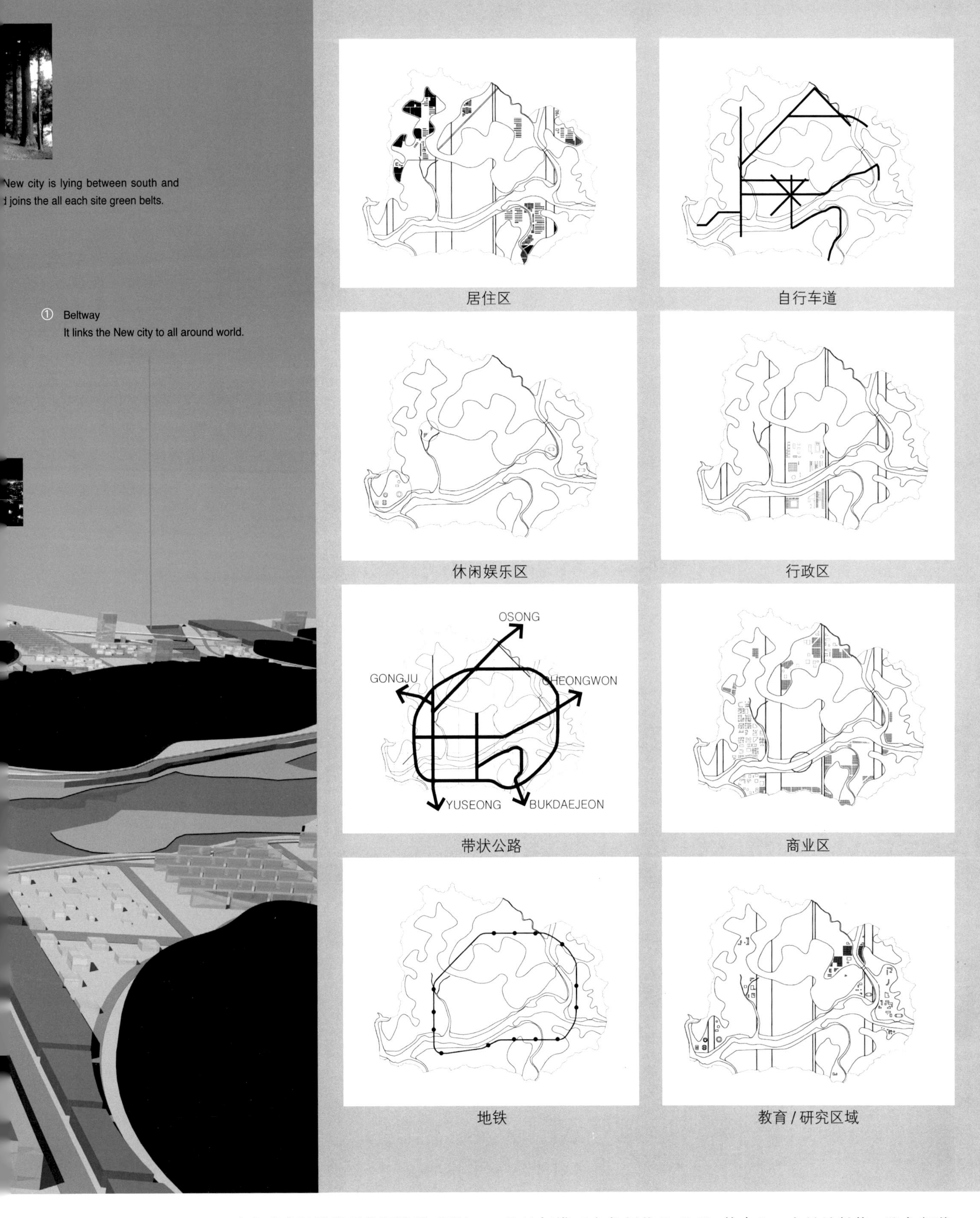

Wonsubong Jeonwol Mt，它们为市民提供了休闲场所，同时，作为绿地，改善了生态环境。

⑥ 江南平原 / 新城的轴线沿用了 Jangnam 平原现有的轴线，它使城市有了很好的朝向。同时，在市区内，可以看到 An-san 的景色。

⑦ 地标塔 / 它象征着 R.O.K. 的中心。它呈放射状，这象征着所有道路都指向中心点。

千岛城

Sumiya, Mamoru (日本)
Hayashida, Toshiko

Sumiya, Mamoru

• qp5共同创始人
• Co-founder member of qp5

■ 评委意见

本方案经过辩证性思维和乌托邦式的构想，借用群岛这一象征性概念，将城市孤岛通过绿地系统彼此相连。方案提供了开辟稻田的途径，来汲取相互依存的自然与人工元素。

■ Jury Comments

Embraced as polemical and utopian, this scheme draws from the metaphor of the archipelago to make urban islands held together by green tissue. The scheme offers ways of constructing the ground of the rice fields, drawing on the ambiguities of nature and man made elements.

■ 摘要

村庄作为一个城市自治单元，在特定的区域内聚集有各种各样的机构和居民，这一密集区便成为了村庄的中心。20 个村庄以群岛的形式散布于整个基地，组成了一个多核心的城市网络。

■ Summary

Village, an autonomous urban unit, has dense space in which various institutions and individuals of a specialized field gather and functions as a core of it. 20 villages dispersed in the site form an archipelago, a city of plural cores' network.

团队
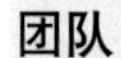
• Hayashida, Toshiko
qp5 共同创始人

village
urban forest
border
alley

村庄

f z 2 8 5 8 3
城市森林

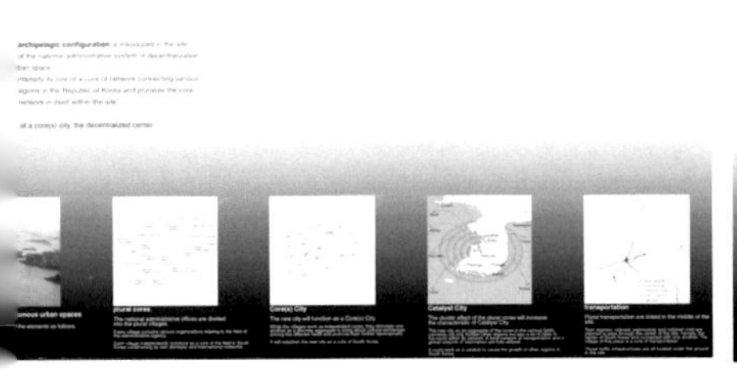
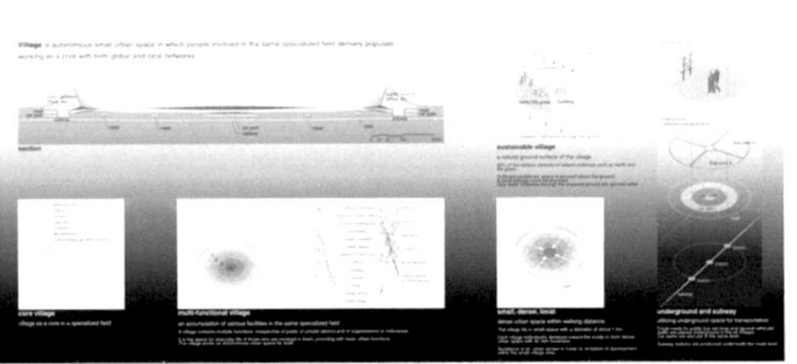
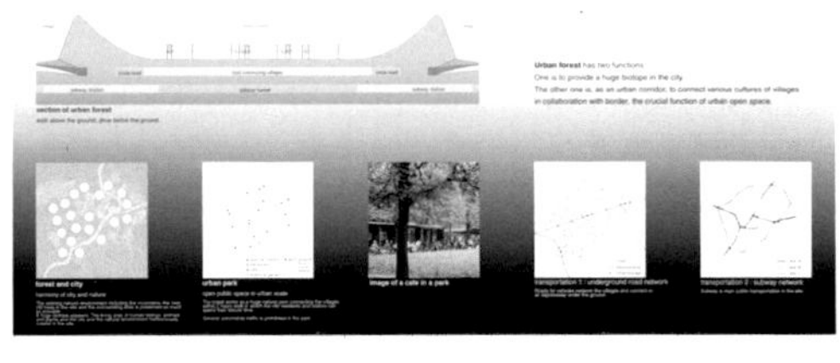

区域

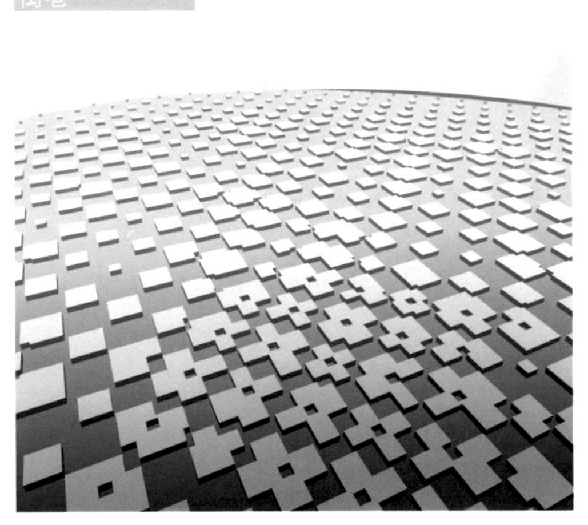
街巷

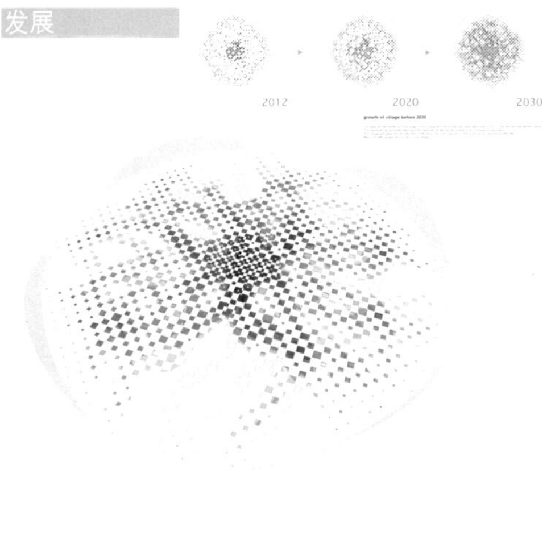
发展
2012
2020
2030

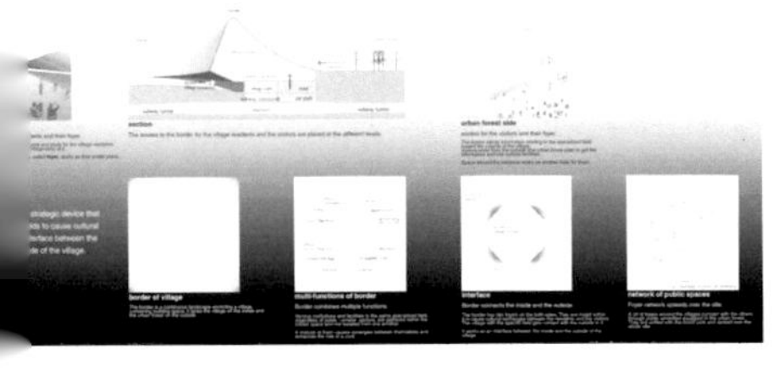
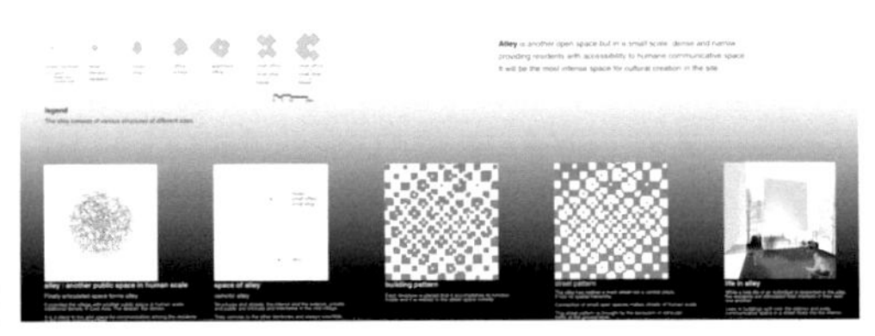
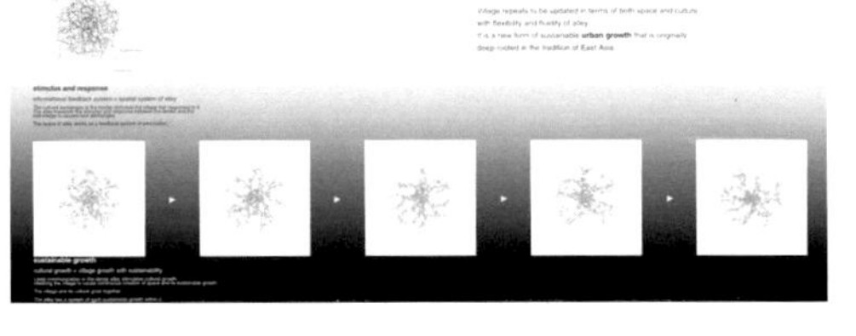

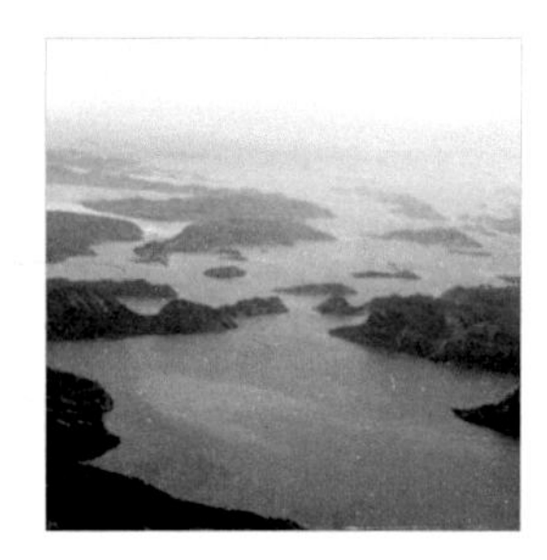

群岛

群岛形的城市结构在基地图中作出了介绍，是将疏散国家行政体系的构想作用于城市空间系统的尝试。新行政中心城市将强化自身作为连接国内不同职能区域的网络核心和本地区内的多核心城市角色。它以中心疏散的形式将核心城市的概念具体化。

Archipelago

The urban structure of archipelagic configuration is introduced on the site.

It is an application of the national administrative system of decentralization, to the system of urban space.

The new city shall intensify its role as the core of the network in connecting various functions of many regions across the Republic of Korea and pluralize the core as being a part of the network within the site.

It is materialization of a core city, as a decentralized center.

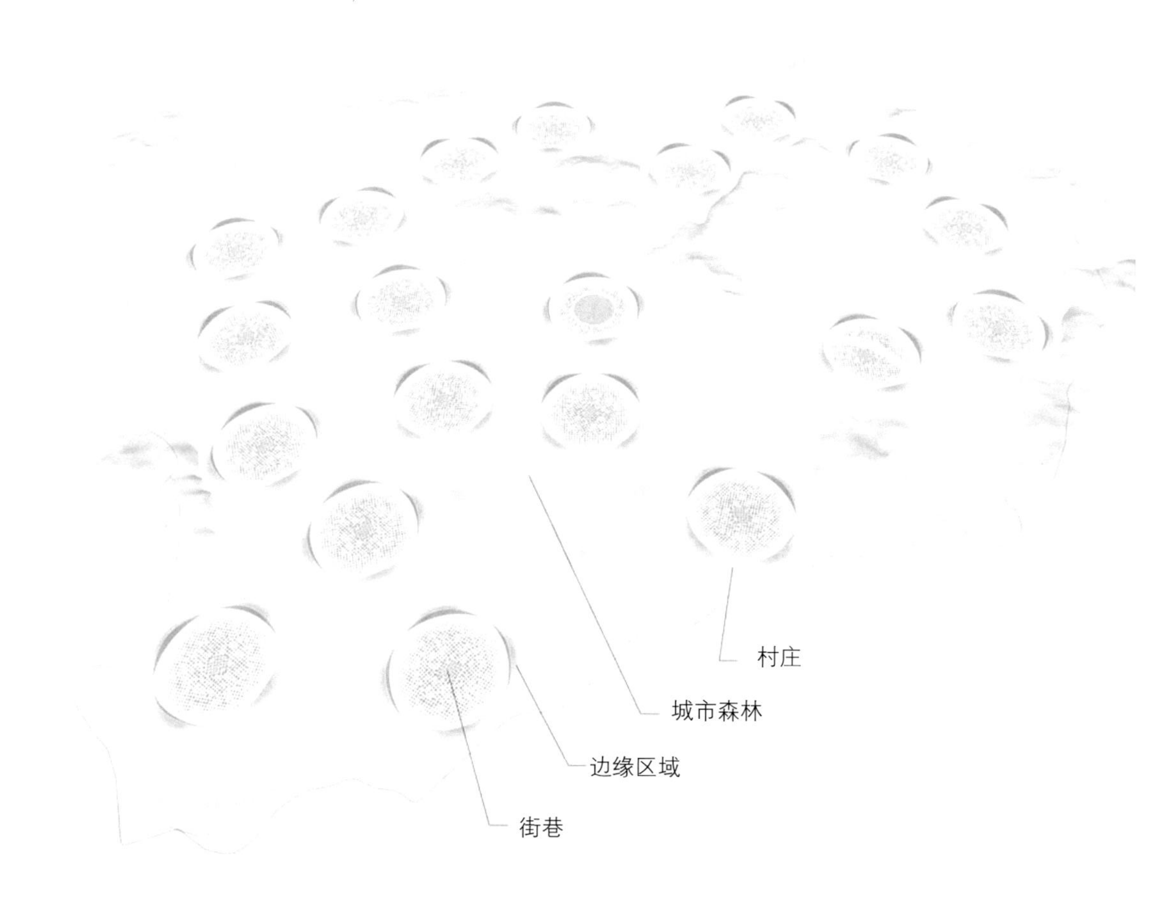

分散自治的城市空间

多核心模式

核心网络

催化城市

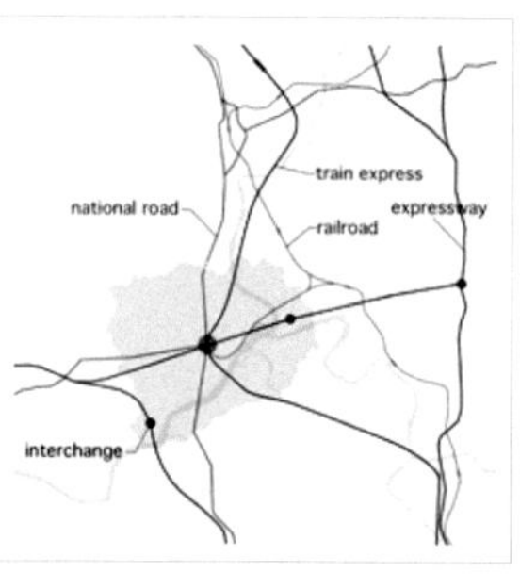

交通网络

■ 村庄

村庄是人口聚集生活在特定区域的小规模自治型城市空间，并作为全球和地方网络中的一个核心。

■ Village

Village is autonomous small urban space in which people involved in the same specialized field densely populated, working as a core with both global and local networks.

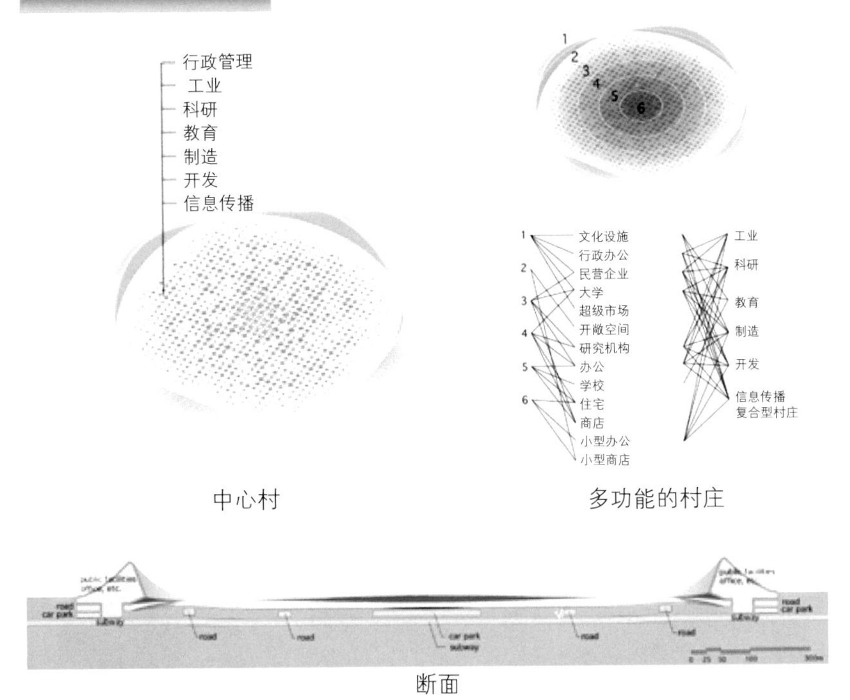

中心村　多功能的村庄

断面

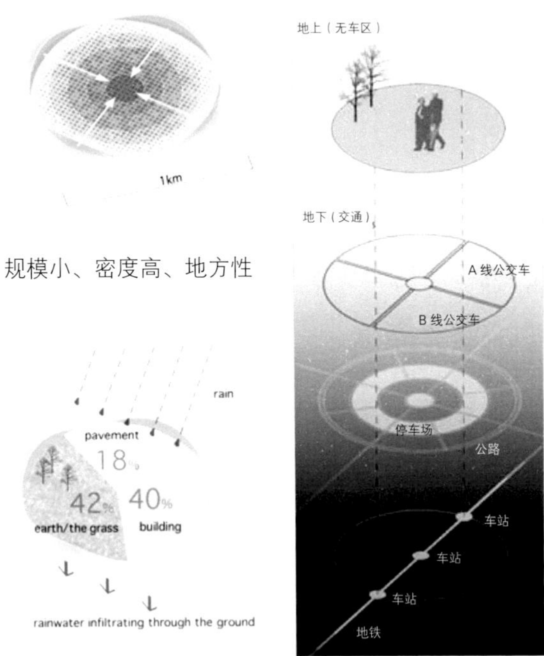

规模小、密度高、地方性

可持续发展村庄　地下空间与地铁

■ 城市森林

城市森林有两个功能：一是作为城市内部的生态群落；二是作为城市走廊，连接了不同文化类型的村庄，并结合村庄的边缘区域共同构成城市开敞空间的主体。

■ Urban forest

Urban forest has two functions.

One is to provide a huge biotope in the city.

The other one is, as an urban corridor, to connect various cultures of villages in collaboration with bordering the crucial functions of urban open space.

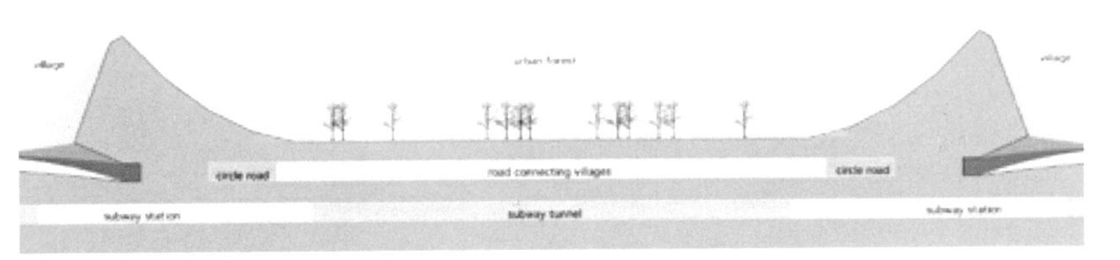

城市森林断面

- 地面步行道
- 地下车行道

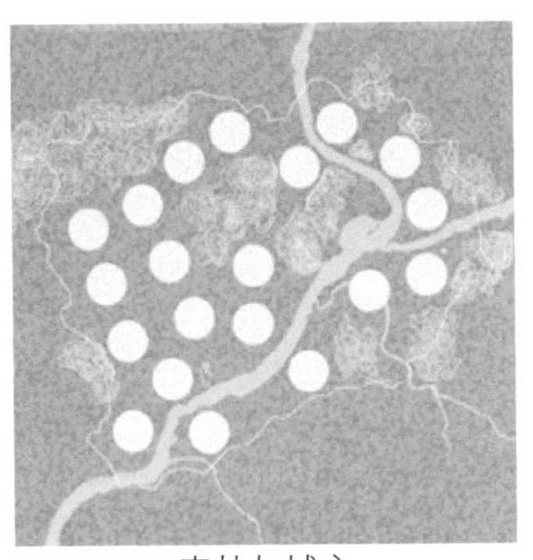

森林与城市

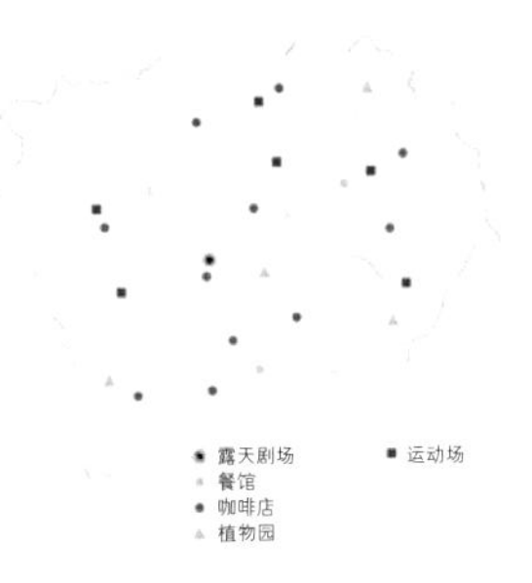

城市公园

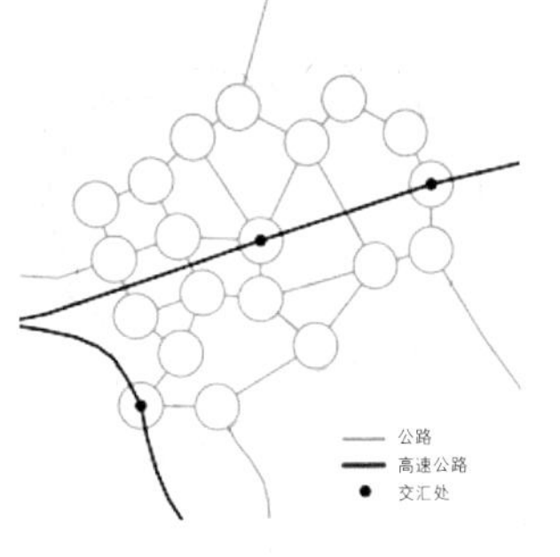

地下道路系统

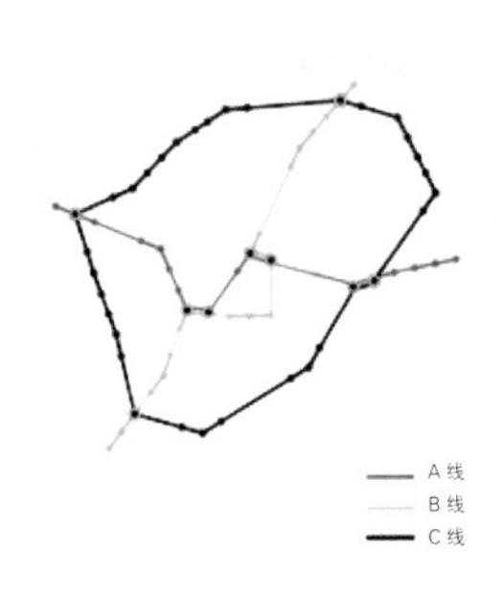

地铁网络

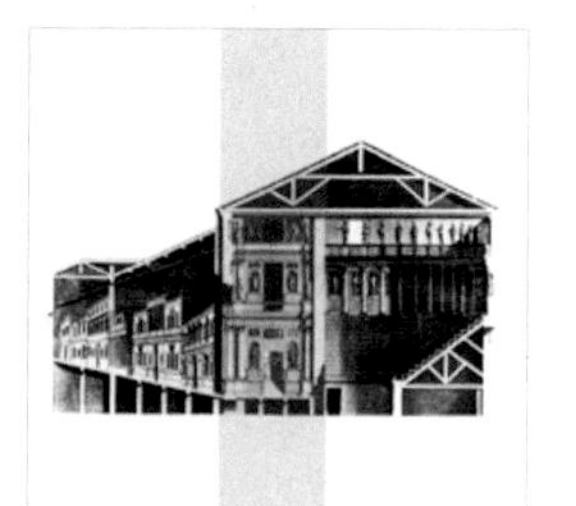

边缘区域

边缘区域是城市的战略资源，在村庄内部和外部之间以临界面的形式连接不同区域，以此来促进文化交流。

Border

Border is an urban strategic device that connect different fields to cause cultural exchanges as an interface between the inside and the outside of the village.

村庄边缘区域

村庄侧景

城市森林侧景

• 从城市战略角度设立的文化交流临界面

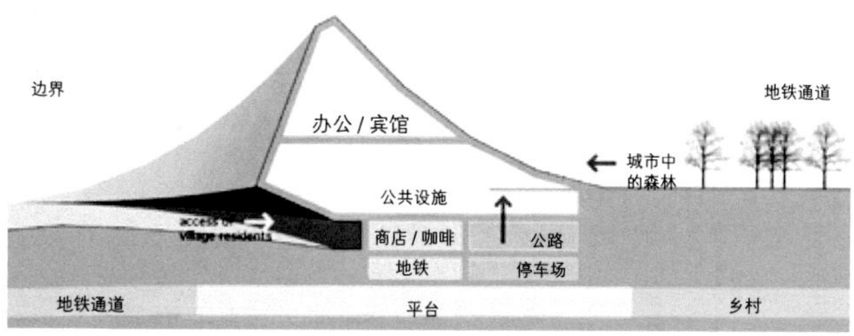

边缘的部分

办公（行政） 图书馆 旅馆 礼堂 体育设施 大学 办公（私人企业） 博物馆 展览馆 购物中心 剧院 电影院

复合型边缘区域

观光者可达 乡村居民可达 为居民提供的开敞空间 为观光者提供的开敞空间

临界面

开敞空间 • 户外交往

公共空间网络

街巷

街巷是一种小型、密集和狭窄的开敞空间，为居民提供就近的的人性化交流空间。它将是城市里文化创造性最强的空间。

Alley

Alley is another open space but in a small scale, dense and narrow, providing residents with accessibility to humane communicative space. It will be the most intense space for cultural creation in the site.

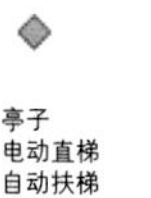

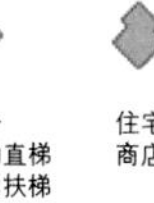

街道家具 e.g. bench flower bed outdoor light | 亭子 电动直梯 自动扶梯 | 住宅 商店 | 办公 学校 | 公寓 办公 | 小型办公 + 小型商店 + 住宅 | 小型办公 + 小型商店 + 住宅

功能和尺寸

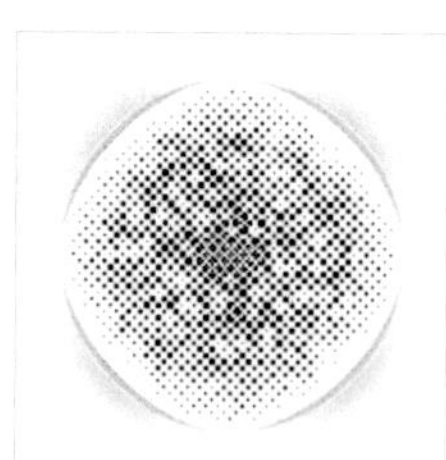

另一种尺度宜人的公共空间

街巷空间

建筑模式

宜人的交往空间

街道模式

街巷生活

■发展

村庄的空间和文化通过街巷的适应性和可变性而得以反复更新。这是一种源于东亚地区传统的新型城市可持续发展模式。

■ Growth

Village repeats to be updated in terms of both space and culture with flexibility and fluidity of alley.
It is a new form of sustainable urban growth that is originally deep-rooted in the tradition of East Asia.

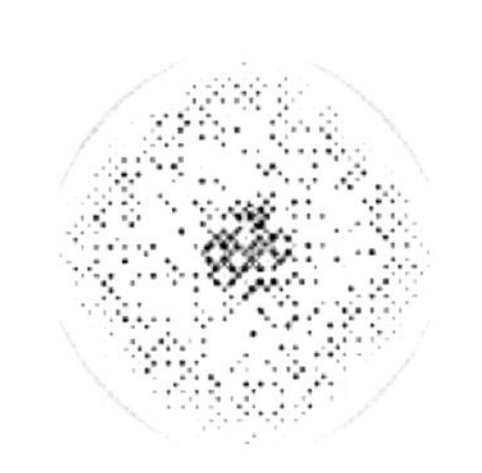

2012　　2020　　2030

• **2030 年之前的村庄发展**

村庄中间部分和边缘区域里的街巷体系将在 2012 年完成，其他地区则保持空白。其他空白地区的街巷将随着人口的增长而逐步形成与繁荣。

2030 年之后，村庄将维持在城市发展到 500000 人口规模时的密度。

一个村庄可以容纳大约 25000 人。

• **Growth of village before 2030**

The alley at the middle and the edge of the village and the border will be built in 2012. But the rest will still remain blank.

The blank will be gradually filled and the alley will be getting denser in accordance with an increase of population.

The village will keep the density after 2030 when the total population of the whole city reaches 500,000.

About 25,000 people will inhabit a single village.

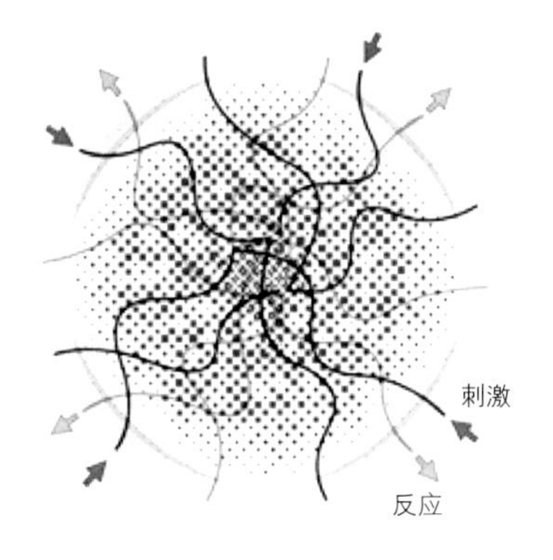

刺激与反应

• **刺激和反应**

信息反馈系统 = 街巷空间系统

边缘区域以临界面的形式连接了村庄内外。街巷空间系统同时也是信息传播系统。边缘区域里的文化交流对村庄产生了刺激。

• **Stimulus and Response**

Informational feedback system = Spatial system of alley.

The border works as an interface connecting the inside and the outside of the village. The system of alley space is also the system of information transmission. The cultural exchange in the border stimulates the village.

可持续发展

适应性，流动性

孕育新都市风格

Undurraga, Cristian(智利)
Allard, Pablo / Lopez, Pablo / Taller Undurraga Deves Arquitectos

Undurraga, Cristian

- 负责人，Undurraga + Deves Arquitectos, Santiago de Chile
- 建筑师，UC Chile '77 。1977年获“国家青年建筑师奖”
- 1991年获“Andrea Palladia国际奖”
- 2004年在基多获得“美洲双年展国际奖”
- 2005年获“迈阿密双年展金奖”
- 2005年获得“Fundacion Futuro创新城市奖”
- 目前负责圣地亚哥行政中心重建工作

- Principal, Undurraga + Deves Arquitectos, Santiago de Chile.
- Architect, UC Chile '77. National Young Architects Award '77.
- Andrea Palladio International Award '91.
- International Award Interamerican Biennale, Quito '04.
- Gold Medal Miami Biennale '05.
- Fundación Futuro Innovation City Award'05.
- Currently in charge of Santiago's Government Center redevelopment.

团队

- Allard, Pablo
- Lopez, Pablo
- Taller Undurraga Deves Arquitectos

■ 评委意见

作品中，拓宽城市中心的河流，这种建议为新城提供一个非常明确的中心。景观围绕着12个滨河的小组群（政府机构）设计，景观和建筑可以相互渗透。

■ Jury Comments

Expanding the river at the core of the city, this proposal offers a clear and unambiguous center for the new administrative city. Organized around 12 pods(12ministries) facing the waterfront, the scheme also offers possibilities of expansion into the landscape, extending each respective grid in their individual trajectories.

“如果我们是真正有思想的人，那我们应当尊重传统。” —Octavio Paz—

“If we want to consider ourselves as true moderns, we should first reconciliate with our tradition”. —Octavio Paz—

摘要

建议整合传统与发展。重新修整金河，修建一个人工湖作为中央公园，由此，赋予了城市性格。在山脉和湖水之间规划了12个区域，规模与传统城市类似；它们还拥有现代化的屋顶花园。功能和所处位置的不同决定了区域的特点，行政区域和商业区位于市中心，混合功能区位于城市的边缘；在这一区域内，建有城市基础设施，包括四条环形的林荫大道、地铁和排水沟。城市的服务设施和四座区域性公园位于各个功能区之间。

Summary

The proposal integrates tradition and development. Recovering the Geum river and defining the character of the city by building a lake as a central park. Between mountain and lake 12 districts are proposed, at a scale that recognizes traditional urbanscape and recovers nature with contemporary rooftop gardens. Each district is characterized by its location and uses, Government and Business areas in the center and mixed uses in the extremes, bringing regional infrastructure into the city, by four rings of local boulevards, subway transport and waterways. The areas between districts host urban services and four Regional parks.

新的复合型行政中心城市
培育一个新的都市，继承传统并得以发展

如果我们认为我们是真正现代的，我们首先应当重新认识我们的传统。

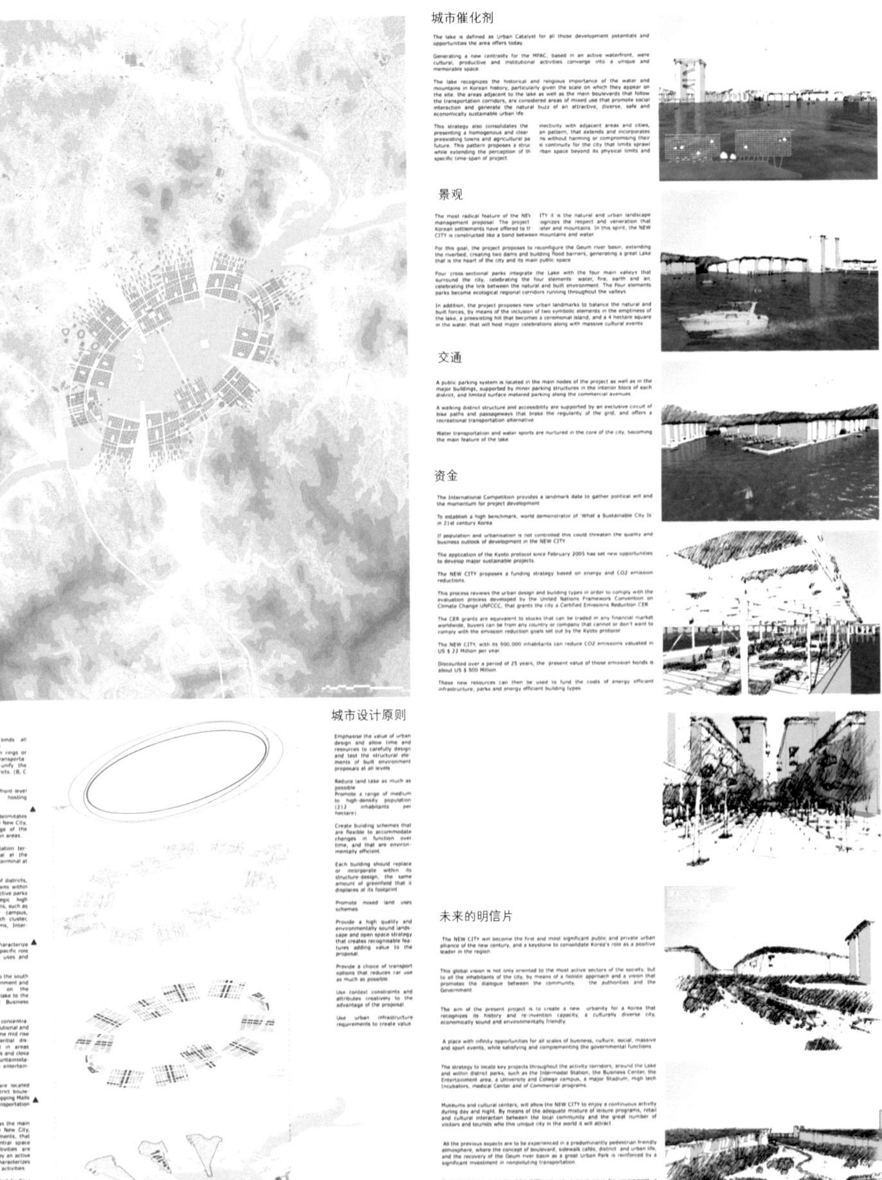

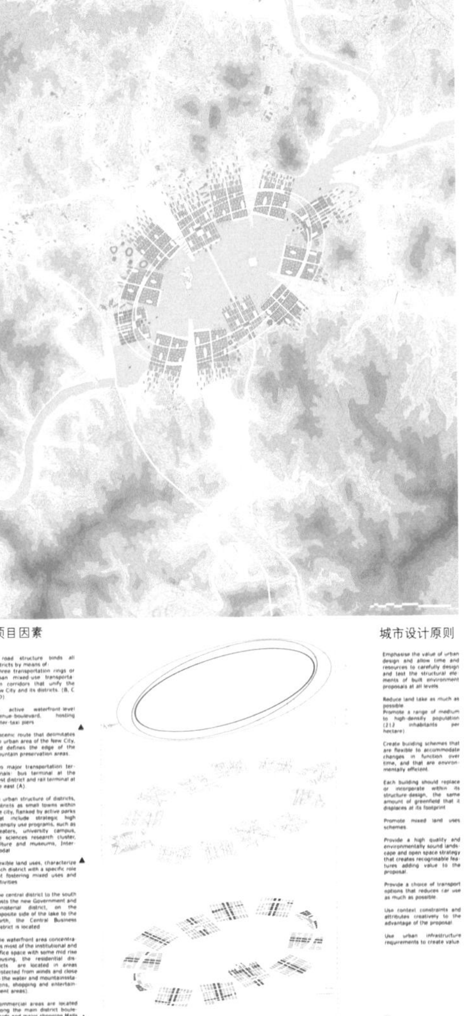

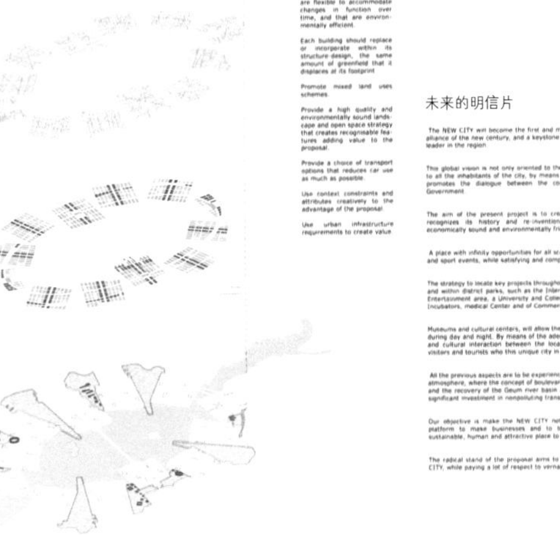

区域整合

孕育一个整合传统与发展的新都市风格

项目致力于营造出一种全民参与建设的社会氛围：居民的生活质量不断提高，环境保护受到重视，经济的可持续发展得以实现，为整个韩国的城市发展作出榜样。

项目为城市形象设定了一个明确的目标：在世界上独一无二的，根植于本土文化的，不朽的城市；能对现有的社会产生积极影响，并能够使周边城市和整个国家受益。

该项目并不希望只是僵化的总体规划，徒有其表；相反，它应该被理解为一个清晰的管理架构，20 年或 25 年以后，在这一架构下，能够形成一个充满活力的统一市场和一个充满活力的政府。

Nurturing a new urbanity that integrates tradition and development

The project aims to guarantee a quality of life suitable to the standards and demands of a growing and leading Korea, promoting environmental protection, sustainable economic development and social integration within a holistic and participative urban environment.

The proposal defines a clear an objective image of the city, a memorable civic experience unique in the world and grounded in local cultural values, with a positive impact in the existing communities, extending its benefits to the surrounding cities and the whole country.

The plan is not intended to be a rigid master plan imposing a particular vision, by the contrary it must be understood as a clear physical management framework that fosters and allows a dynamic collaboration of market and government actions for the next 20 or 25 years.

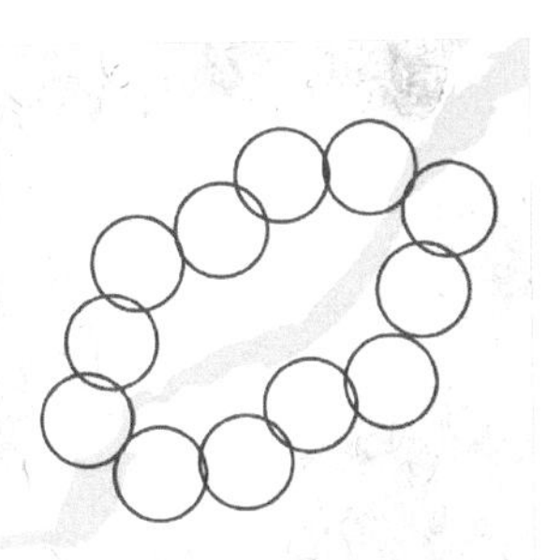

基地由12个直径为500m的圆形地块组成。这些地块的步行半径均为15分钟，因此，服务设施的布置和城市活动的发生就会比较集中。

The site is defined by twelve 500 meter diameter areas. These areas determine a walking distance radius of 15 minutes, where all activities and services should converge.

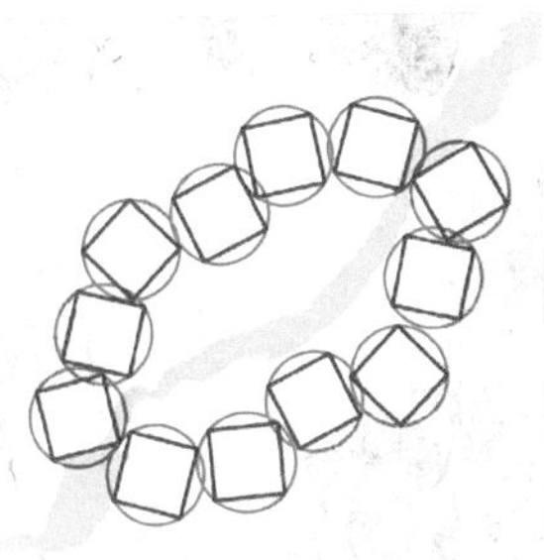

步行半径决定了城市街区的规模；并且地理特征决定了街区的位置，同时也要利于风向和自然景观。

The walking distance ratios define the size of each city district and their orientation is determined by the geography, the protection from the winds and the views.

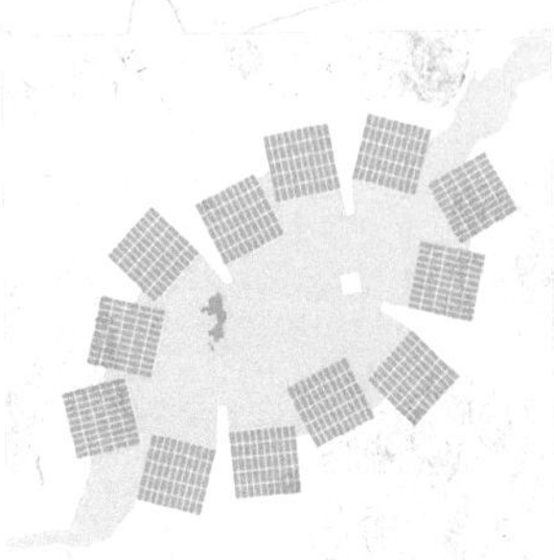

12个功能区中央是一片由河堤和河滨围成的水域，山与水的关系决定了每一个区域中道路的类型。

The area surrounded by the 12 districts is filled with water by the construction of a river dam and a new waterfront. The street pattern of each district follows the relation between water and mountain.

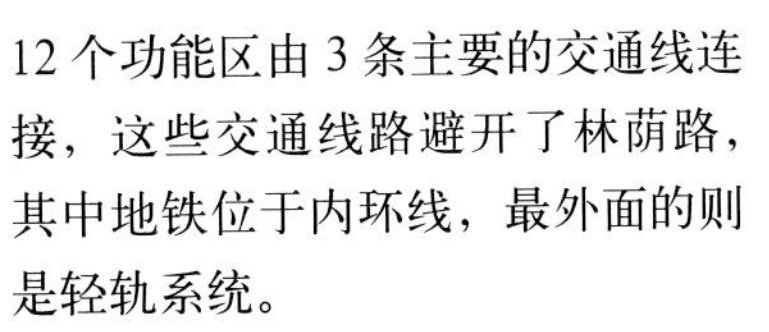

12个功能区由3条主要的交通线连接，这些交通线路避开了林荫路，其中地铁位于内环线，最外面的则是轻轨系统。

Three major transportation corridors connect the 12 districts. These corridors compromise superficial boulevards, a subway line in the inner circle and a light rail system in the outermost.

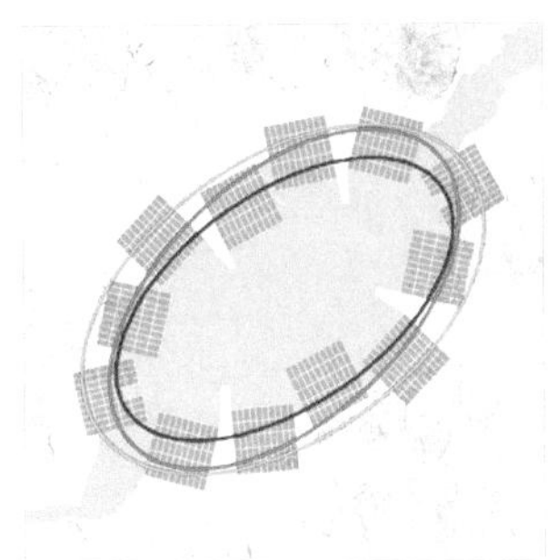

城市中建有主要的区域性的基础设施，区域性的林荫道、排水沟和绿化景观带成为城市可以生长的边界，很好地保护了现有的农田和自然景观。

Major Regional infrastructure is brought into the city, local boulevards and waterways are defined, and a scenic ring route becomes the growth boundary for the city, protecting existing agricultural and natural values.

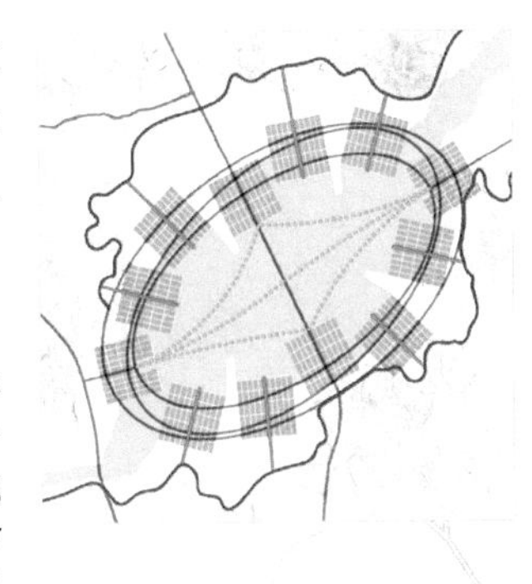

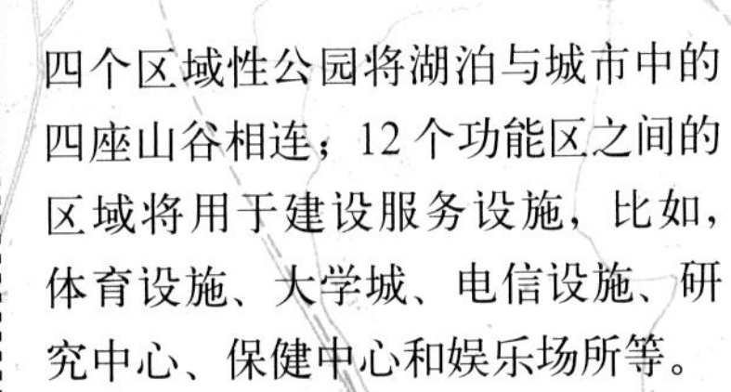

四个区域性公园将湖泊与城市中的四座山谷相连；12个功能区之间的区域将用于建设服务设施，比如，体育设施、大学城、电信设施、研究中心、保健中心和娱乐场所等。

Four regional parks connect the lake to the four valleys that converge to the city. The areas between districts host urban services such as sport facilities, a university campus, a telecom cluster, a research center, a health center and a recreational marina.

■ 区域整合 / Region Integration

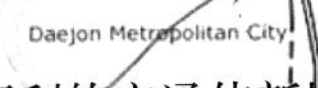

作为韩国经济发展的新动力，新城拥有很多商业机遇。位于Jochiwon-eup，gongju-si和Daejeon-si沿线，便利的交通使新城可以大力发展旅游业，教育产业和技术服务型产业，它们可以为政府提供充足的财政收入。这些产业将会沿主干道集中布置。

Located as the new economic driver for Korea generating multiple business sector opportunities. Proximity to Jochiwon-eup, Gongju-si and Daejeon-si, - opportunity for tourism, education and technology-based service industries supporting government and financial activities. The proposal crosses and is inserted in the convergence point of some of the most significant transportation infrastructure corridors for the region.

规划设计元素 / Project Elements

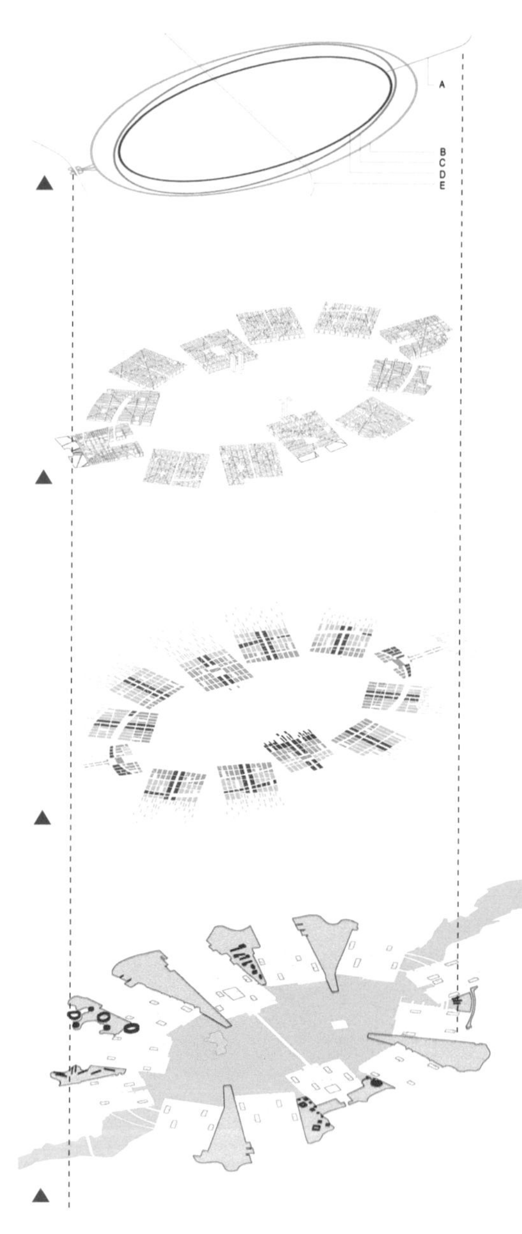

• 通过以下的方式，道路系统分布于所有的功能区中：

三条交通环线和城市混合交通线将新城以及各个城区联系了起来（B、C 和 D）。充满活力的滨河区，林荫道和水上巴士码头。道路沿线景色优美，道路不但划分了城市，也成为山地保护区的边界。

城市中有两个交通枢纽：位于城市西部的公交终点站和位于城市东部的轨道交通终点站。

• 区域的城市结构，区域像城市中的小镇，周围有活力四射的公园，镇中心有剧院、大学城、生命科学研究中心、文化建筑、博物馆、联运站、商铺和娱乐中心。

• 根据功能需要，灵活确定土地的用途，增加混合功能区的活力。

• 新城中的公共空间没有等级之分，市中心的湖泊成为市民休闲娱乐的好去处，滨水区各具特色，充满活力。

四个区域性生态公园将湖泊与山脉连成一体，这四座公园分别代表了四种自然元素：空气、水、土和火。

• A road structure binds all districts by means of:

Three transportation rings or urban mixed-use transportation corridors that unify the New City and its districts. (B, C & D). An active waterfront-level avenue-boulevard, hosting water-taxi piers. A scenic route that delimitates the urban area of the New City, and defines the edge of the mountain preservation areas.

Two major transportation terminals: bus terminal at the west district and rail terminal at the east (A).

• An urban structure of districts, districts as small towns within the city, flanked by active parks that include strategic high intensity use programs, such as theaters, university campus, life sciences research cluster, culture and museums, Intermodal stations, shopping and entertainment areas.

• Flexible land uses, characterize each district with a specific role but fostering mixed uses and activities.

• A hierarchic void as the main public space of the New City, the lake of the elements, that serves as the central space where all civic activities are focused, animated by an active waterfront that characterizes each district and its activities.

The lake is penetrated by Four Regional ecological reserve parks that connect the lake with the mountains. These parks are related to the four elements: Air, water, earth and fire.

城市设计原理 / Urban design principles

• 重视城市设计价值，在所有的设计中，我们需要投入时间和精力认真研究和测试那些影响建造环境的因素。

• 尽量减少占用土地，倡导适度的高密度模式（212 人 /hm^2）。

• 制订建设规划，这样既能满足未来城市功能变化的需要，又可以充分利用环境。

• 在设计建筑时，应当考虑设计同等面积的绿化景观。

• 倡导土地使用的综合性的方案。

• 提供一种高品质的，对环境无破坏的景观设计；提供一种开放的空间设计策略，并使之易于识别。

• 提供多种交通方式供市民选择，最大限度地减少汽车的使用。

• 利用文脉延续，使之成为项目的优势所在。

• 利用城市基础设施创造价值。

• Emphasize the value of urban design and allow time and resources to carefully design and test the structural elements of built environment proposals at all levels.

• Reduce land take as much as possible. Promote a range of medium to high-density population (212 inhabitants per hectare).

• Create building schemes that are flexible to accommodate changes in function over time, and that are environmentally efficient.

• Each building should replace or incorporate within its structure-design, the same amount of greenfield that it displaces at its footprint

• Promote mixed land uses schemes.

• Provide a high quality and environmentally sound landscape and open space strategy that creates recognizable features adding value to the proposal.

• Provide a choice of transport options that reduces car use as much as possible.

• Use context constraints and attributes creatively to the advantage of the proposal.

• Use urban infrastructure requirements to create value.

城市催化剂 / Urban Catalyst

- 湖泊成为城市催化剂，它为今天整个区域的发展提供了机遇。
- 以滨水区为基础，作品为城市设计了一个新的中心，一个独一无二的，令人难忘的空间：在文化、生产和制度方面，它都充满活力。
- 经过规划的混合功能区位于湖边，也紧邻林荫大道，这样的设计可以使这个区域成为具有吸引力的、多样化的、安全的和经济上可持续发展的城市中心。
- The lake is defined as Urban Catalyst for all those development potentials and opportunities the area offers today.
- Generating a new centrality for the city, based in an active waterfront, were cultural, productive and institutional activities converge into a unique and memorable space.
- The areas adjacent to the lake as well as the main boulevards that follow the transportation corridors, are considered areas of mixed use that promote social interaction and generate the natural buzz of an attractive, diverse, safe and economically sustainable urban life.

交通 / Transportation

- 大型公共停车场位于城市主要节点或重要建筑里，另外，每个街区内部还有较小的停车场，沿商业街，采取计时收费的地面停车方式。
- 步行系统是一条封闭的环线，由人行道和自行车道组成，它为市民提供了一种休闲方式。
- 水上交通和水上运动在湖面上展开，成为湖泊的特色。

- A public parking system is located in the main nodes of the project as well as in the major buildings, supported by minor parking structures in the interior blocks of each district, and limited surface metered parking along the commercial avenues.
- A walking district structure and accessibility are supported by an exclusive circuit of bike paths and passageways that brake the regularity of the grid, and offers a recreational transportation alternative.
- Water transportation and water sports are nurtured in the core of the city, becoming the main feature of the lake.

景观 / Landscape

项目计划重整金河，延伸河床，建造两座河堤和防洪坝，从而形成一个湖，这个湖会成为城市中心和主要的公共场所。

而且，项目计划建设新的城市地标，新地标将与自然环境和建筑物相统一。做法是在湖中引入两个象征性元素，一个是由现有的一座小山改建成的岛屿，另一个是面积在 $4hm^2$ 的水上广场，重大的活动可以在这里举行。

The project proposes to reconfigure the Geum river basin, extending the river-bed, creating two dams and building flood barriers, generating a great Lake that is the heart of the city and its main public space.

In addition, the project proposes new urban landmarks to balance the natural and built forces, by means of the inclusion of two symbolic elements in the emptiness of the lake, a preexisting hill that becomes a ceremonial island, and a 4 hectare square in the water, that will host major celebrations along with massive cultural events.

储备 / Funding

新城的战略储备计划建立在能源和降低二氧化碳排放量的基础上。拥有 50 万居民的新城一年中，通过降低二氧化碳排放量可以创造 2200 万美元的价值。25 年之后，这一数据将变成 5 亿美元。新能源可用于基础设施建设，公园建造即为建筑物提供能量。

The new city proposes a funding strategy based on energy and CO_2 emission reductions. The new city, with its 500,000 inhabitants can reduce CO_2 emissions valuated in US $22 Million per year. Discounted over a period of 25 years, the present value of those emission bonds is about US $500 Million. These new resources can then be used to fund the costs of energy efficient infrastructure, parks and energy efficient building types.

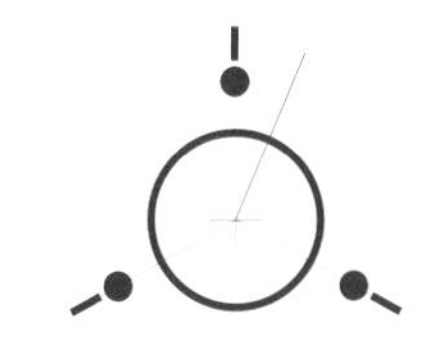

入围作品
Entry Works

无题

Kurokawa, Kisho (日本)
Kumazawa, Akira / Hashimoto, Kenichi / Jin, Guanyian / Koike, Daisuke

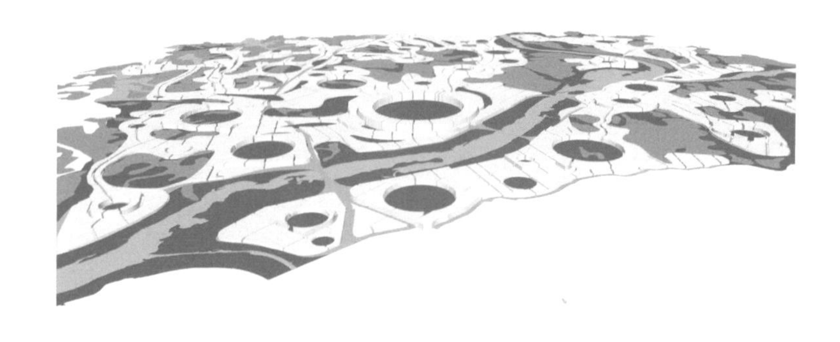

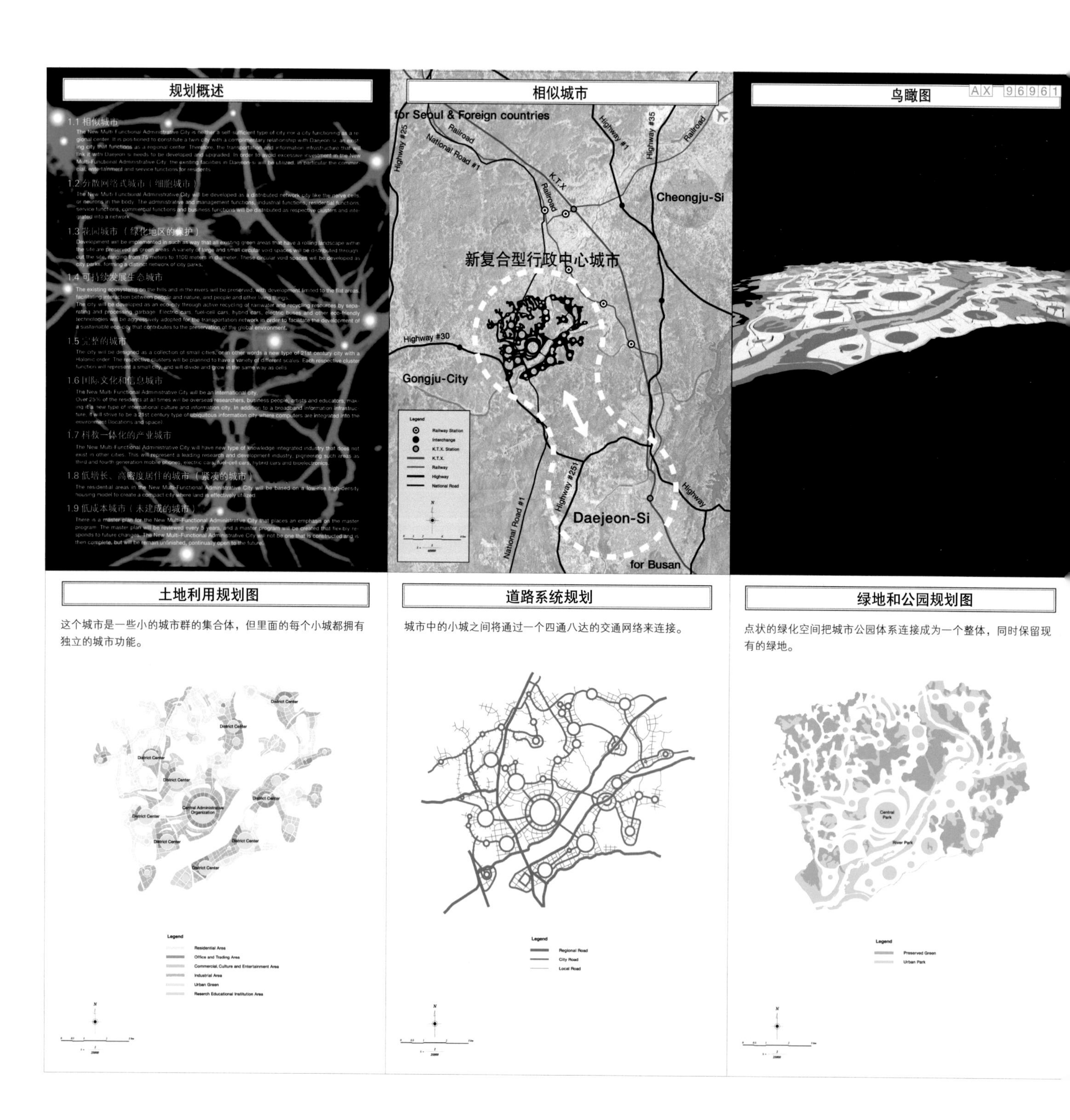

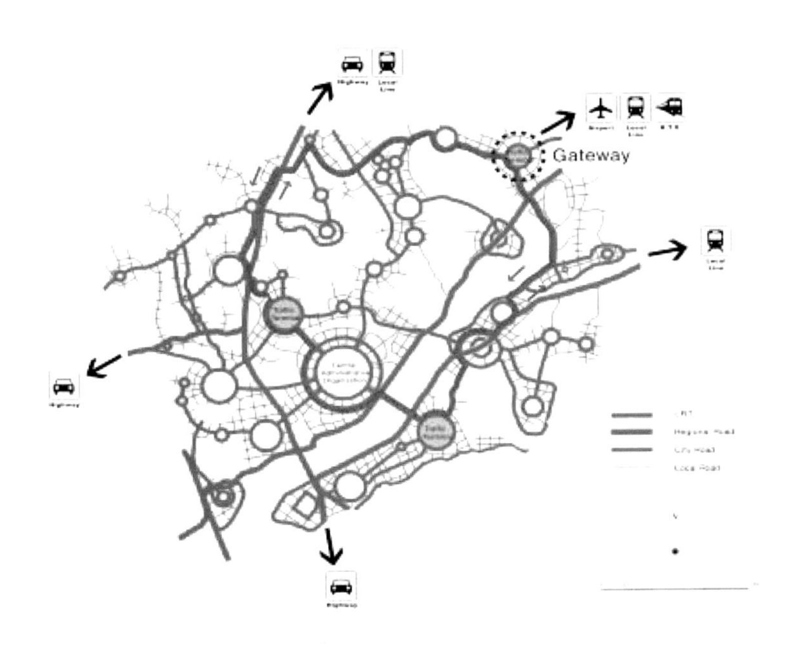

交通规划

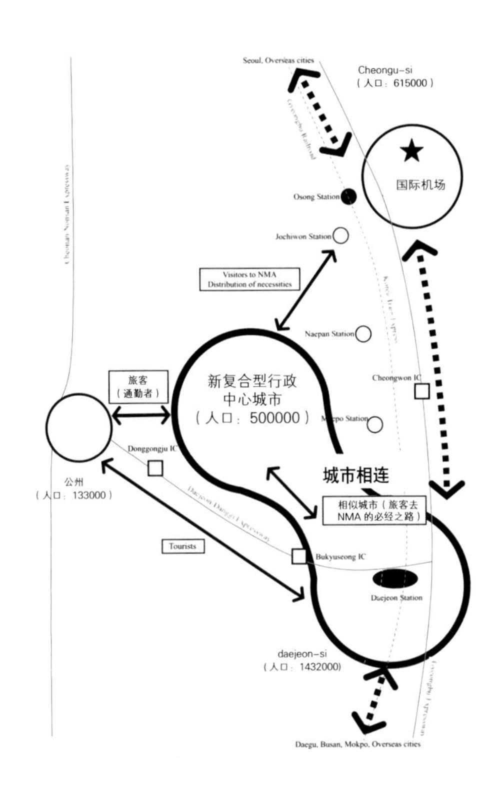

- 通过发达的交通网络来连接新城和周边地区。
- Transportation Network Linking New Multi-functional Administrative City with Surround-ing Areas

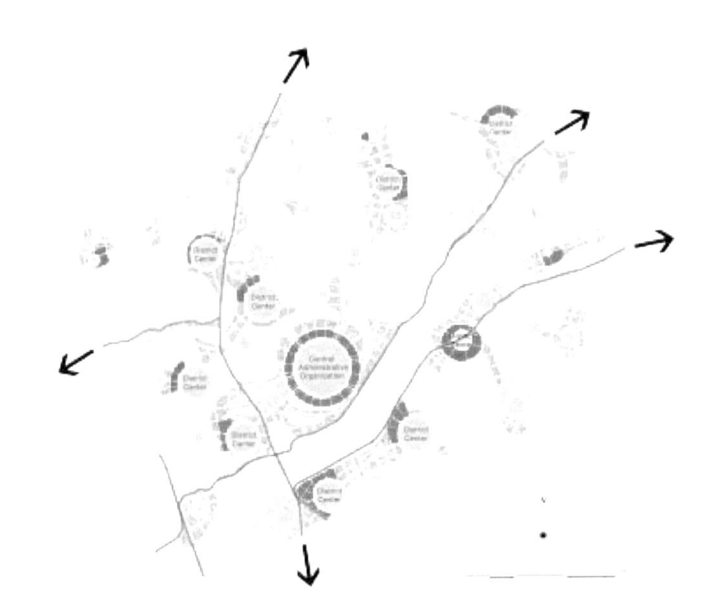

图例	•	面积（hm^2）	比例（%）
总计	•	7314	-
水域	•	469	-
山岭和绿地	•	2078	-
附加绿地	•	1240	-
道路	•	730	-
城市用地	•	2797	100
•	居住用地	1025	37
•	办公和贸易用地	138	5
•	商业、文化和娱乐用地	189	7
•	工业用地	722	26
•	城市公园用地	483	17
•	研究和发展用地	240	9

土地利用规划

第一阶段

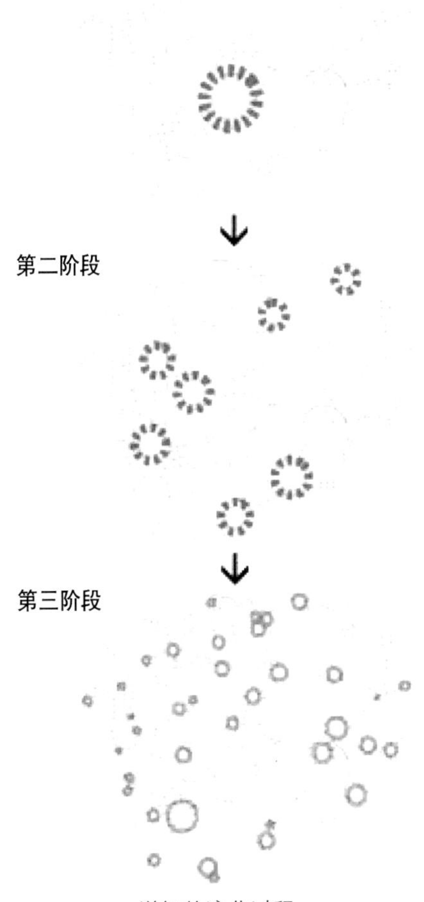

详细的演化过程

快速发展的城市

Arbanas, Magraret（美国）
Krarmuk, Uenal / Kuo, Jeannette

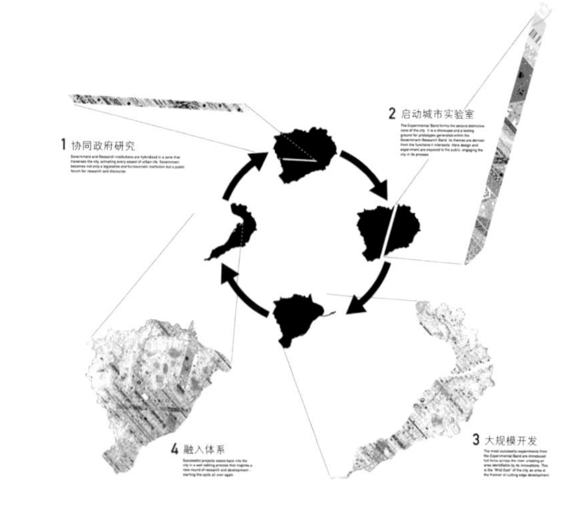

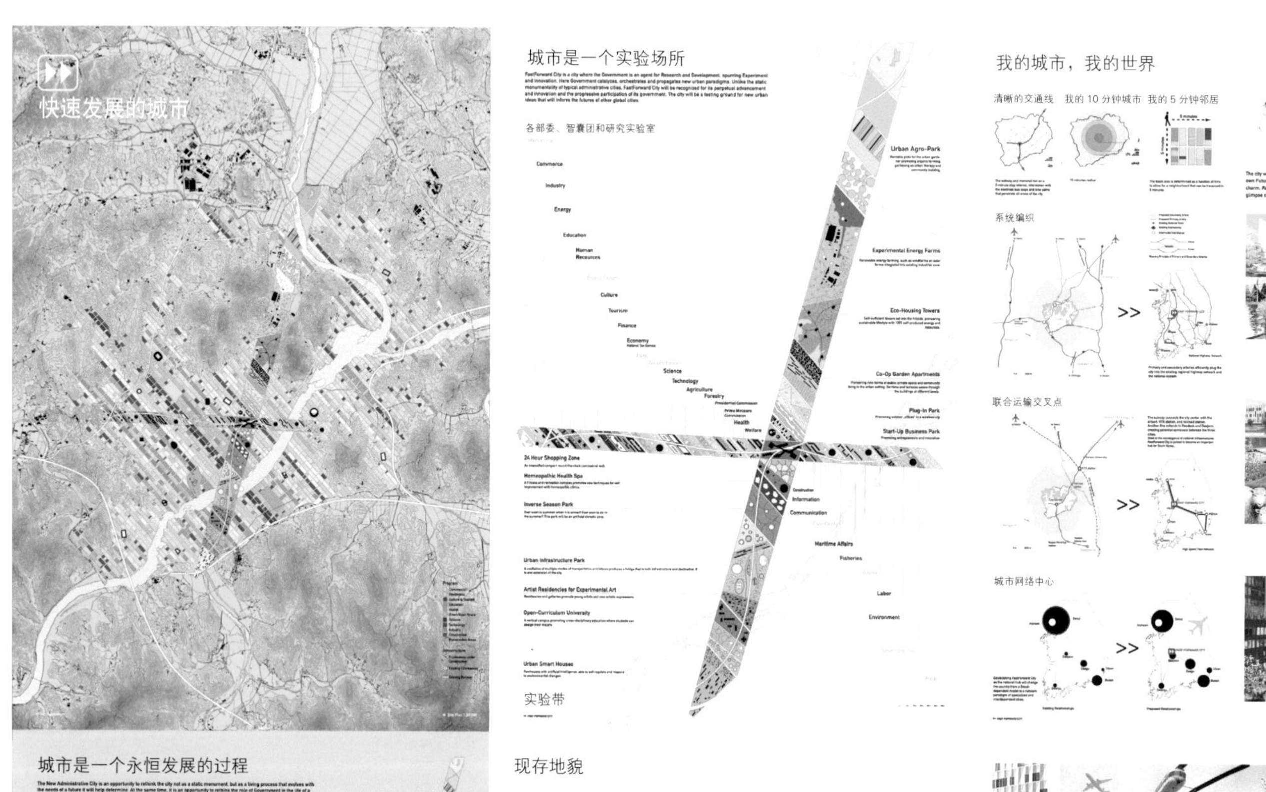

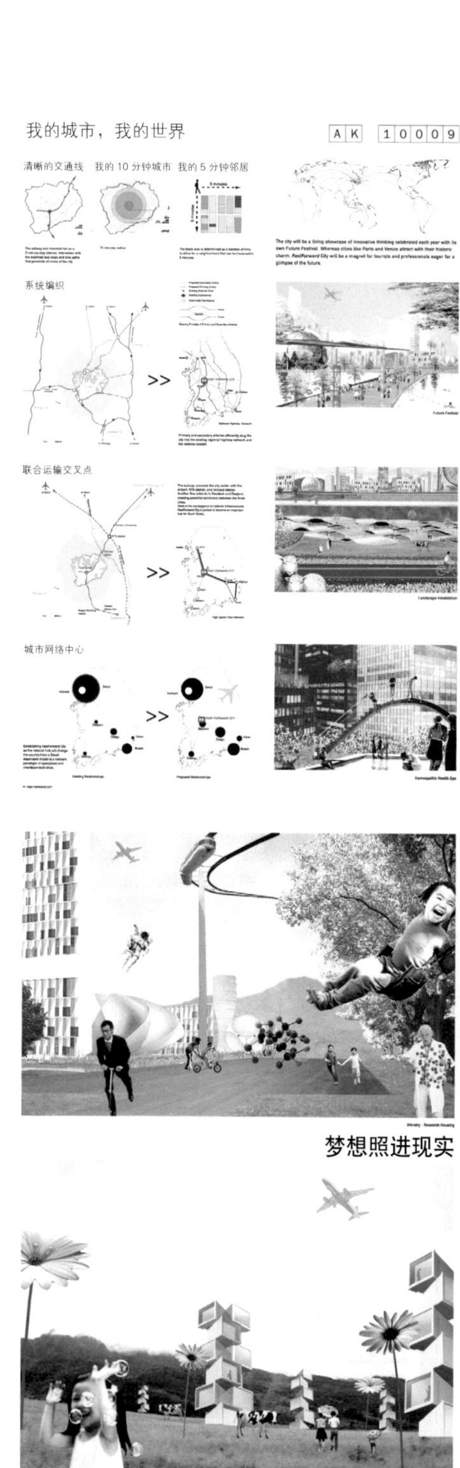

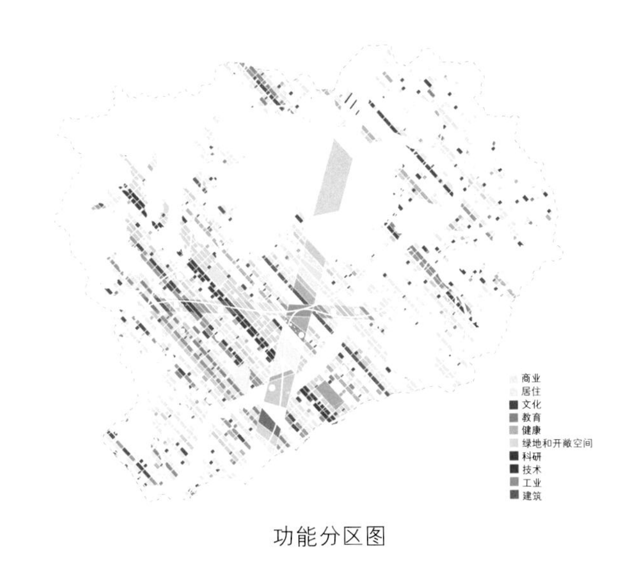

功能分区图

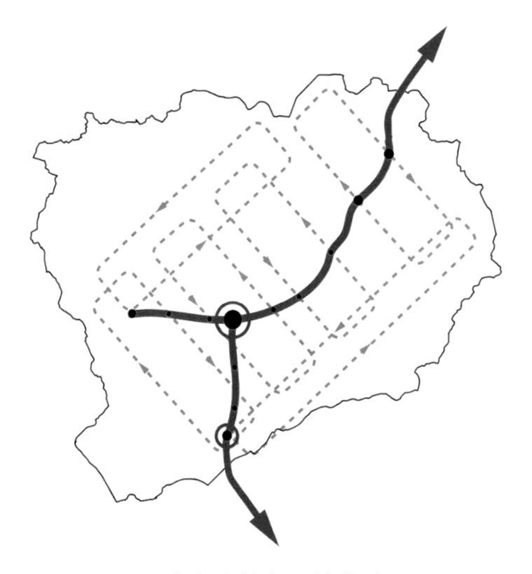

清晰的城市运输线路

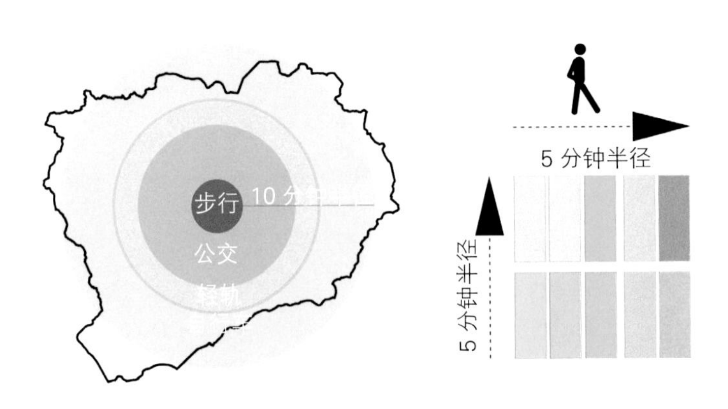

我的 10 分钟城市，5 分钟近邻

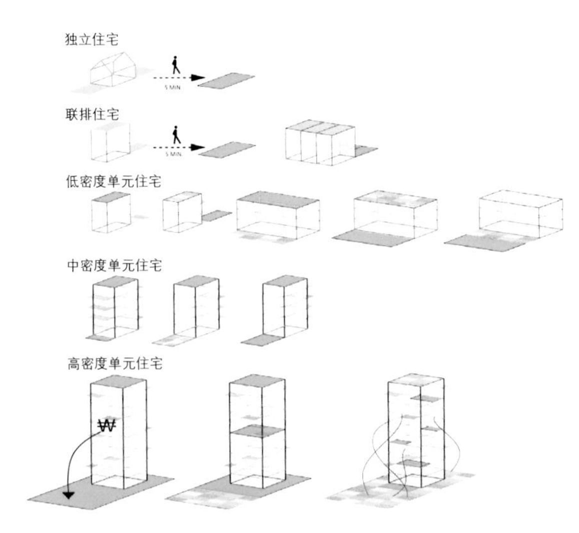

开敞空间的流通

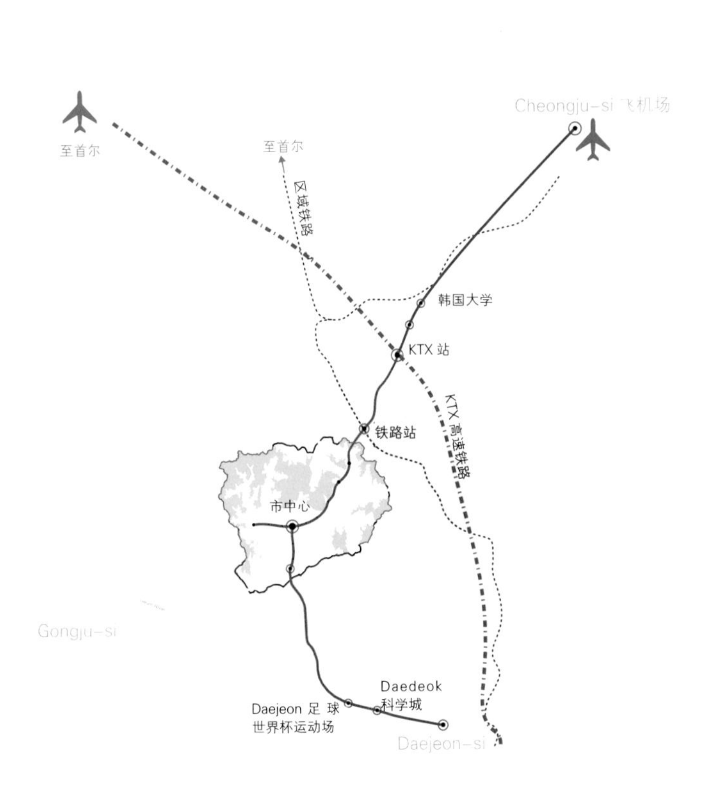

15 分钟可到达目的地的韩国快速列车，30 分钟可到达机场。

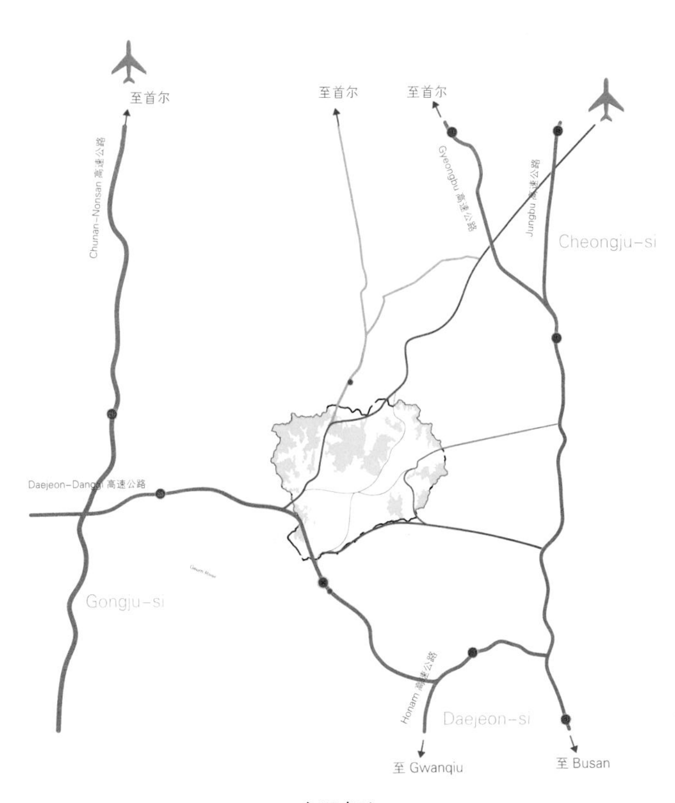

交通枢纽

无题

Schenk, Leonhard（德国）
Lehan Drei, Architects+Urban Planners
Feketics Schenk Schuster / Muller, Felix /
Flury, Christina / Witulski, Stephanie

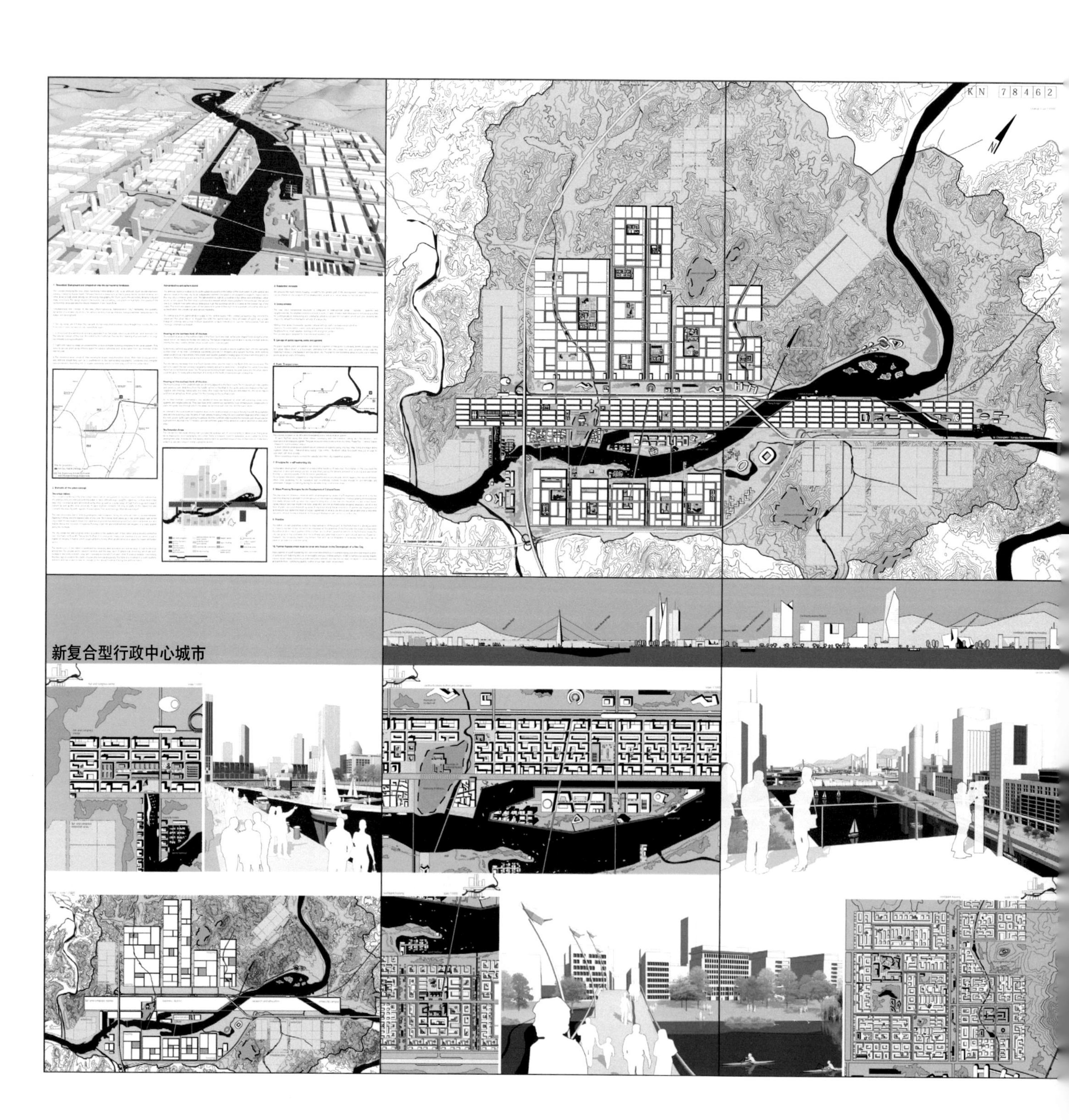

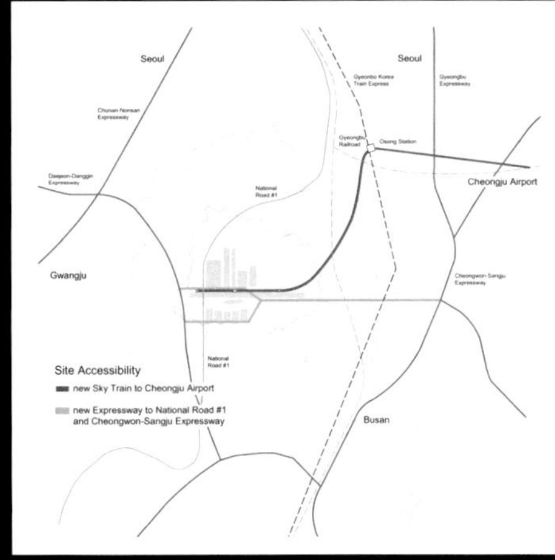

可通达的场所：机场、铁路、高速路、国道

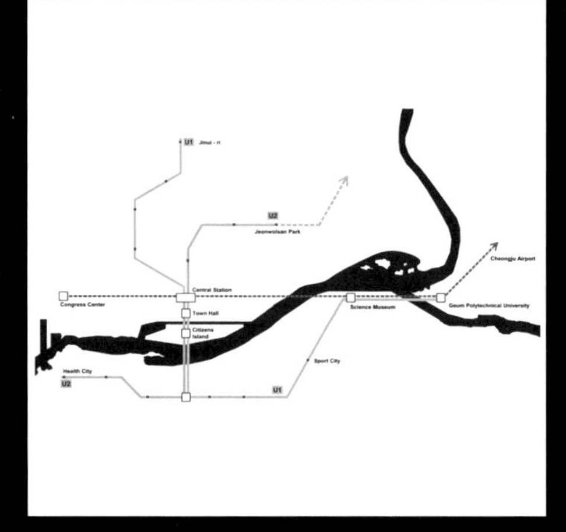

公共运输

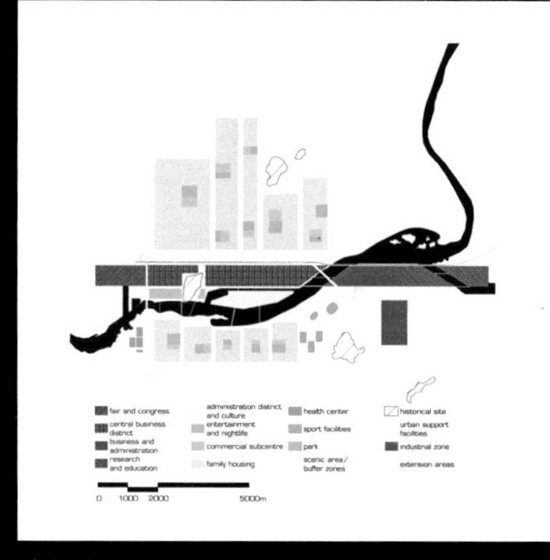

功能分区

行政和文化岛

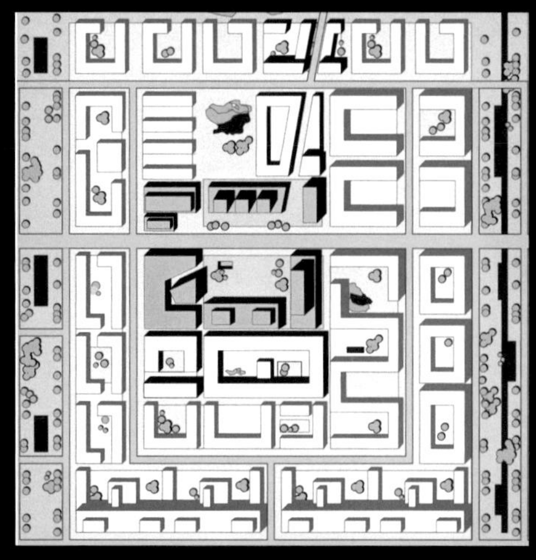

在南岸的住宅

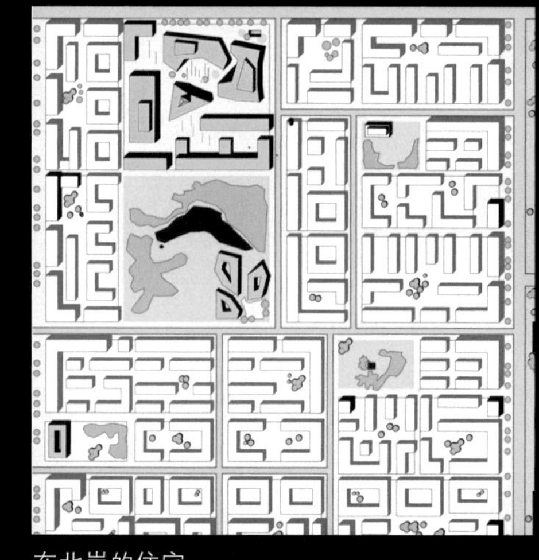

在北岸的住宅

平和的生态政策

Ock, Han Suk(韩国)
Seo, Tae Yeol / Lee, Sang Yong /
Huh, Tai Yong / Lee, Sang Kyoo

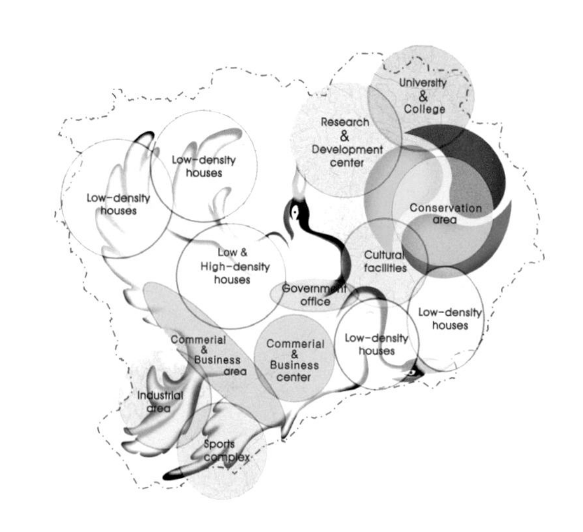

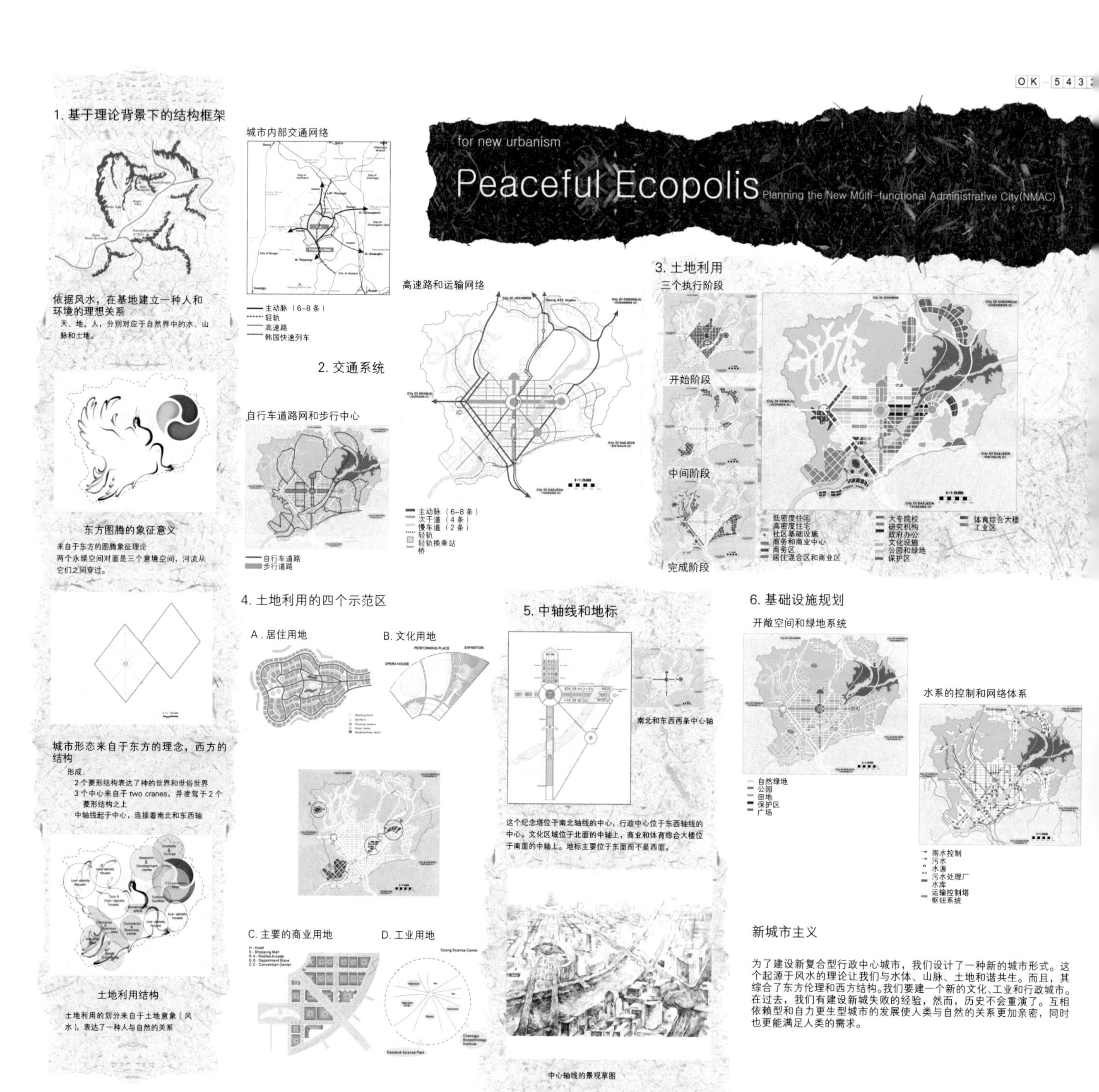

居住用地的一个示例（低密度）

H– 旅馆
S– 购物中心
R.A– 商业步行街
D.S– 百货中心
C.C– 会议中心

S S S S S H
R.A
S S S S S H
S
S D.S D.S C.C
S H
S H
H

商业用地的一个示例

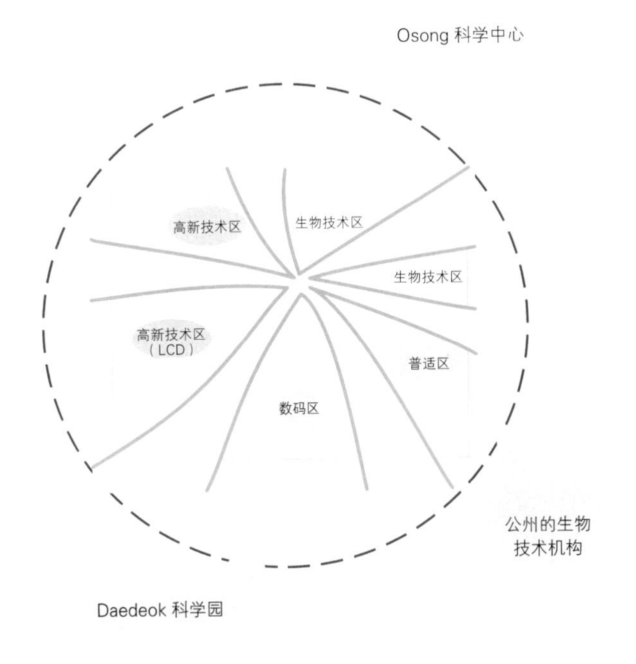

工业用地的一个示例

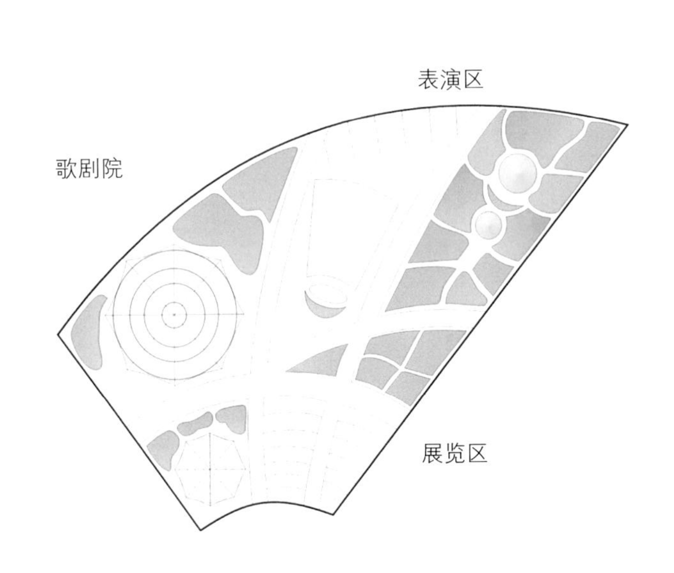

文化用地的一个示例

■ 执行阶段

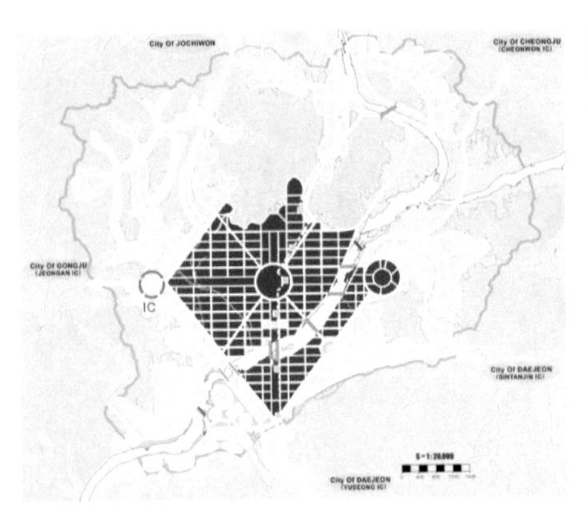

前期

City Of JOCHIWON
City Of CHEONGJU
(CHEONWON IC)
City Of GONGJU
(JEONGAN IC)
IC
City Of DAEJEON
(SINTANJIN IC)
City Of DAEJEON
(YUSEONG IC)
S=1:20,000

中期

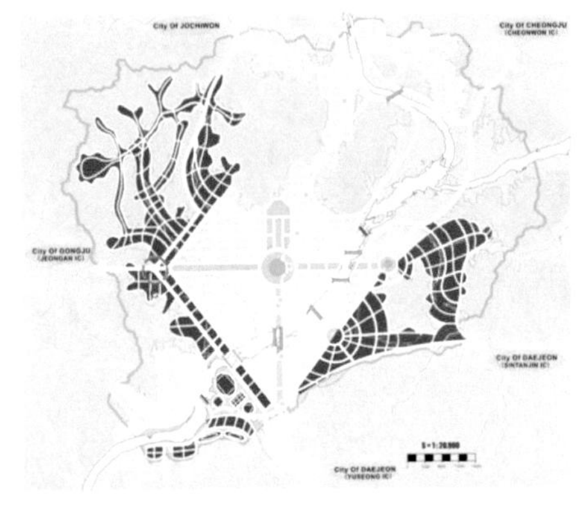

后期

开园

Lee, In Won(韩国)
Mooyoung Architects&Engineers /
Michell, Anthony / Kim, Tae Kyung

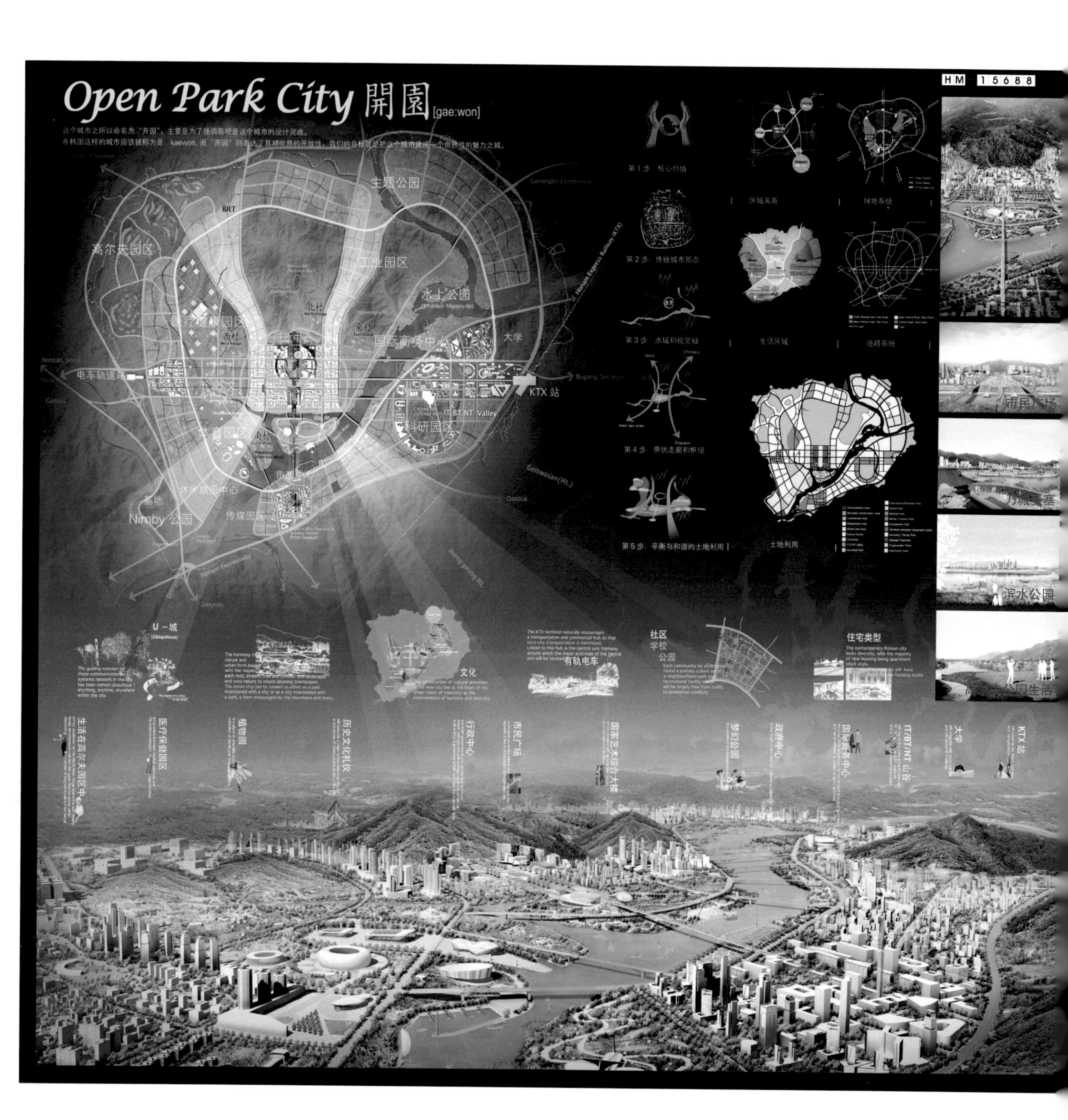

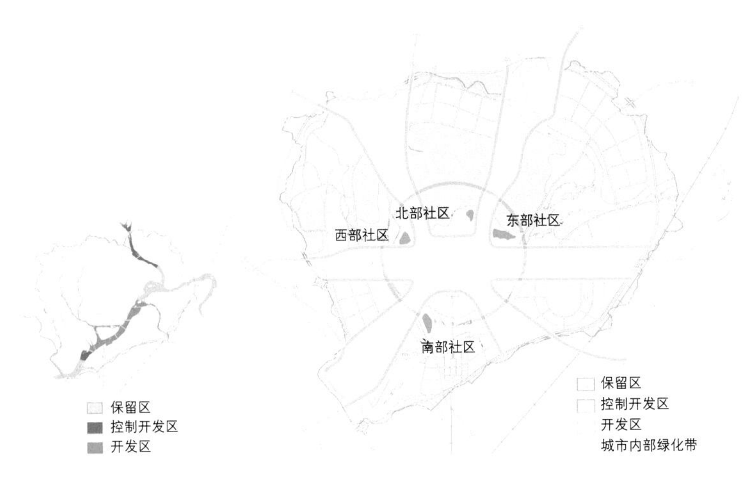

以带状形式把城市划分为 3 个基本的区域

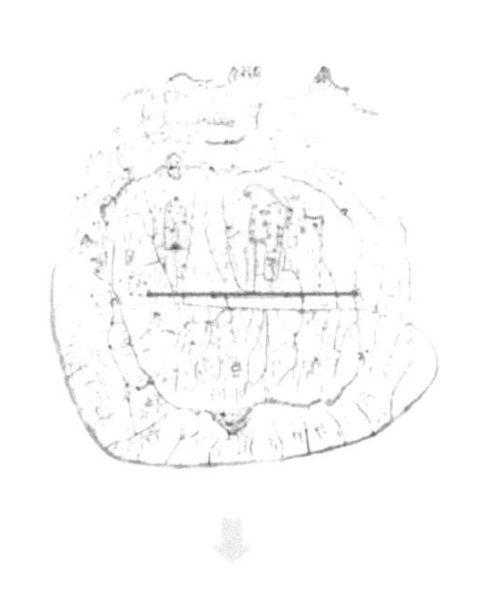

土地利用概念规划图

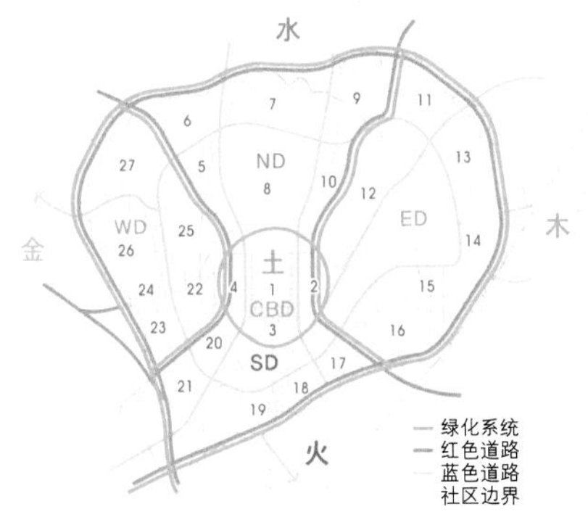

把城市划分为 5 个居住区，27 个社区

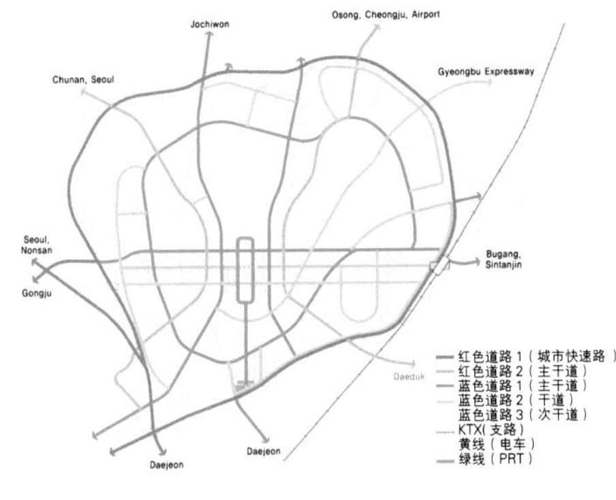

主要的交通体系和动脉连接

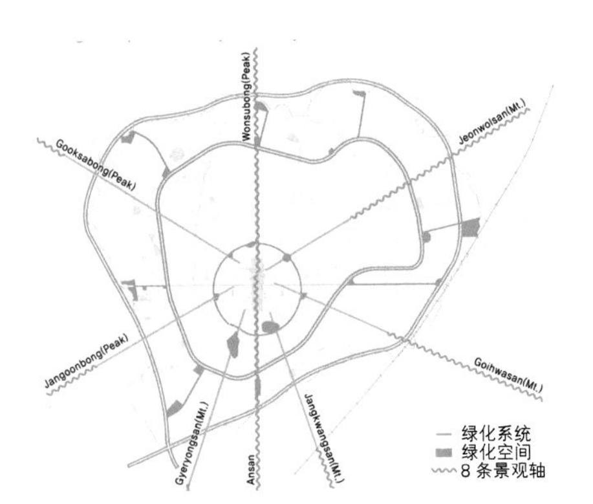

依据主路网形成的绿化系统

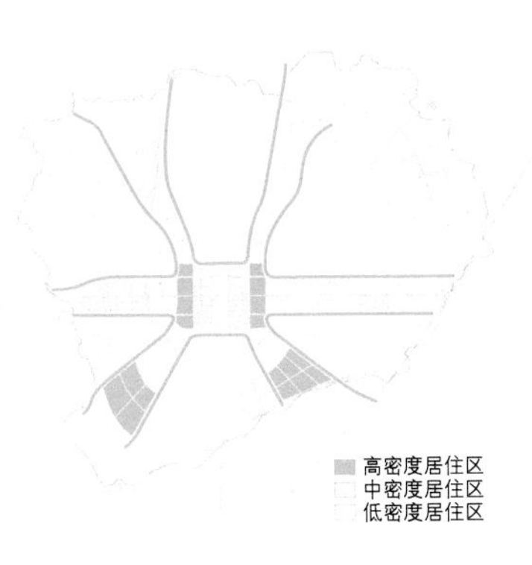

居住区的发展模式

新城的形成过程

无题

Lee, Houng Chul (韩国)
Eric, Strauss / Shin, Kyungsik Irene /
KunHwa Engineering CO., LTD. /
Urban and Regional Plannning Program
Michiegan State University

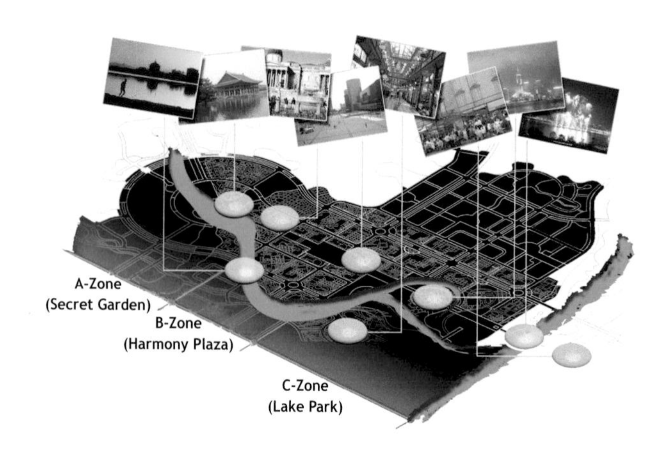

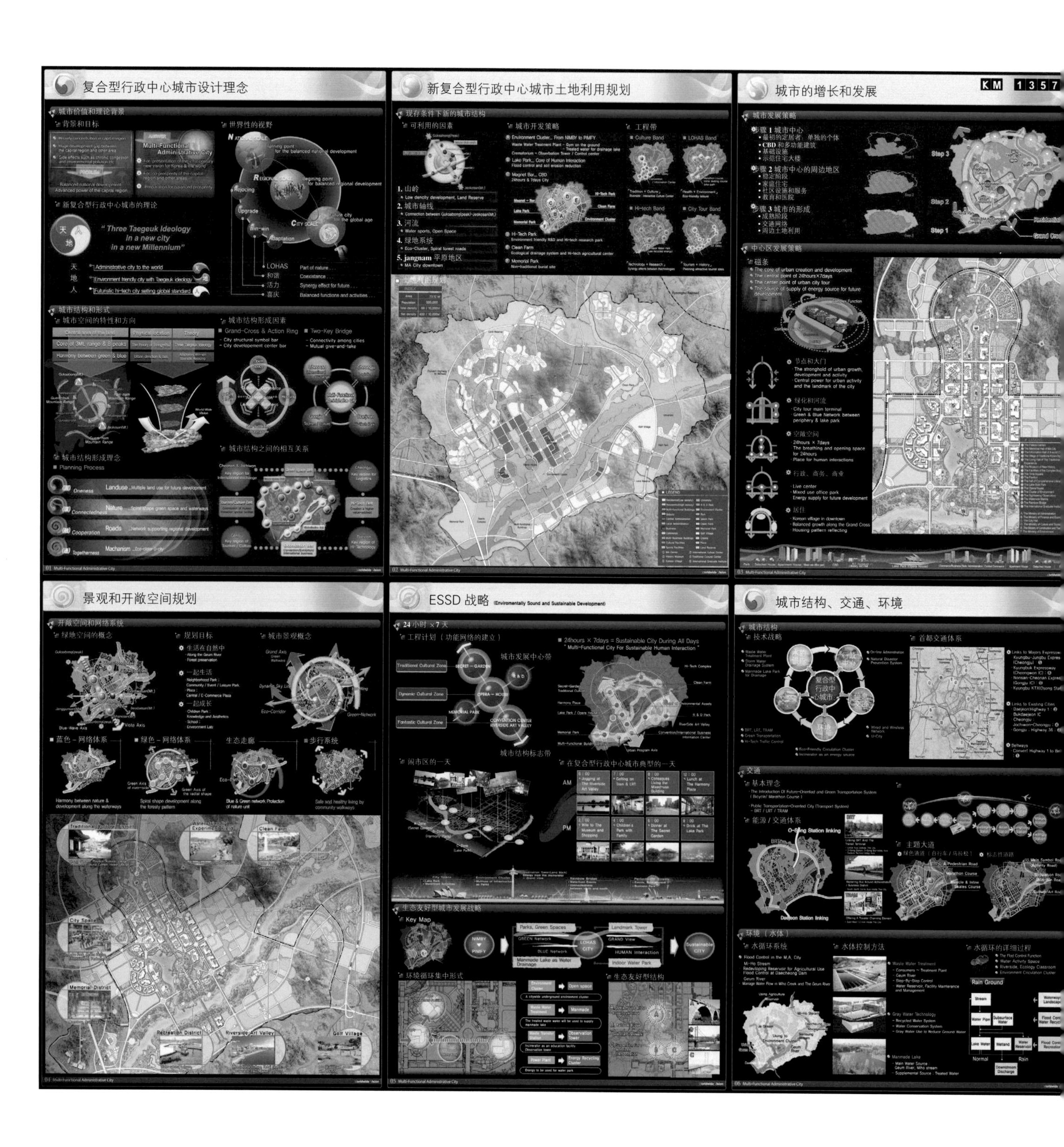

总体规划图

闹市区航拍图

景观总体规划图

混合巨尺城市

Space Group（韩国）
Lee, Sang Leem / Purini, Franco

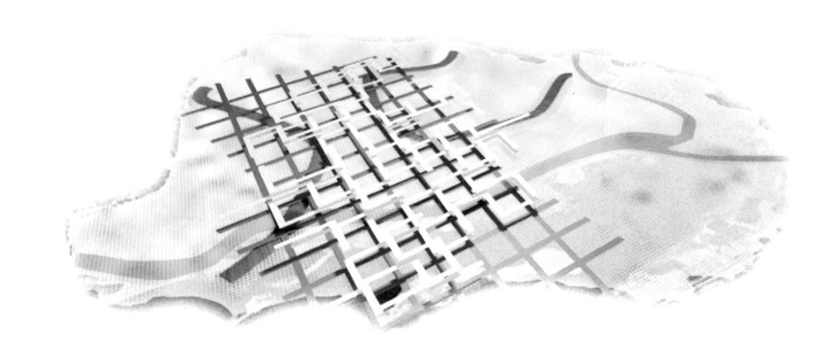

MM 02030

混合巨尺城市

Urban activity chases the flow of street. We suggest to expand CBD (central business district), which was overly centralized in the past, through the entire city. The expanded CBD will be spread along the city streets in type of the flow of network. We call it as "Urban Band".
The Urban Band will be a space, filled with various urban activities without automobiles. In the Urban Band, new structures and characteristics of the new city will be introduced through applying the concepts of Upside-down the Urban Fabric, MIX the Urban Fabric in 3-dimensional way, and MIX Program & MAX Urbanity.

We suggest a city where the coexistent of health from rural life and culture from the crowded city life is possible. We call the domain, surrounded by Urban Bands spread through the entire city, as "Ruban" (urban+rural) Block. In each Ruban Block, the dwellers can enjoy living in ecological environments, and they can connect anytime and anywhere to the CBD within 10 minutes simply by walking. The Ruban Blocks will provide rural (natural) environments to the city dwellers in pursuit of MIX Uniqueness.

We pursue the coexistent of the extreme attraction of the city life and the extreme health of rural life in the MIXMAX city through mixing characteristics between rural and city, and growth between unplanned low-rise Urban Fabric and high-rise Urban Fabric.

混合城市＋田园风光＝最大的城市

Environmentalists including many people say the density of a city should be lower than today. However, the efficiency and the true beauty of a city come from a principal of the hustle and bustle of the city streets. There are some people who hope for the pleasure of a rural life, and there are other people who want to enjoy the busy and crowded city life. A huge number of people who inhabit the city might want to enjoy both sides.

Through the entire city, the Urban band and Ruban block surrounded by the Urban Band will be spread out. The system will provide the coexistent of the extreme attraction of the city life and the extreme health of rural life in the MIXMAX City to the city dwellers.

混合城市结构和最大城市

The Urban Band is 110mX110m X600m as possible development space, and it has to be encouraged to develop with block units not pieces, joint building development, and connecting buildings through its many incentive programs. Alternatively, MIX the Urban Fabric in connected 3-Dimensional way will be possible.

混合城市和乡村

示范城的结构

扩张的 CBD

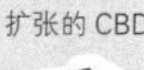

连接到城市的任何时间和任何地点

生活在城市旁边的一个公园里

混合的独特性

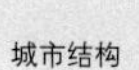

城市结构

BEFORE　AFTER

混合城市结构

颠倒的城市构造

In traditional Korean society, urban events and activities were often held along the main cities' roads. However, cars occupy the wide roads, and people walk through the alleyways in the modern Korean cities. We propose some ideas to reverse this situation. The center of Urban Band is designed only for pedestrians and mass transportations while other car roads will be situated around the edge.

城市结构在三维空间中混合

New Urban Fabric will be created through mixing the characteristics of unplanned low-story and high-tech multi-story Urban Fabric. Buildings which were isolated from the past will be connected to each other through low, middle and high stories. Through introducing a new way of connecting buildings and then creating spaces between them, it will lead the city to be flourished where possibilities are endless.

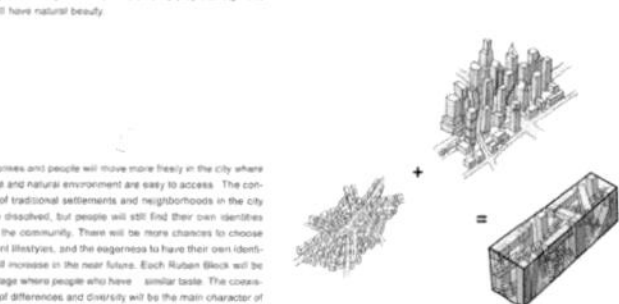

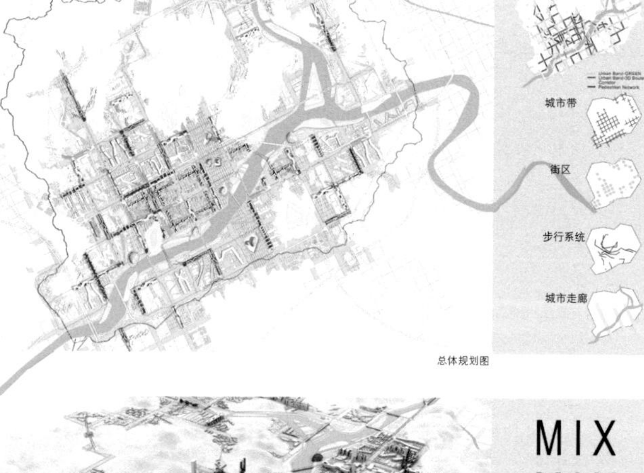

总体规划图

城市带

流动

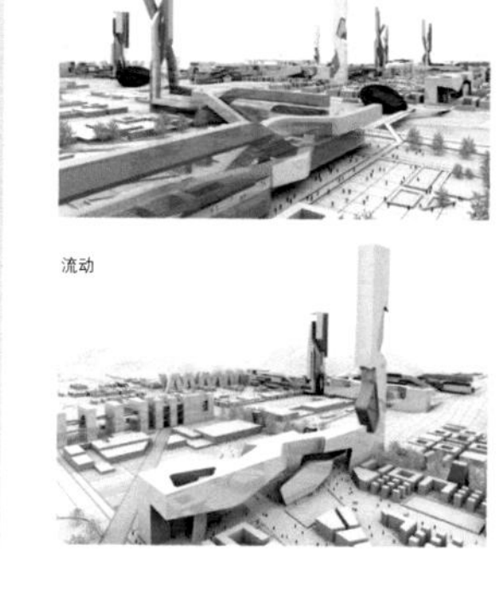

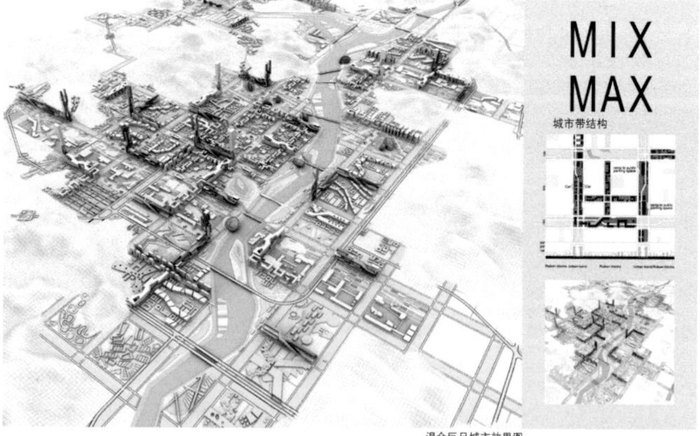

混合巨尺城市效果图

街道 1

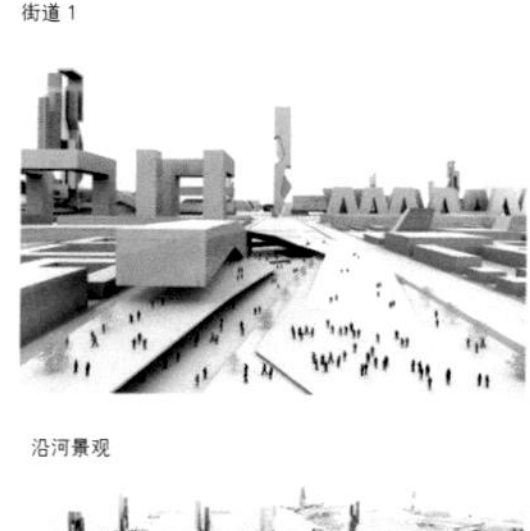

沿河景观

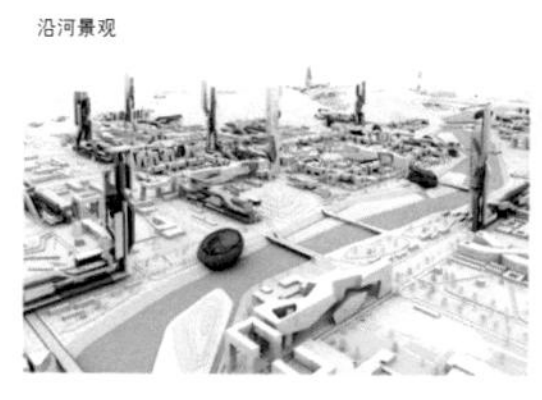

街区

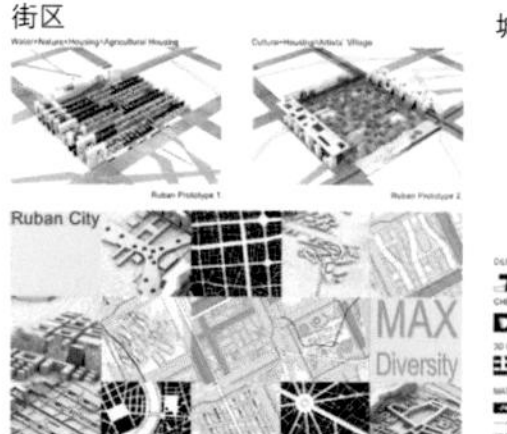

城市带

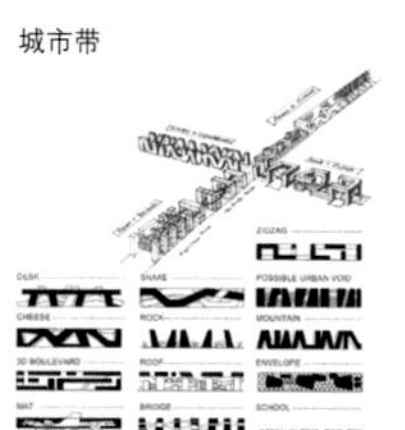

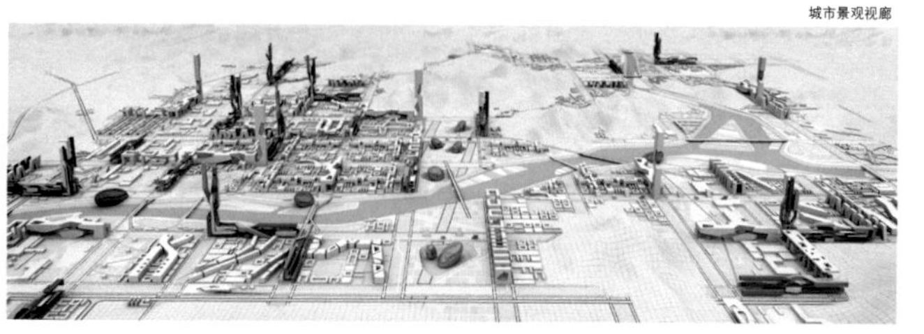

城市景观视廊

夜景观

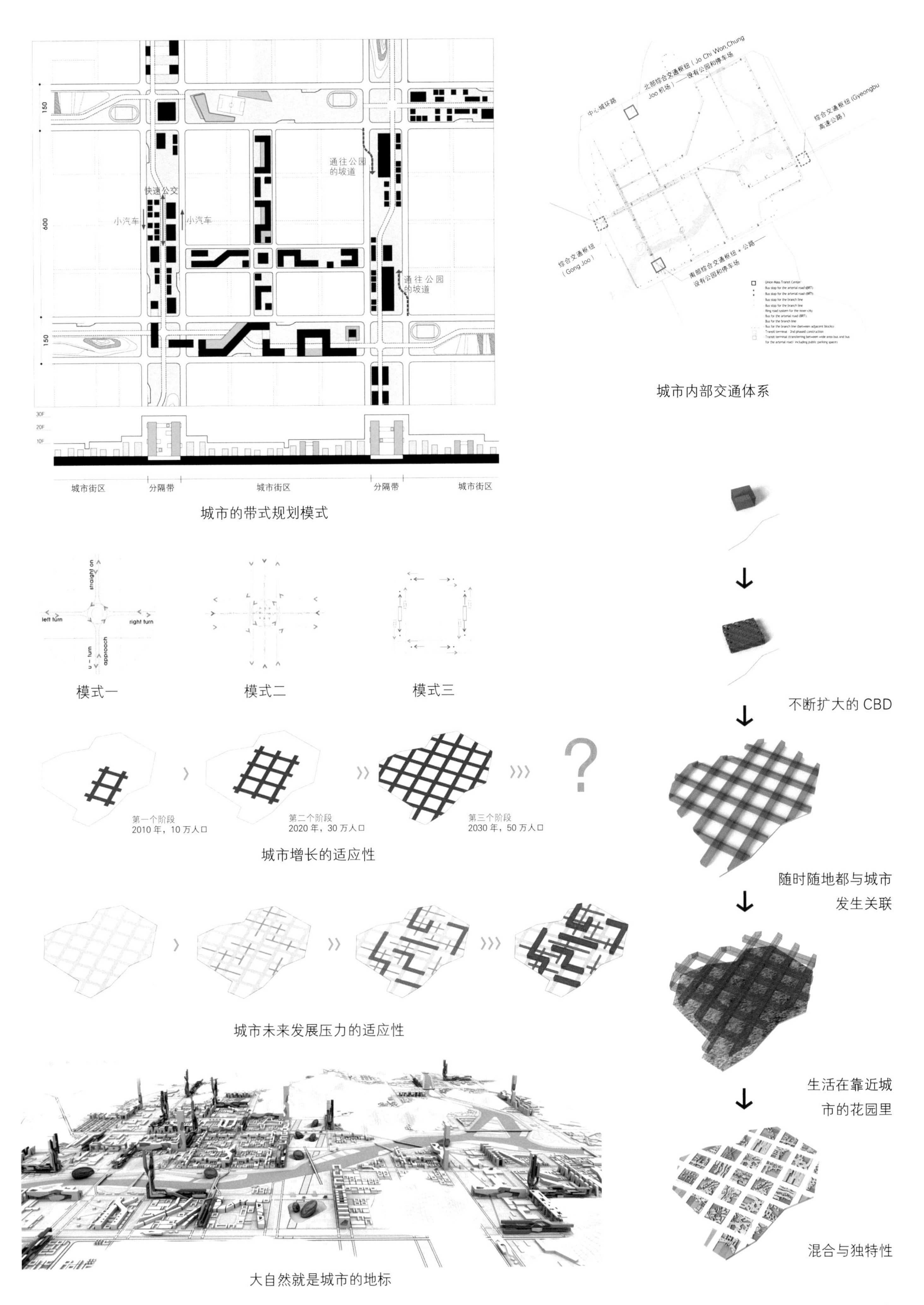

城市的带式规划模式

城市内部交通体系

模式一

模式二

模式三

城市增长的适应性

城市未来发展压力的适应性

大自然就是城市的地标

不断扩大的 CBD

随时随地都与城市发生关联

生活在靠近城市的花园里

混合与独特性

无题

Pak, Hun Young (韩国)
A. rum Architects / Post Media /
Go, Eun Tae / Kim, Han Jun

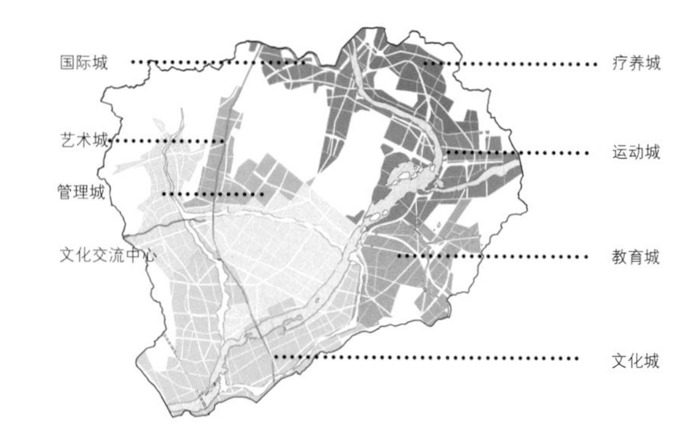

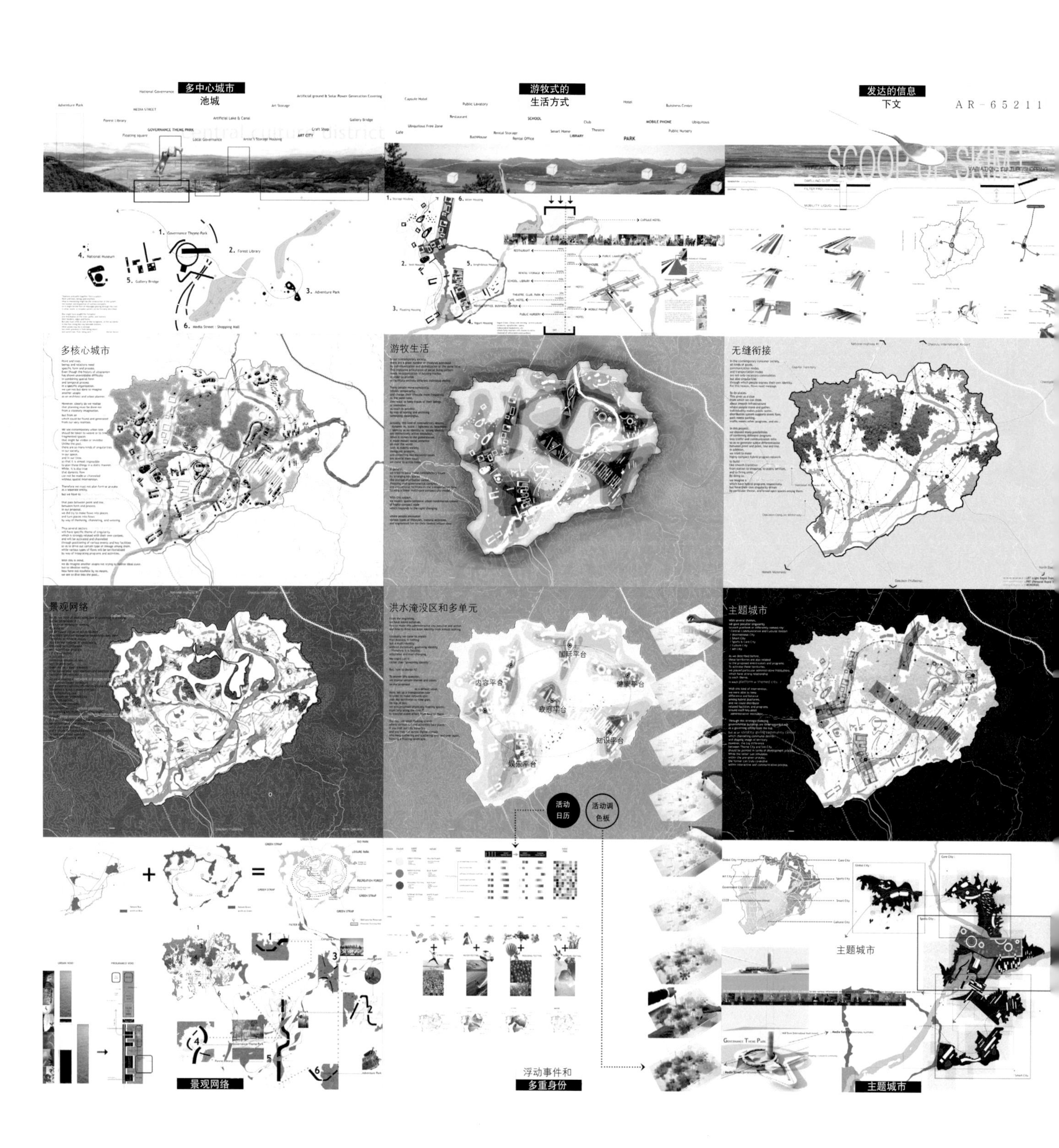

管理城

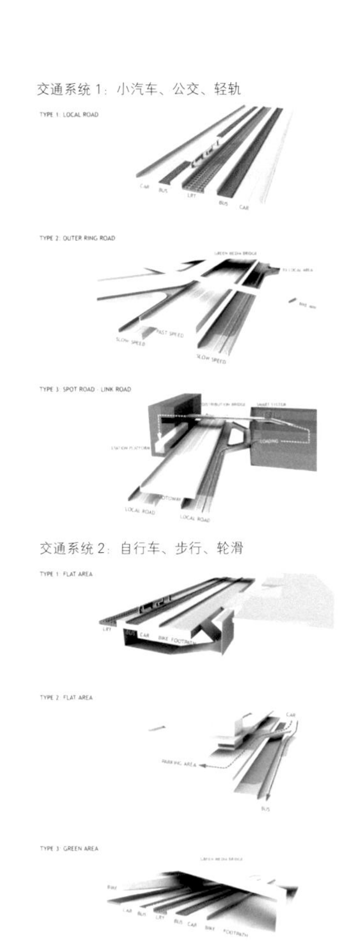

道路系统

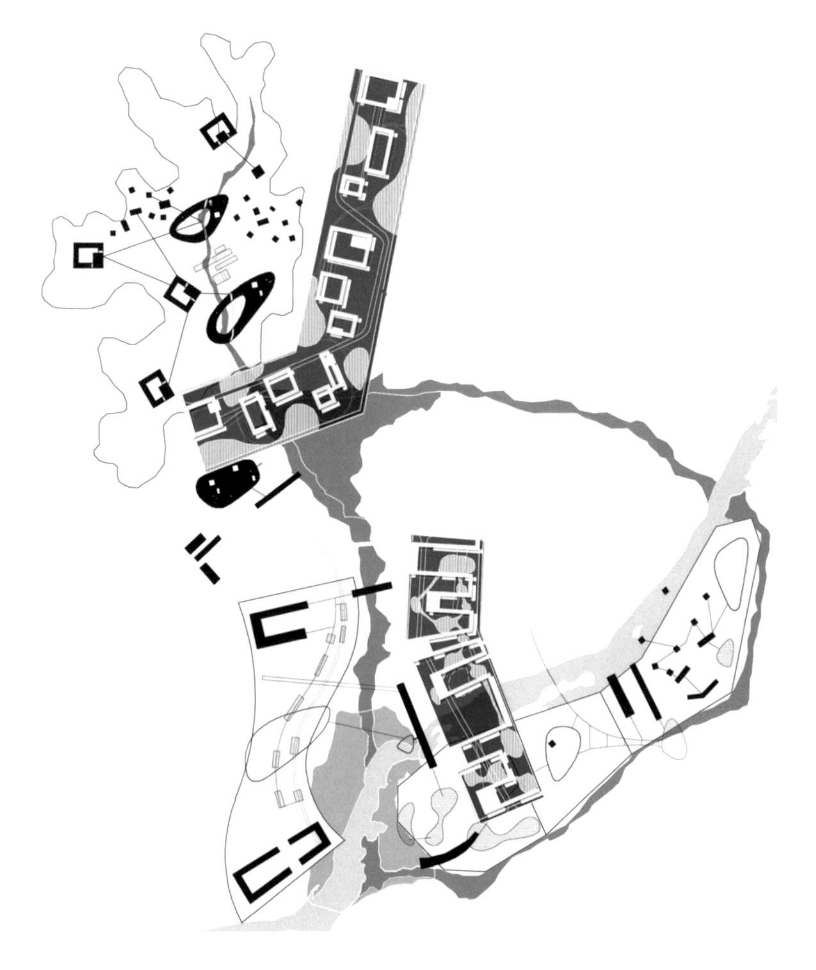

艺术和文化城

每一个阶段的发展密度

国际城

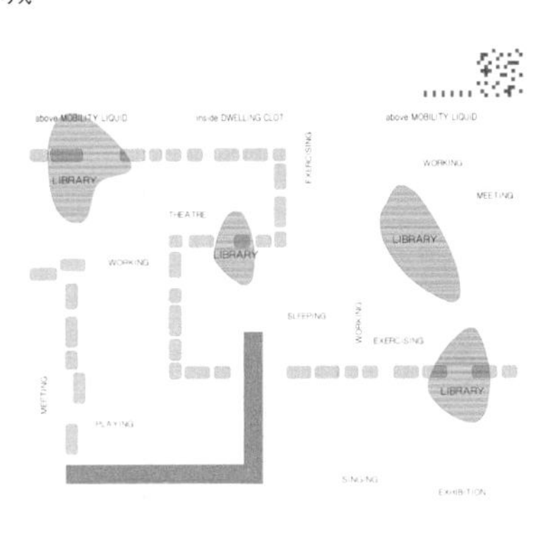

社区共享设施

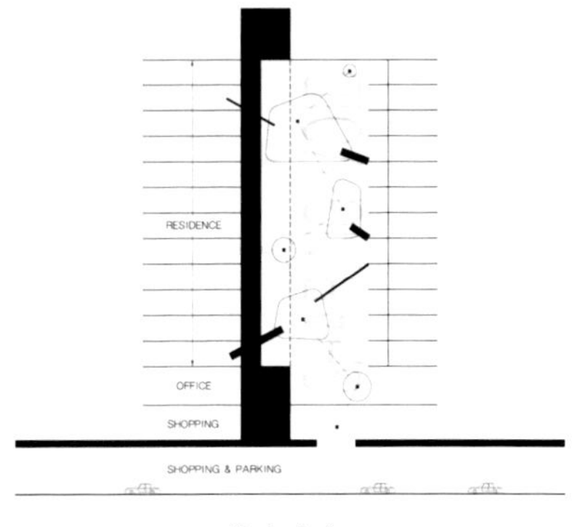

共享空间

数码城堡

Hwang, Doo Jin（韩国）
Hong, Soon Jae

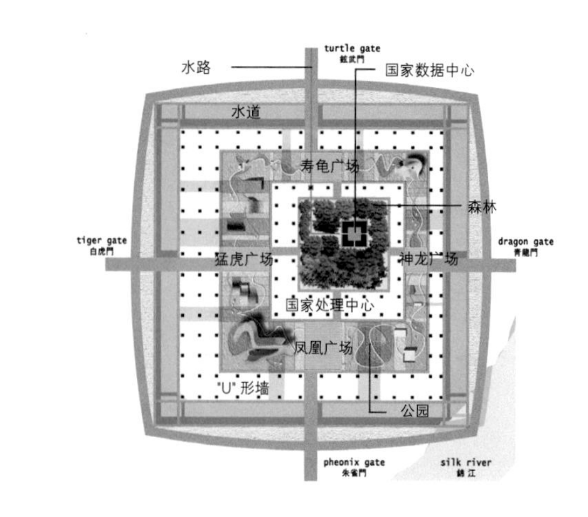

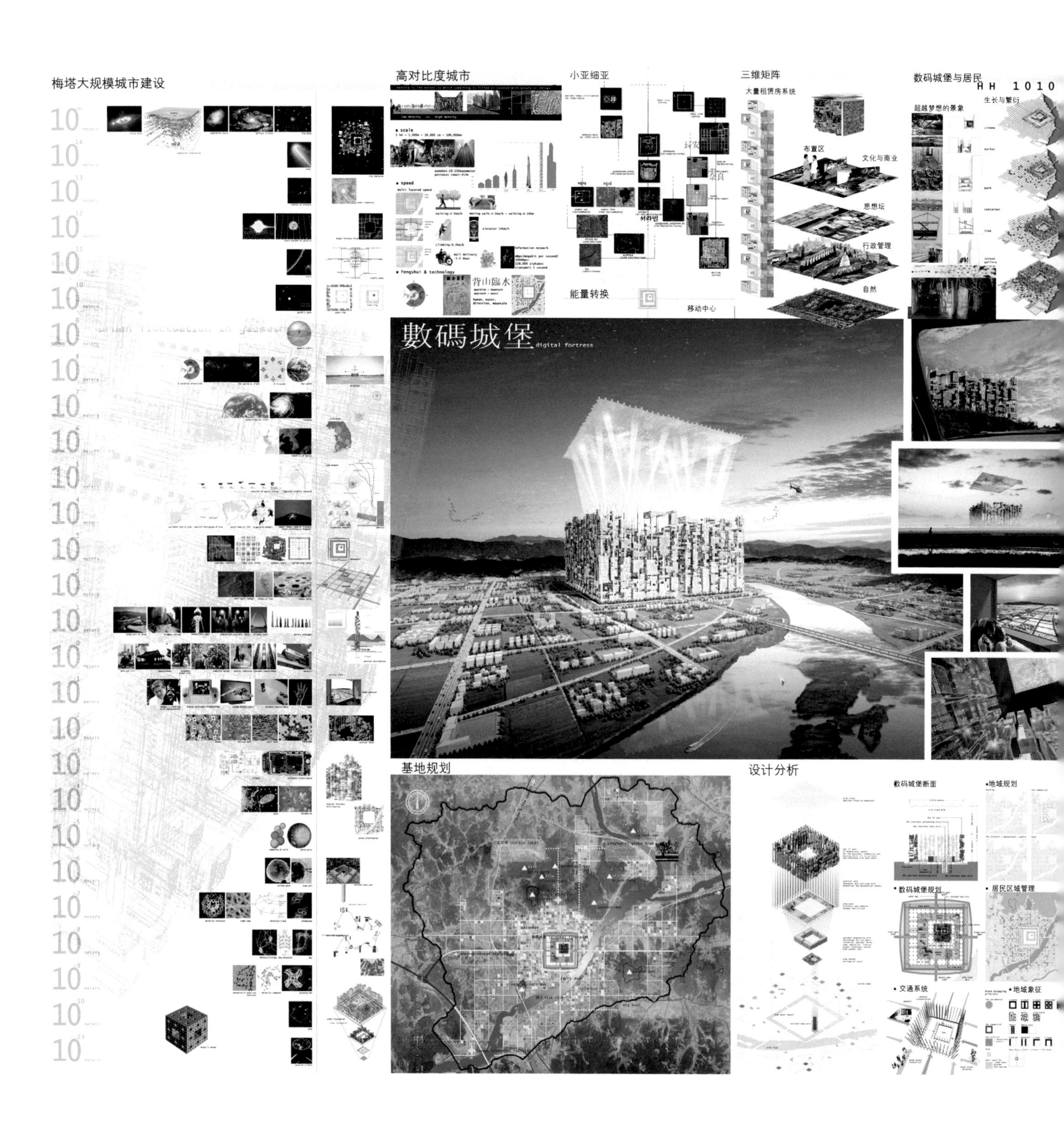

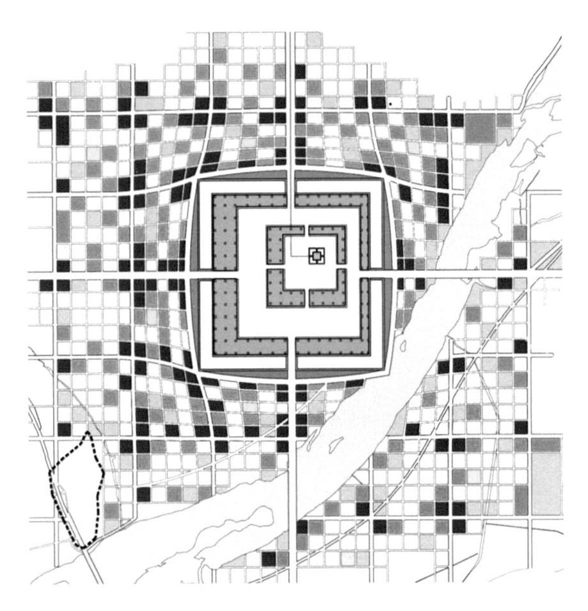

内部尺度的城市生活对照三位数字矩阵图

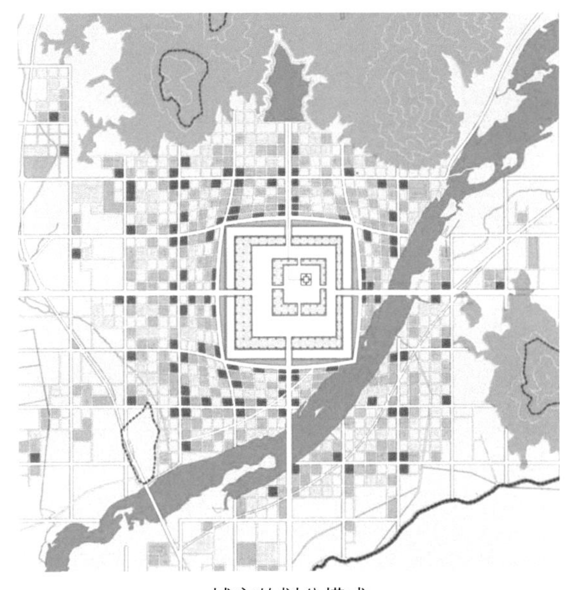

城市的划分模式

青龙 白虎 朱雀 玄武

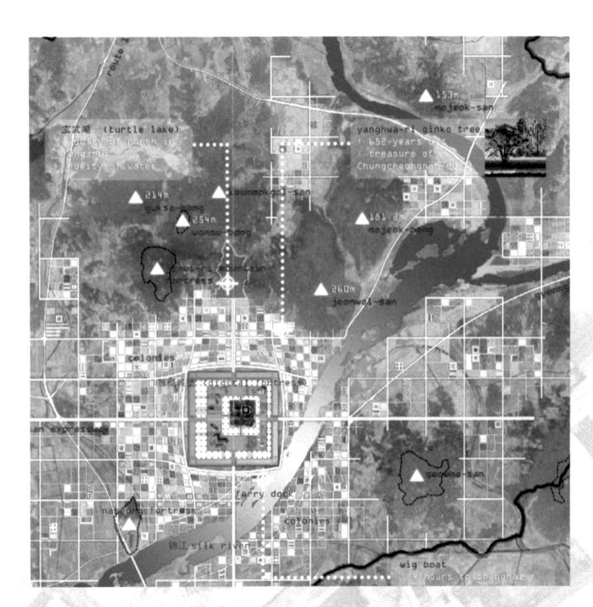

山脉和水域系统

无题

Yi, Eun Young（韩国）

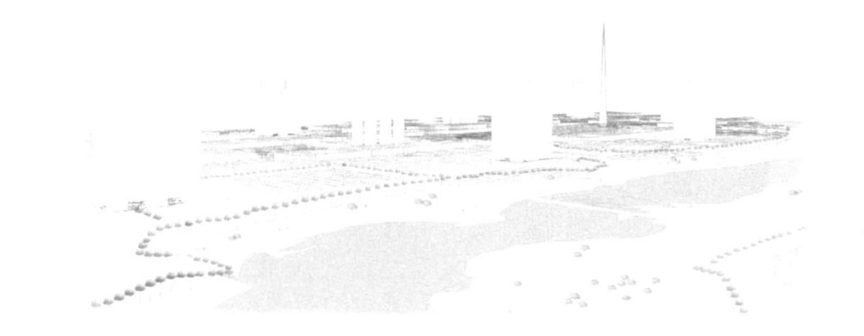

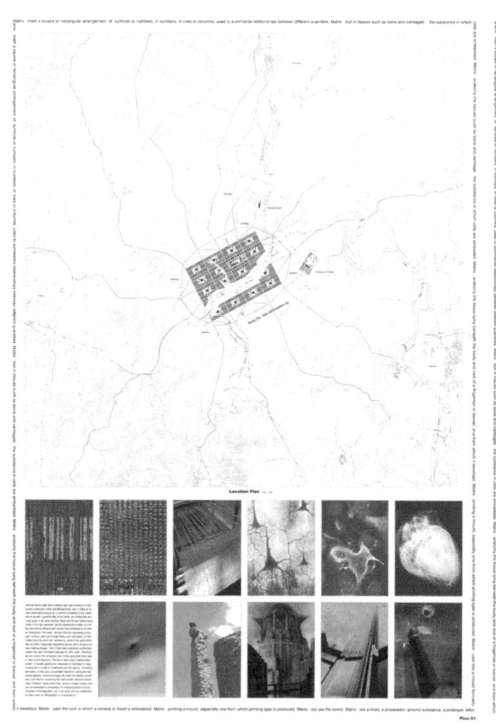

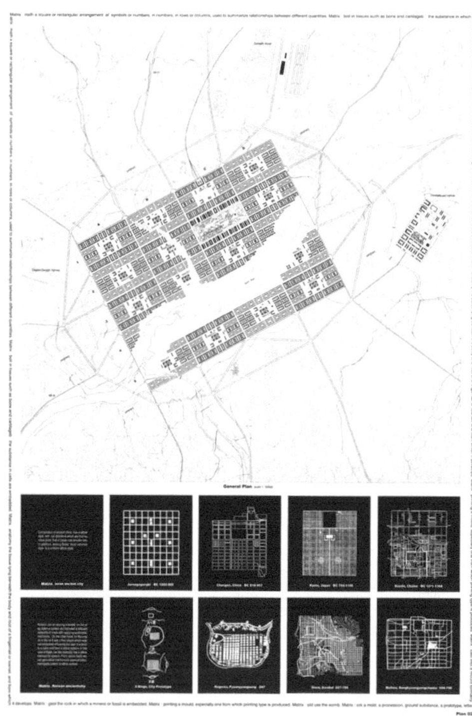

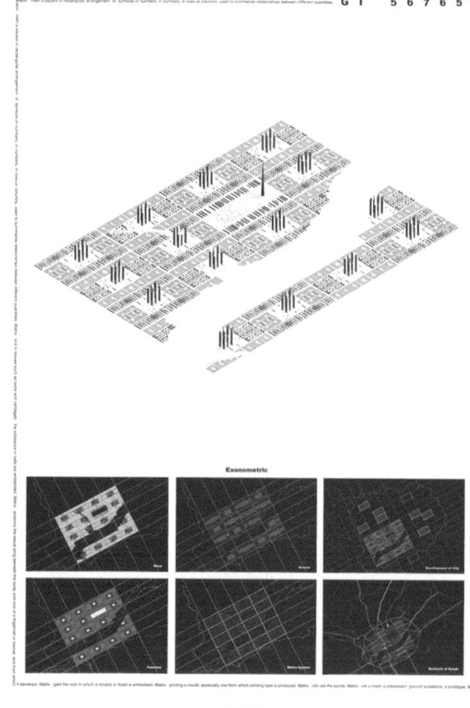

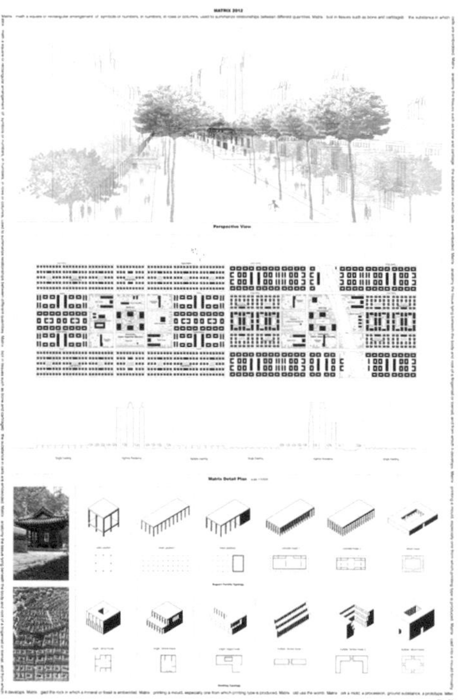

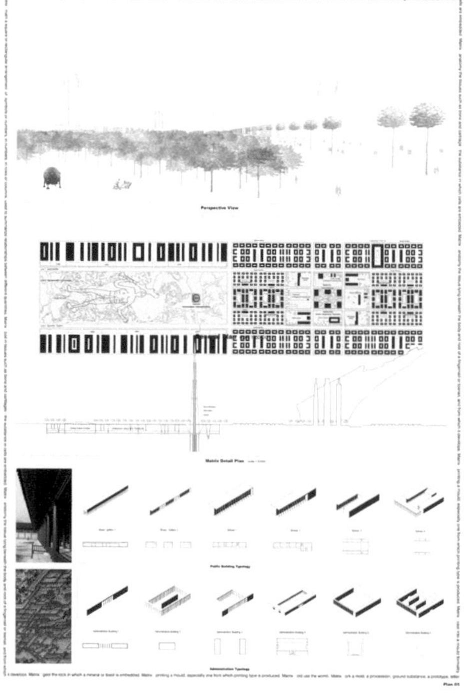

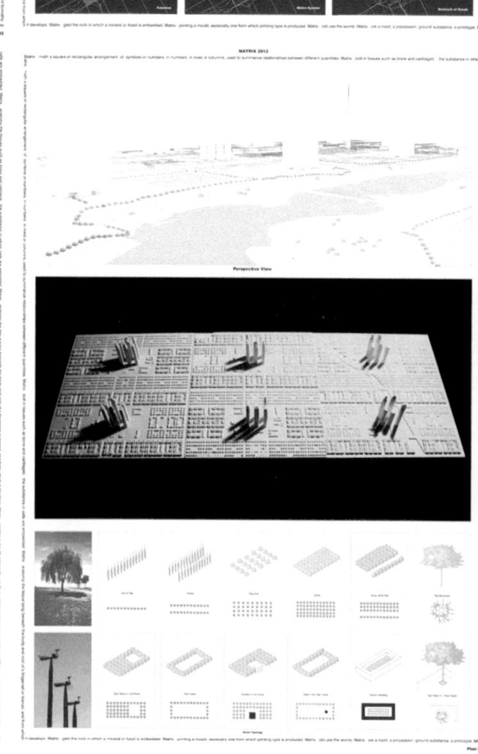

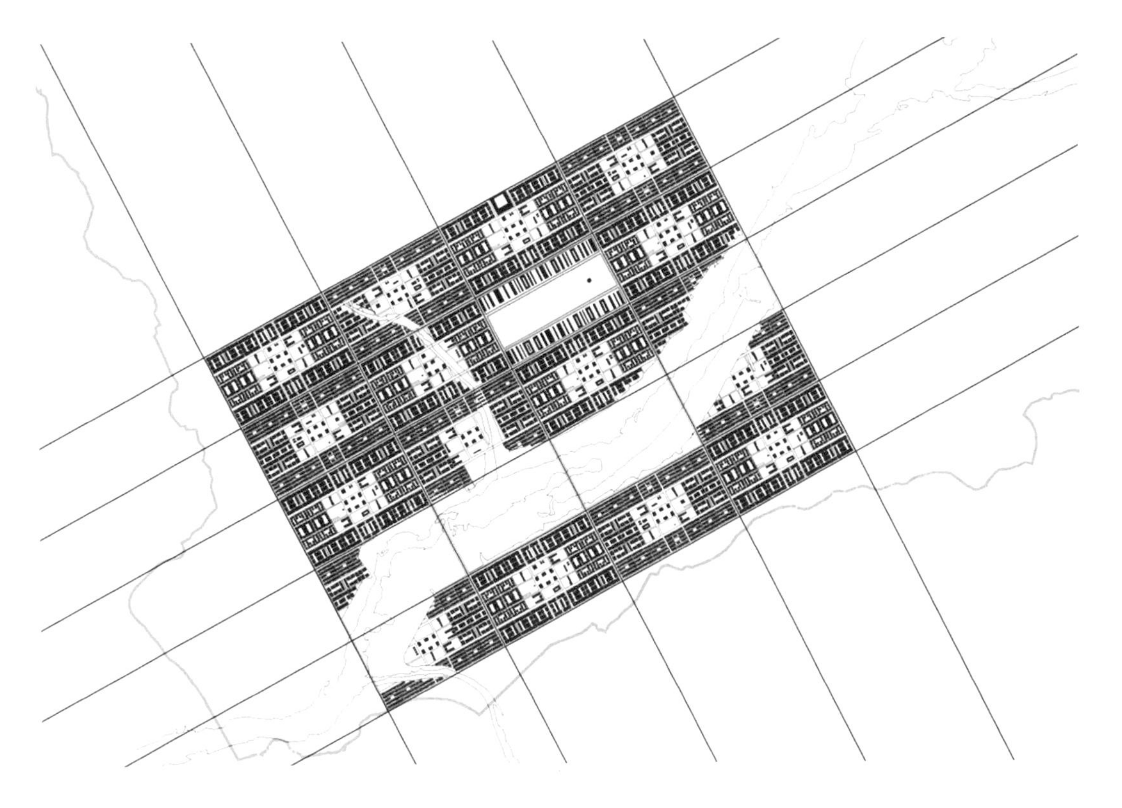

概念性图表 / 总平图

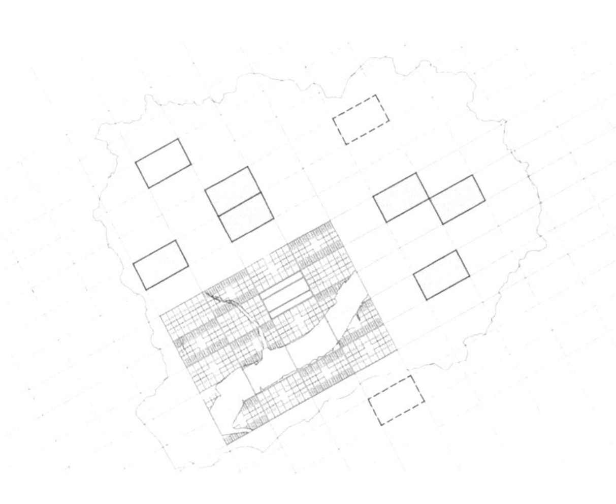

概念性图表 / 城市发展

概念性图表 / 道路系统

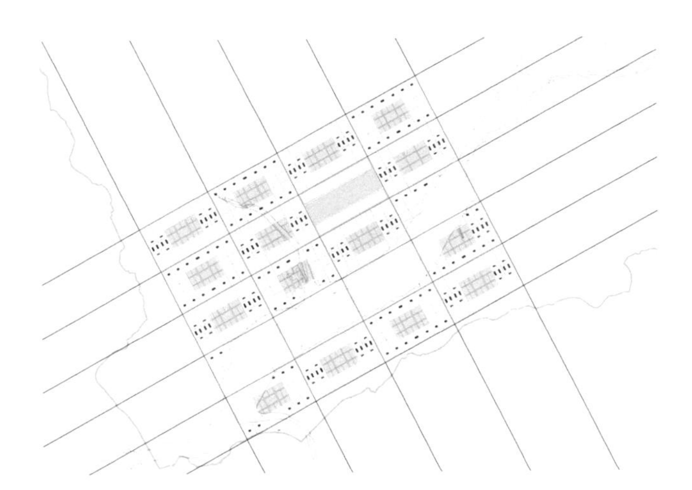

概念性图表 / 绿化

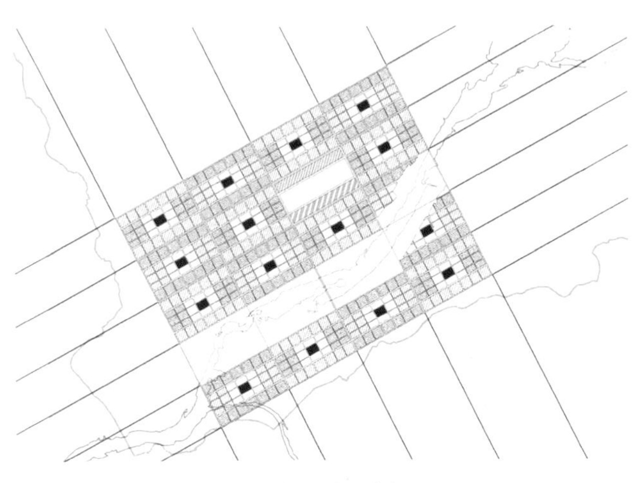

概念性图表 / 功能

无题

Treuttel, Jerome (法国)
Treuttel, Jean Jacques / Garcias, Jean Claude /
Pourrier, Stephane / Fichou Torres, Laurent

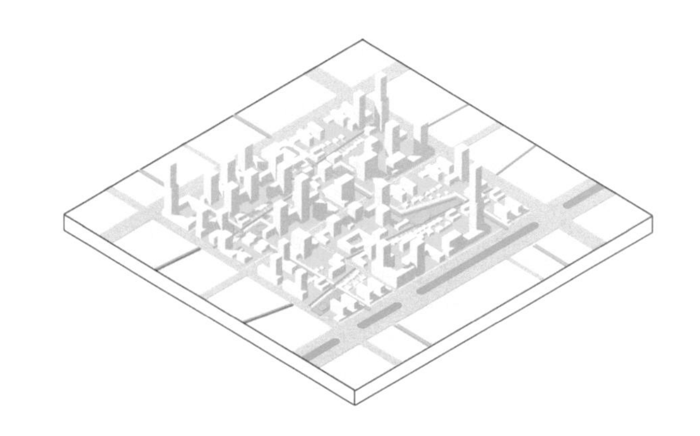

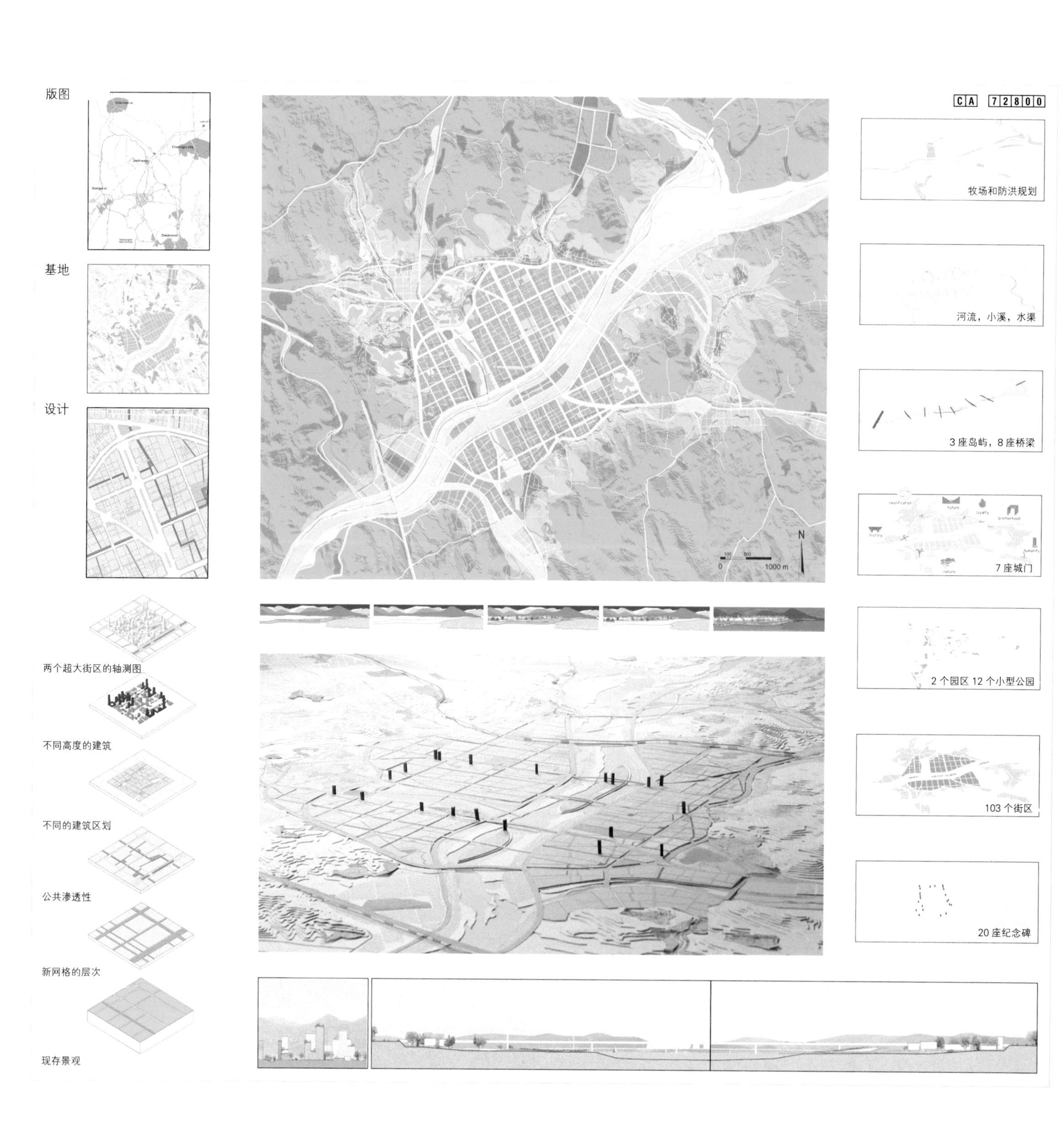

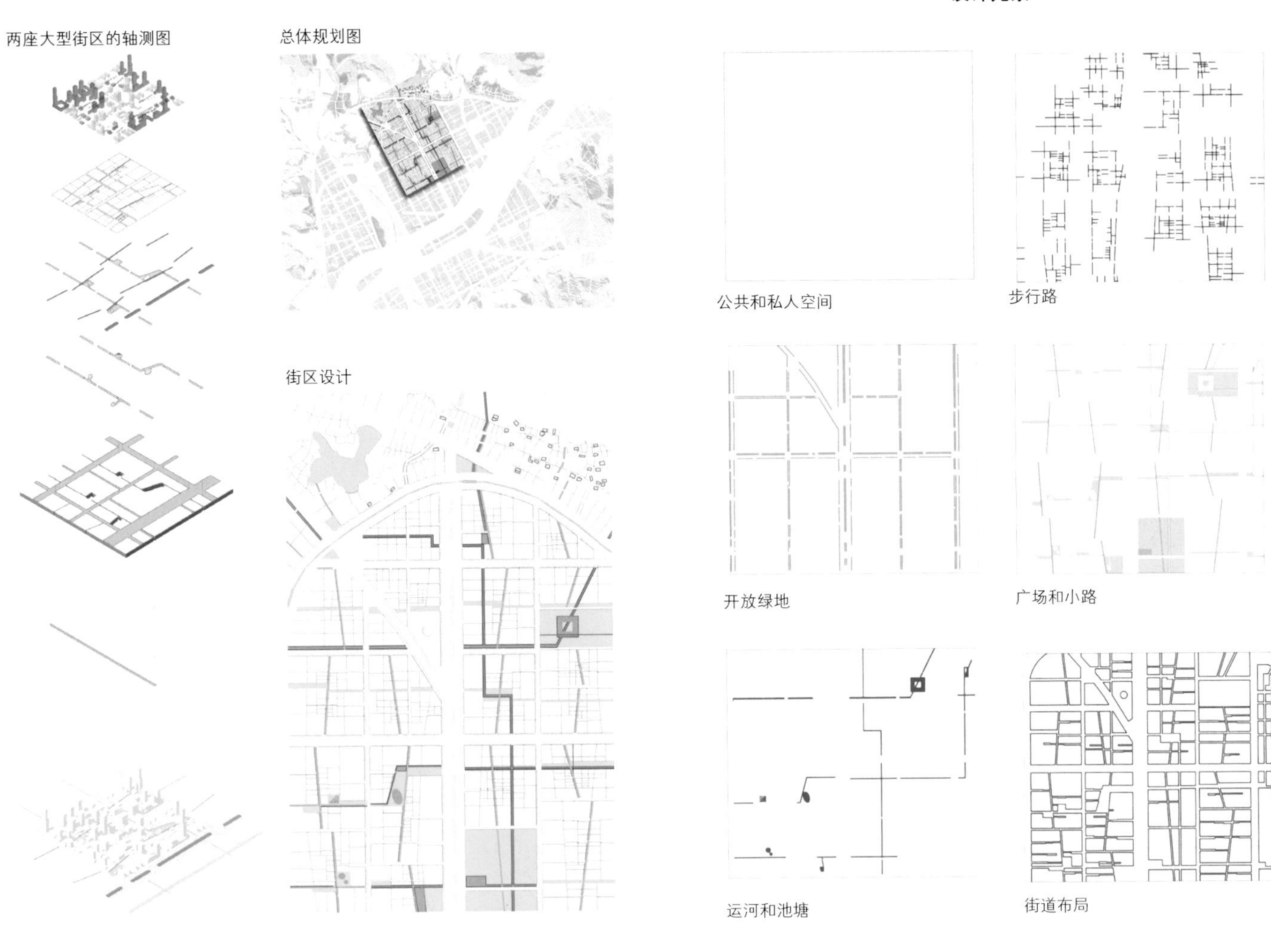

自然轴线和生物圈

Cheonan-si
Cheongju-si
ECO - Ring
Gongju-si
National Park
Daejeon-si

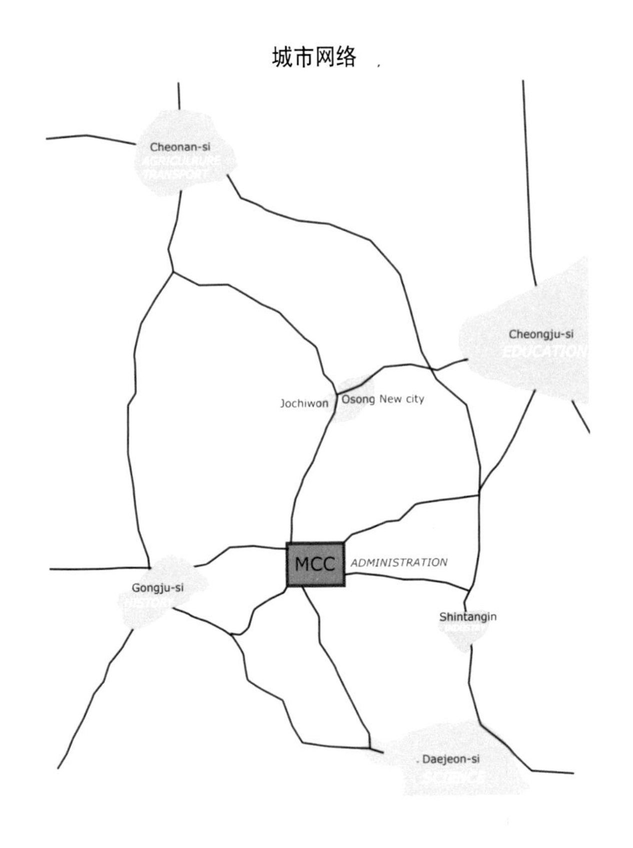

无题

Jonathan D. Solomon（美国）
Kutan A. Ayata / Aleksandr Y. Mergold

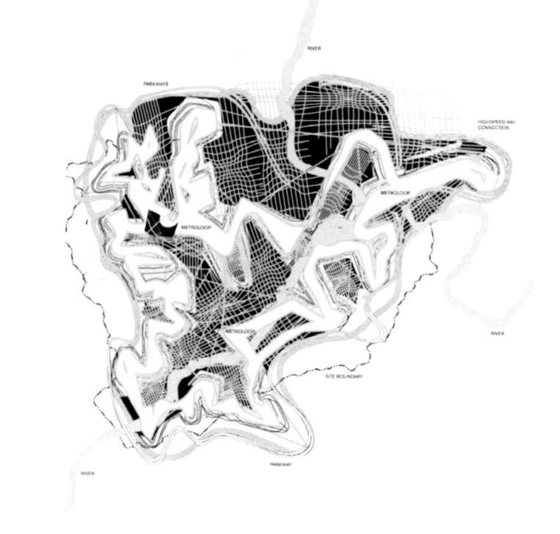

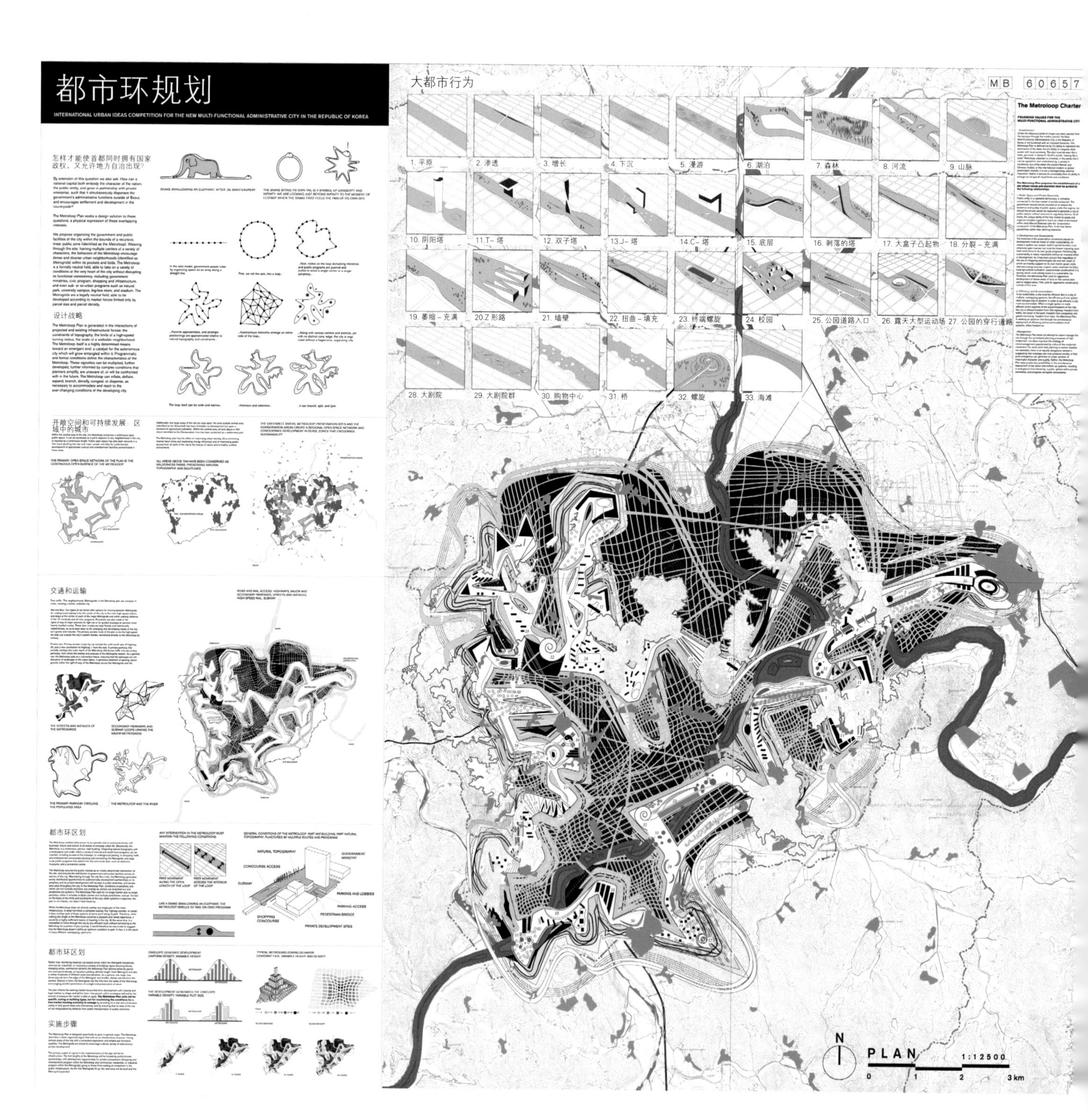

密集交错的街道和林荫道

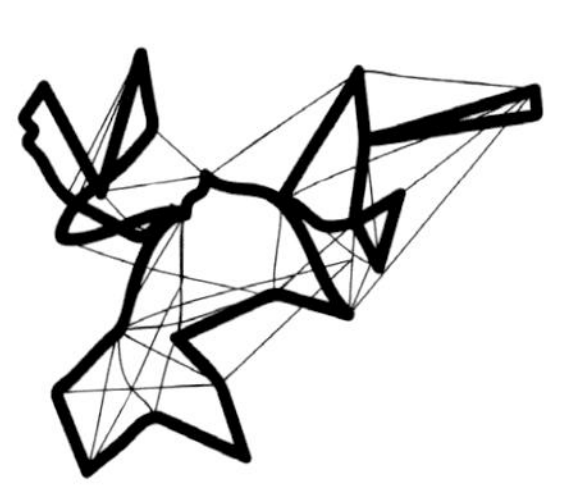
二级道路和地铁环路连接主要的道路网

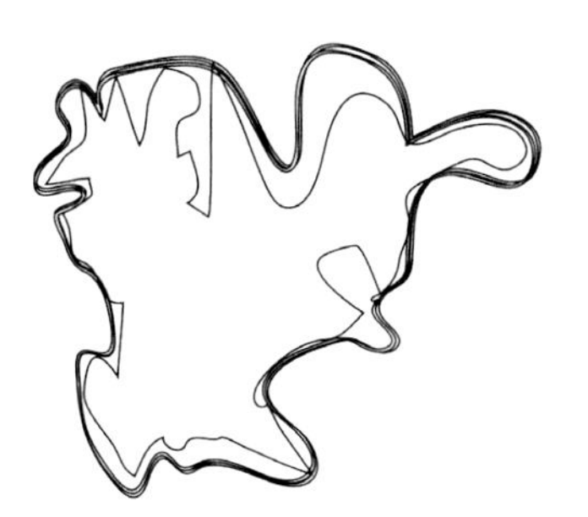
初级道路环绕连接着居住地区

都市环与水系

封套式发展：统一密度，不同高度

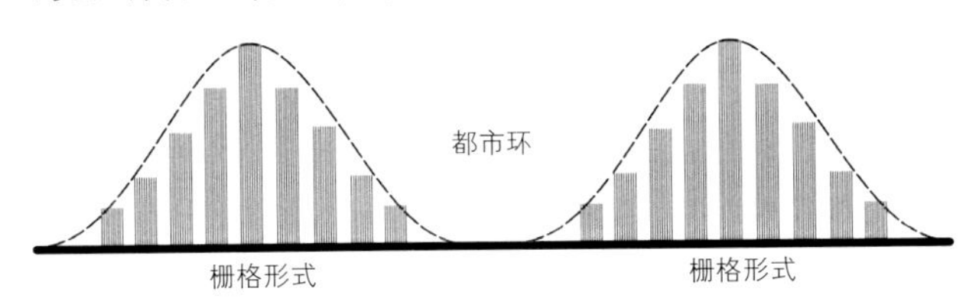

发展导致了封套模式：不同密度，不同尺寸

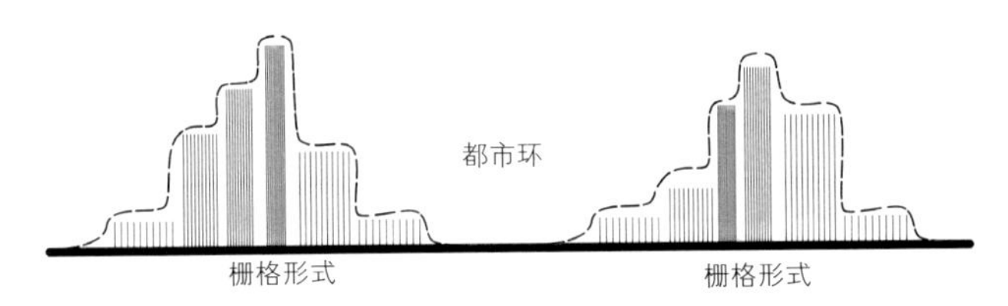

对都市环的任意的干涉都必须遵循以下的条件：

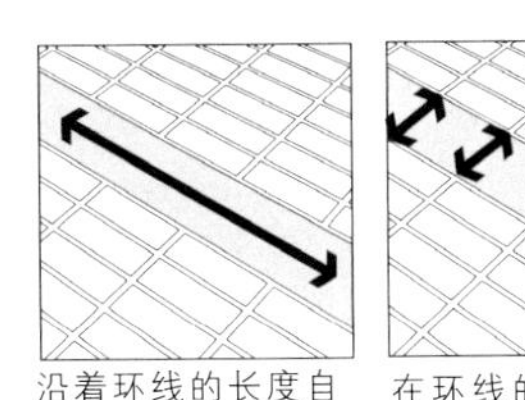
沿着环线的长度自由的移动

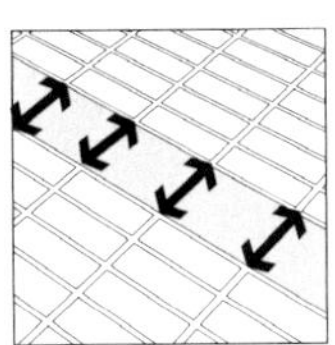
在环线的内部自由的穿越

像一条蛇一样吞下一头大象。都市环膨胀以容纳整个城市

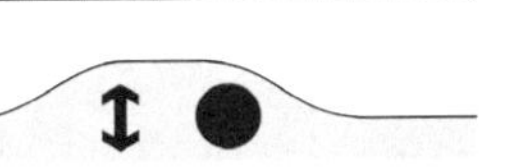

典型的栅格式区划模式，不变的F.A.R，不同的高度和密度

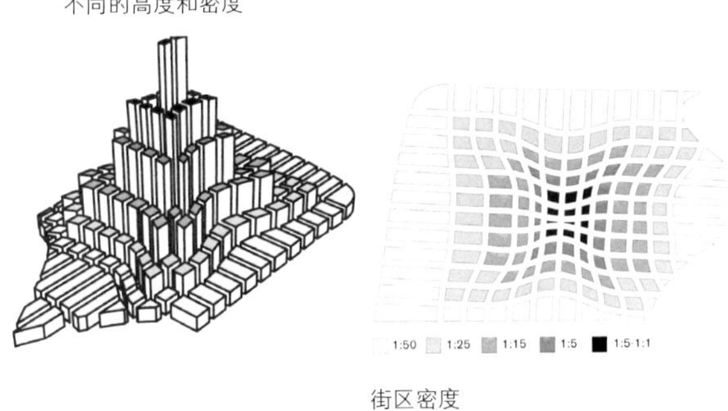

街区密度

都市环的普通状态：一部分是建筑的基础，另一部分是自然地形，线路和纲领

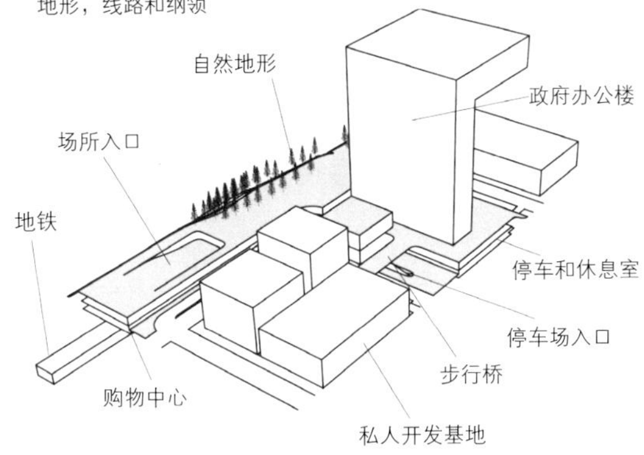

5年

10年

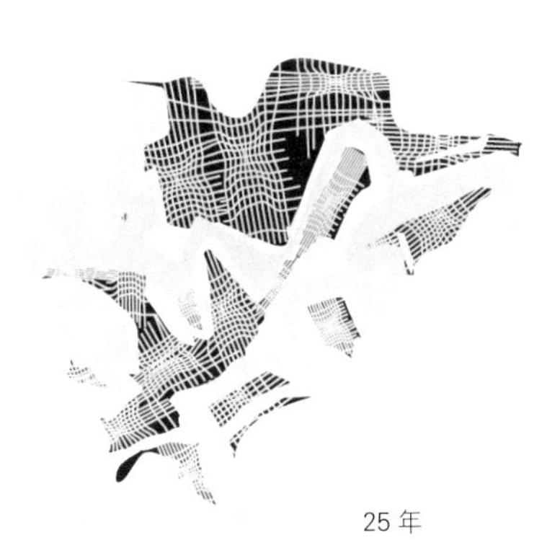
25年

都市环规划阶段

无题

Choi, John（韩国）
Choi, Ropiha / Mcgregor and Partners /
Garbutt, Michael

阴： 自然 开放 深沉的 安静的 保守的 山脉 河流 洪水 平原 呼吸 旅行 女性 守望 新鲜 捕鱼 散步 支持 运动 游泳 缓慢

阴阳：

阴阳是相反的两个方面。阴的一部分是阳，阳的一部分是阴。阴和阳是相辅相成的。

阳： 城市 强烈 内敛 喧闹 工作 浓厚 男性 社会 消费 宴会 痛快 政治 粗犷 冒险 创新 金钱 炫耀 迅速

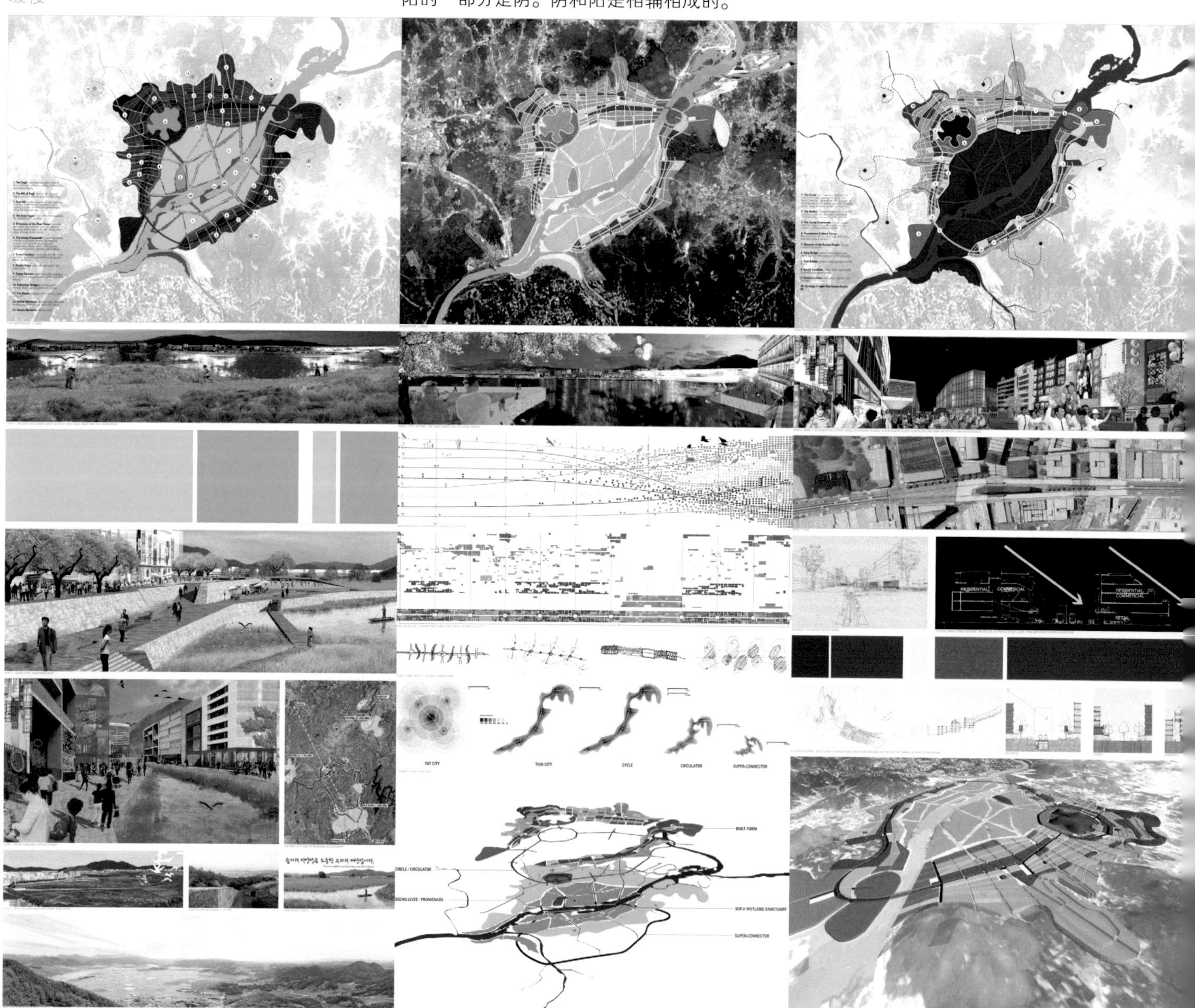

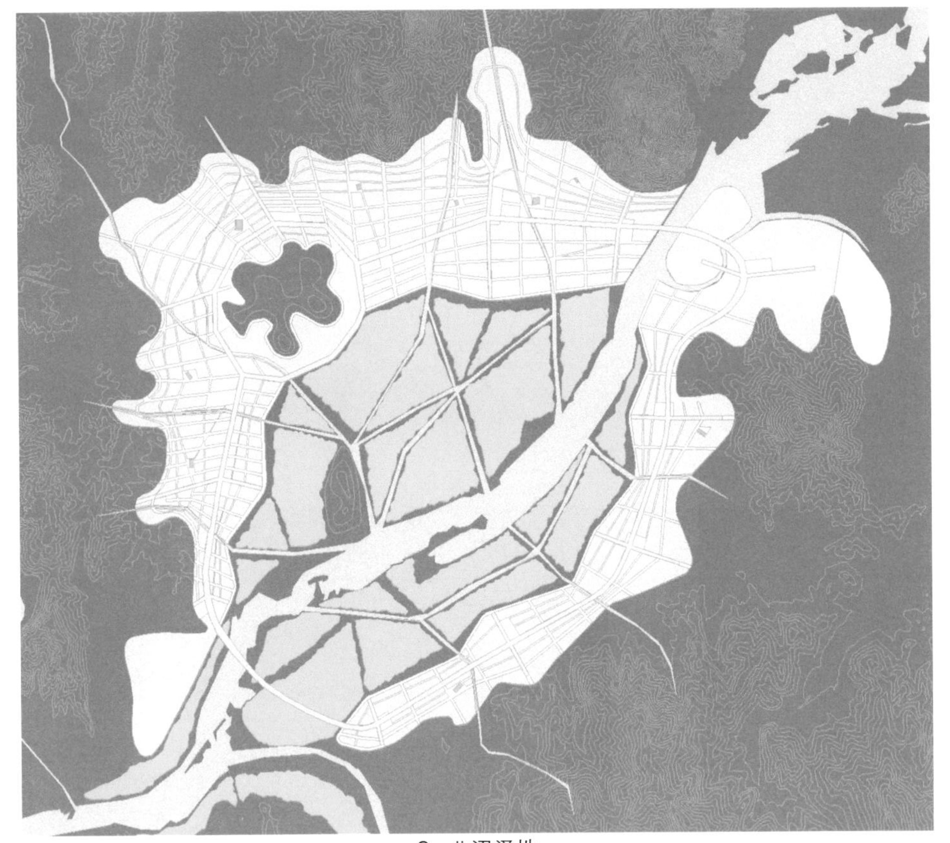
Supji 沼泽地

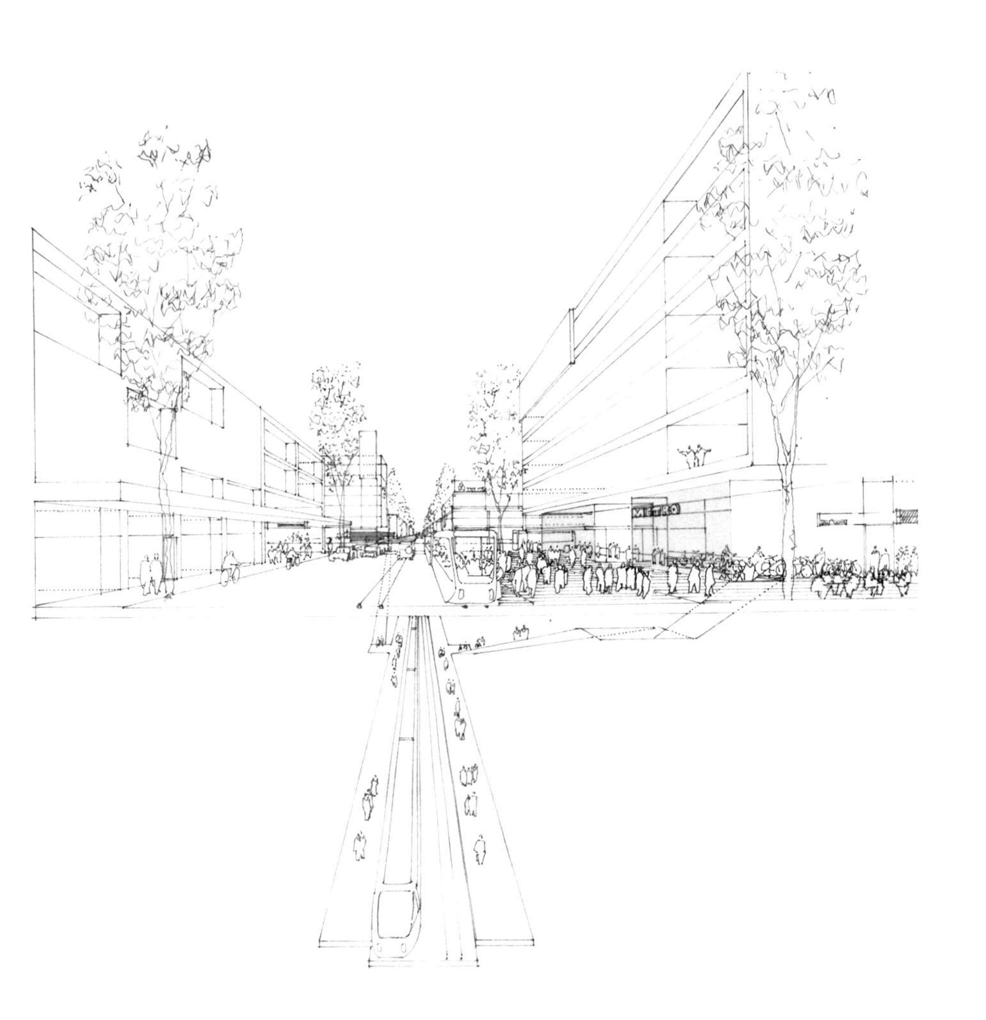

街道绿化和建筑形式

Seong 的防洪堤和散步道是新城的一个标志

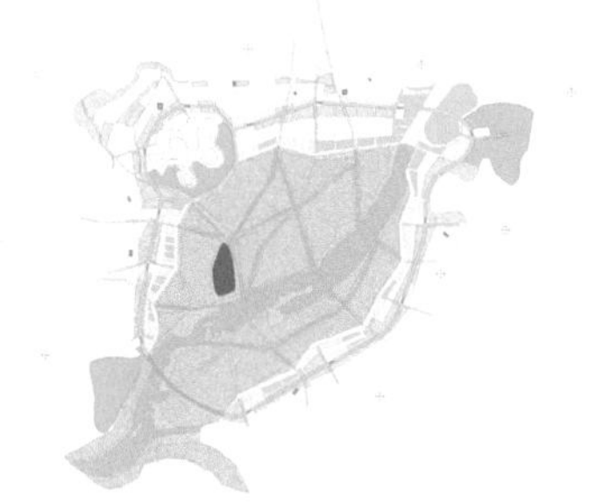
Supji 和 Supji 的小山

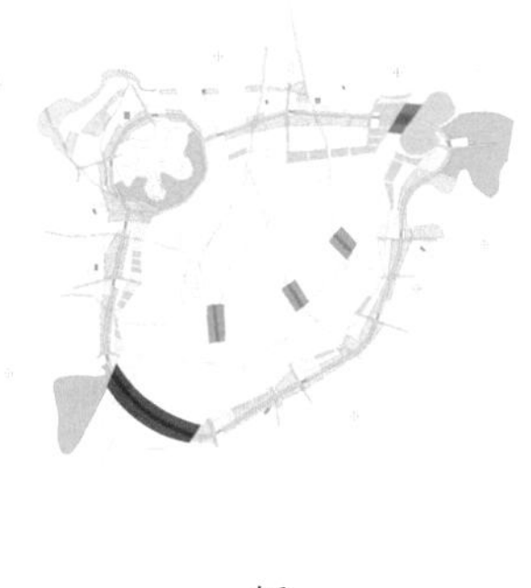
桥

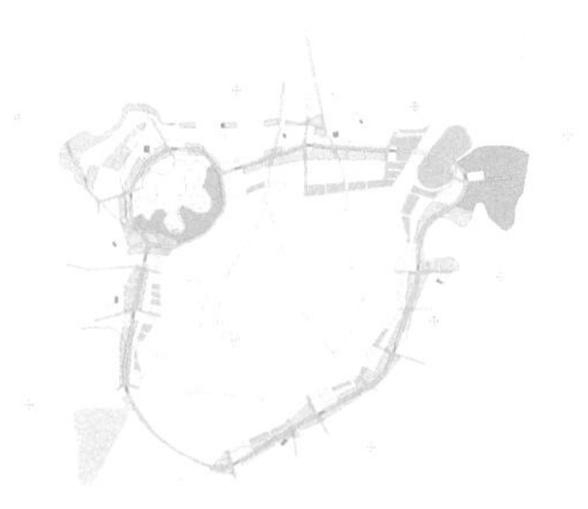
韩国人民的博物馆

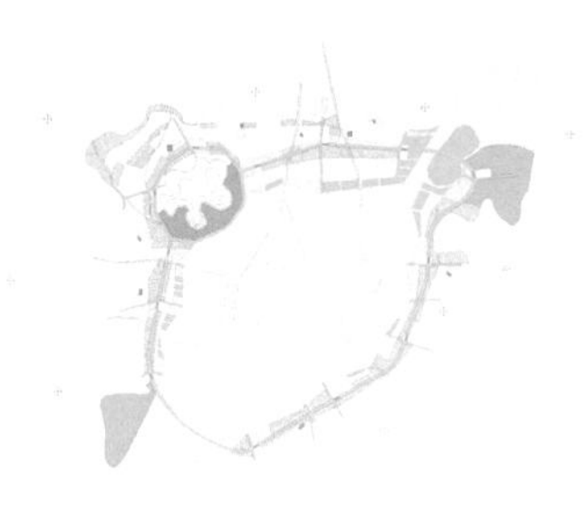
政府

花园城市

Stienon, Christopher（美国）
Kim, Eung Soo / Huang, Shin Yau

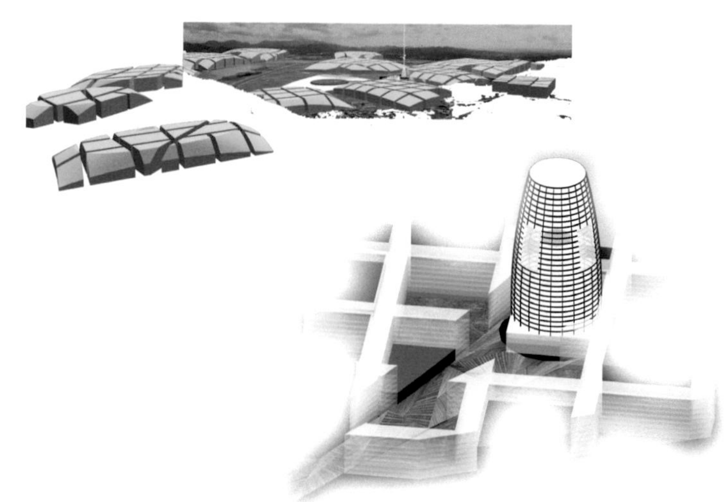

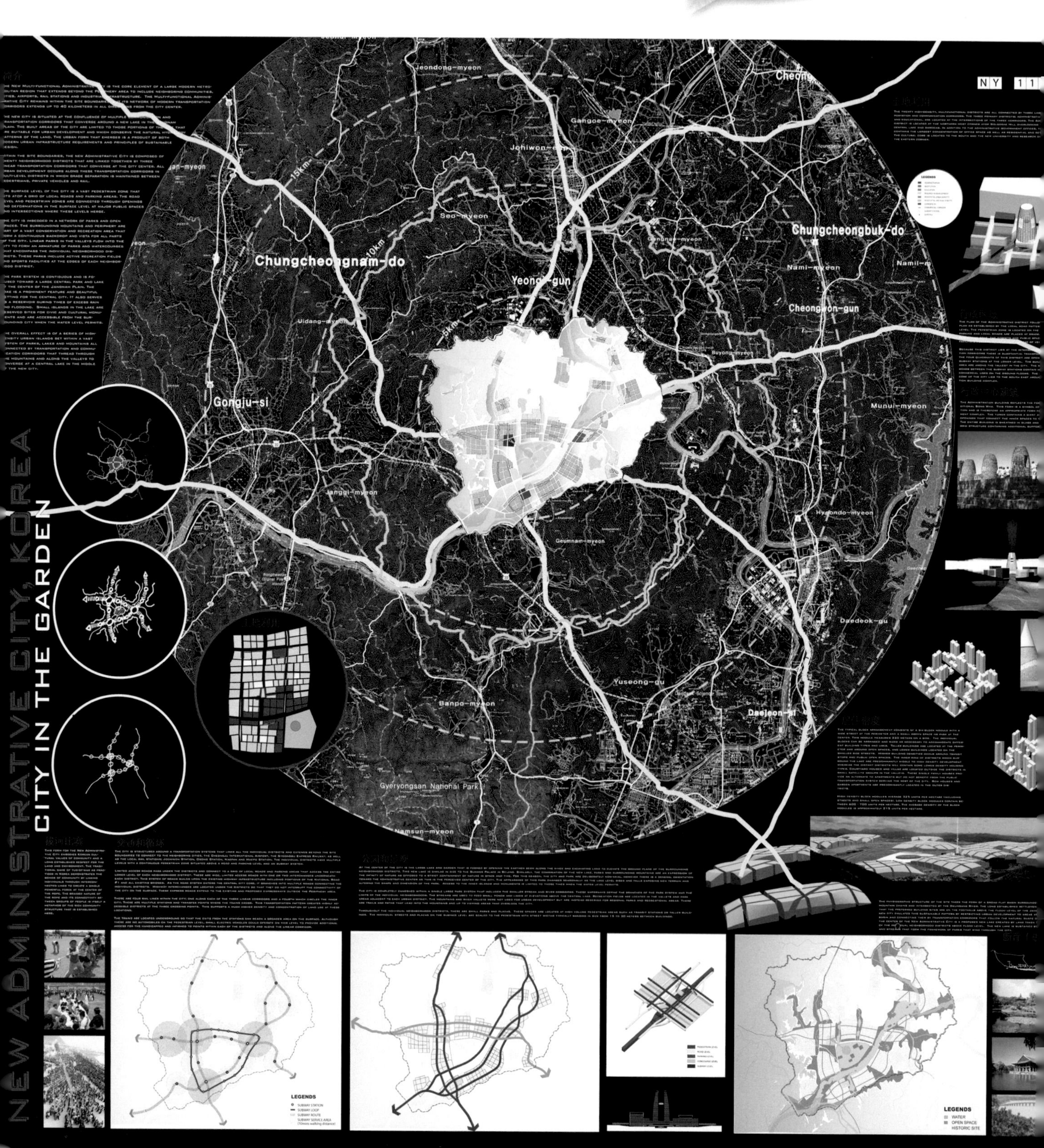

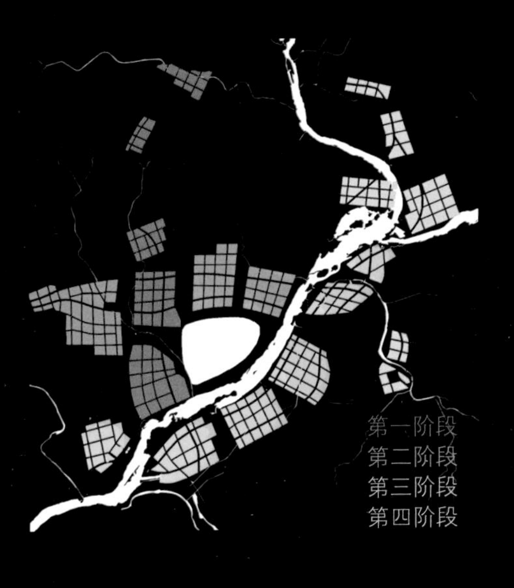

概念性示意图：三个交通走廊

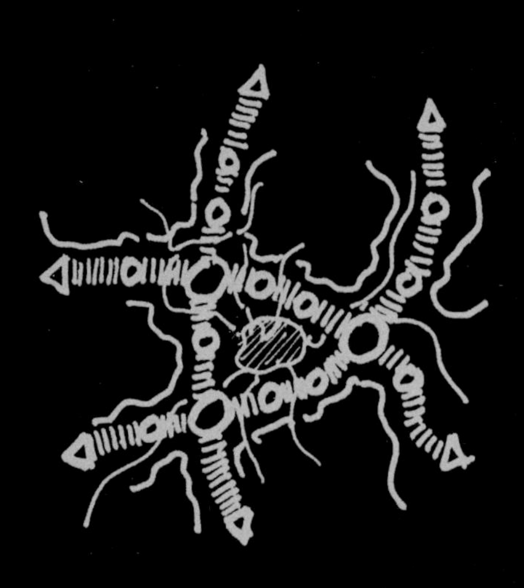

概念性示意图：处于花园中的城市

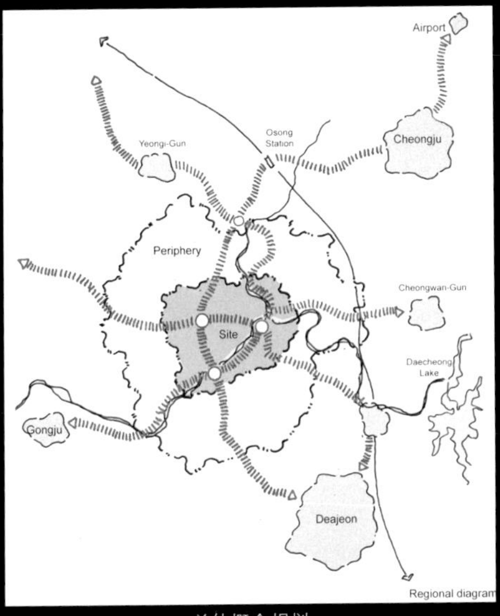

总体概念规划

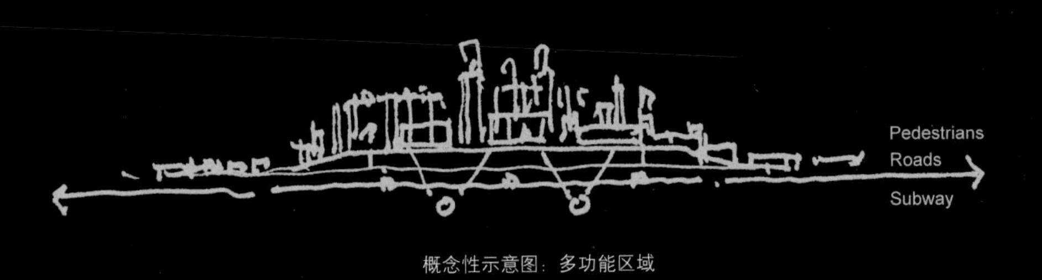

概念性示意图：多功能区域

城市建于山脚下，山下湖泊的周围是一个大池塘

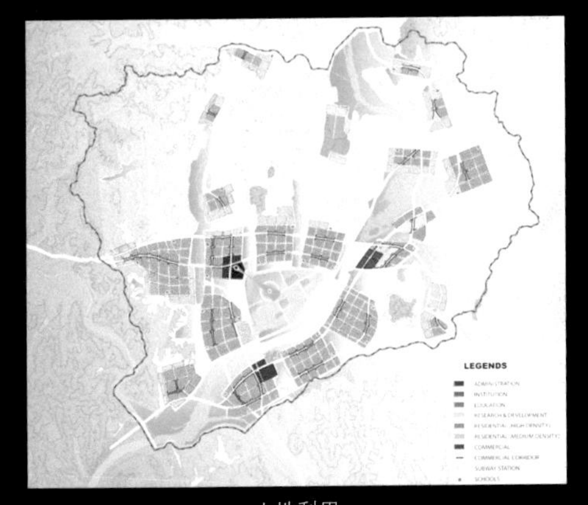

土地利用

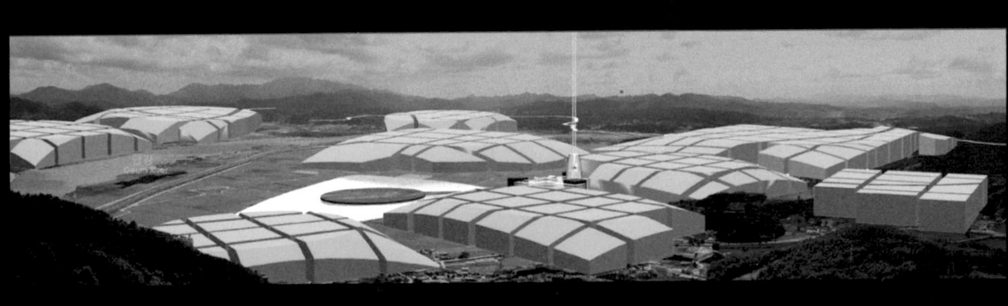

这个概念性示意图展示了城市的大致风貌

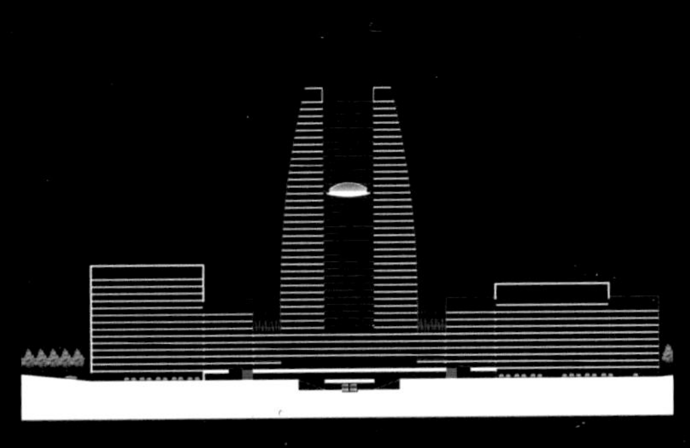

行政大楼立面图

湖面视角的行政大楼效果图

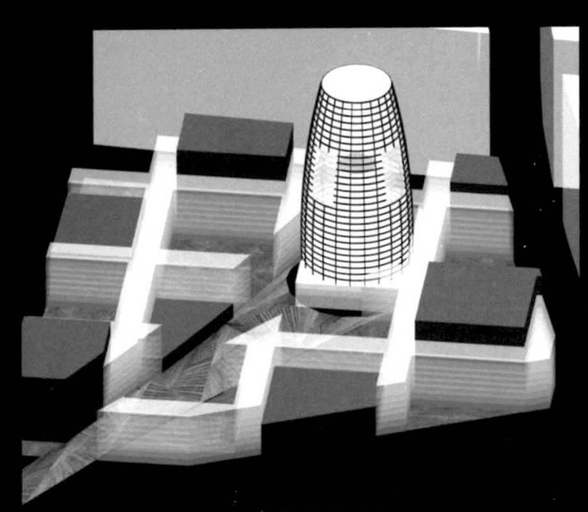

行政大楼的轴测图

衍生城市

BaritoAdi Bulden Raya Ganda Rito（印度尼西亚）

Ilya Fadjar, Maharika / Revianto Budi, Santosa / Ariadi, Susanto / Prihatmaji, Yulianto Purwono

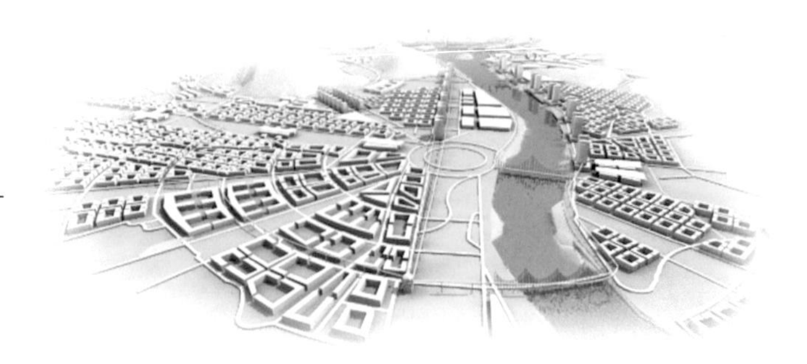

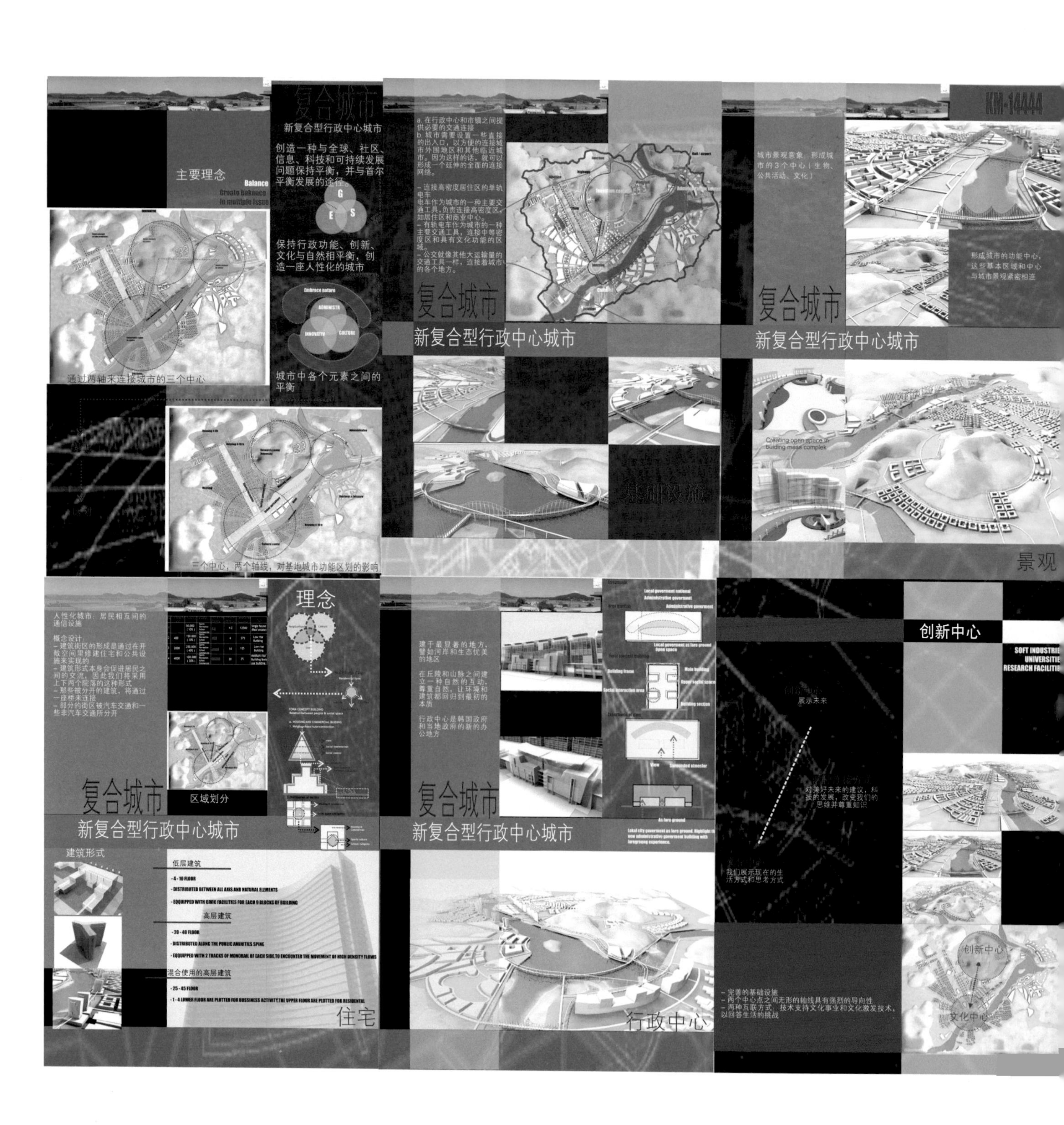

国际的—多元化，社区，可交换的身份

重新理解对于场所的感知

可持续发展—资源的匮乏，过度的消耗，与自然的分裂

重新建立与自然的关系

信息和技术—交流模式、交通模式、社会关系模式

重新形成一种模式

国际的理念 本土的城市

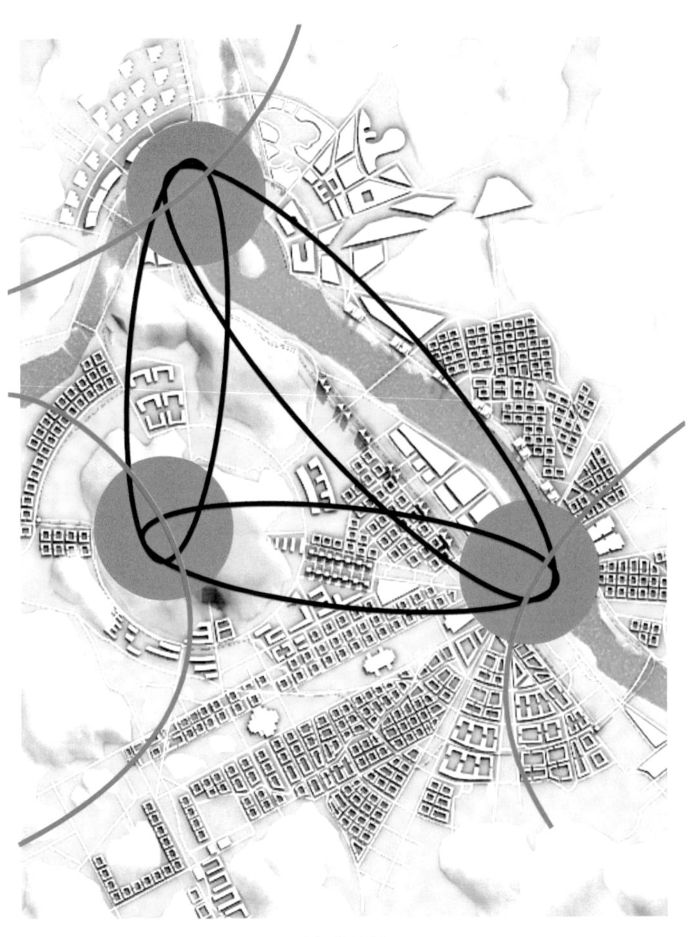

基础设施

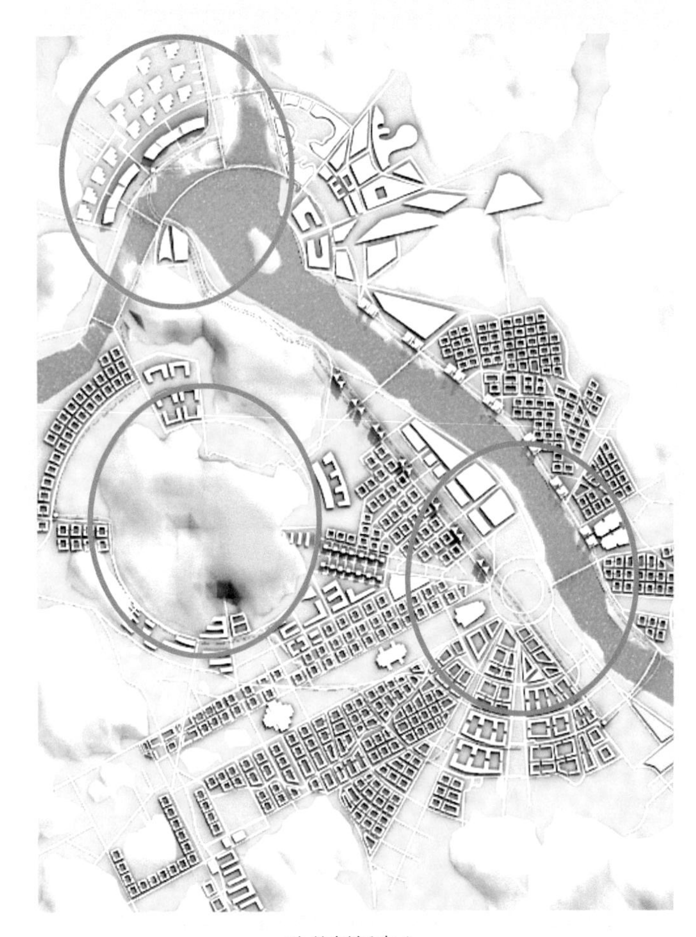

改革创新中心

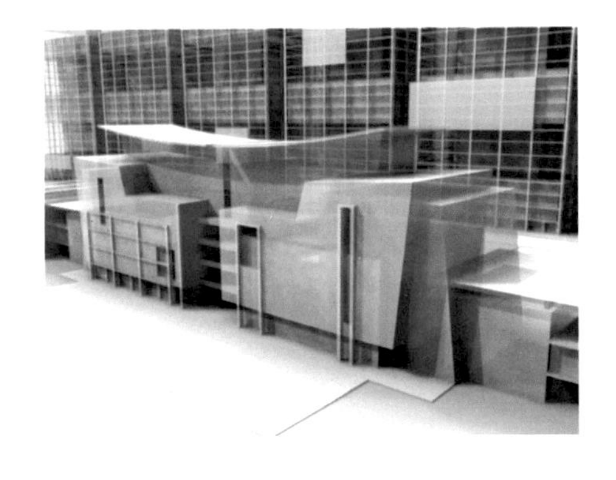

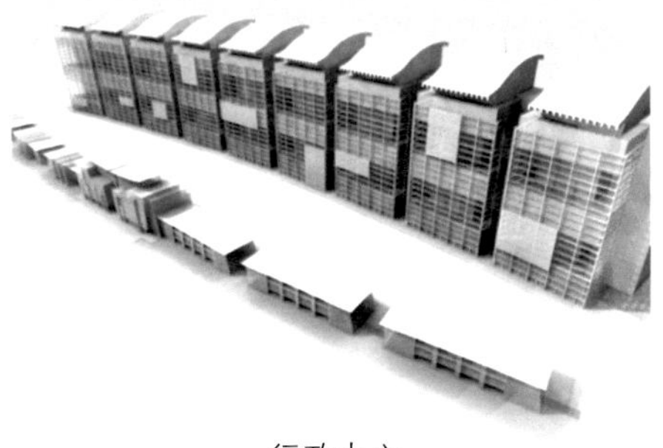

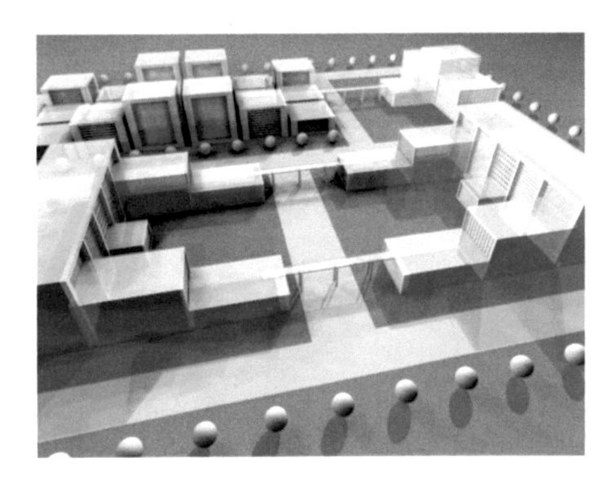

行政中心

无题

Grosch, Leonard（德国）
Wessendorf, Joerg

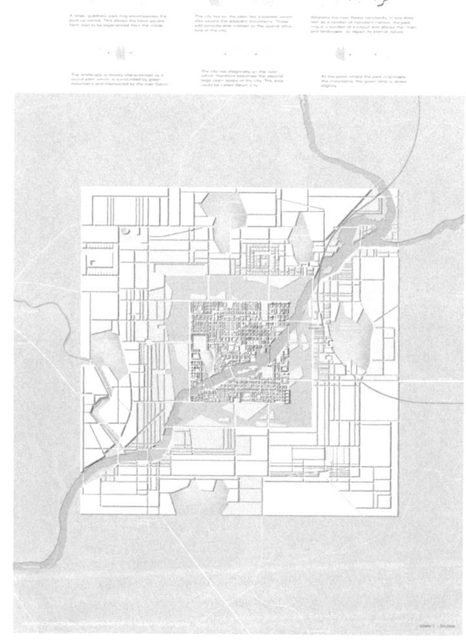

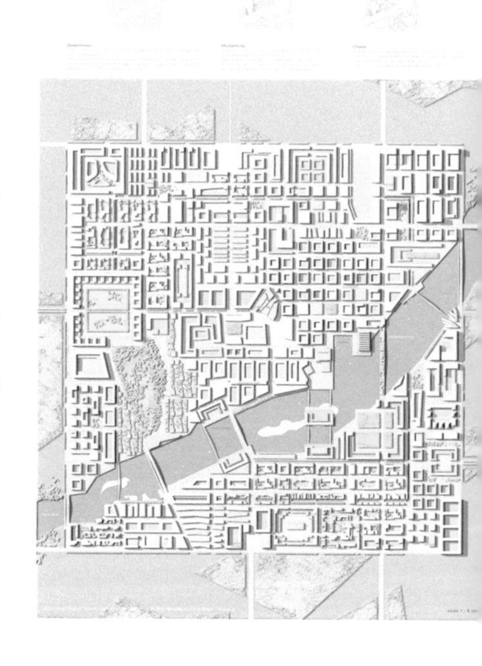

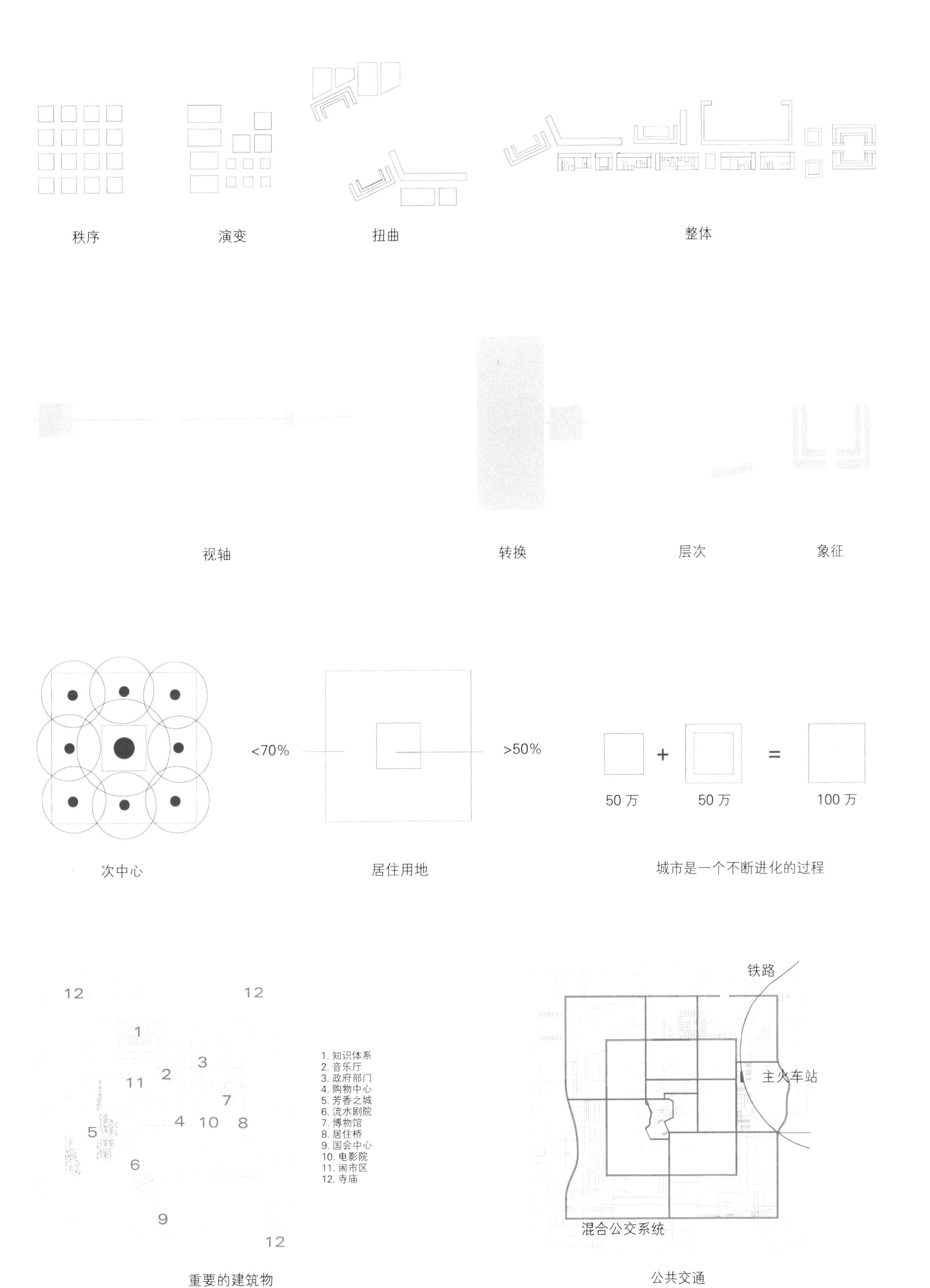
秩序
演变
扭曲
整体
视轴
转换
层次
象征
<70%
>50%
50 万
50 万
100 万
次中心
居住用地
城市是一个不断进化的过程
12
12
1
3
2
11
7
4
10
8
5
6
9
12
1. 知识体系
2. 音乐厅
3. 政府部门
4. 购物中心
5. 芳香之城
6. 流水剧院
7. 博物馆
8. 居住桥
9. 国会中心
10. 电影院
11. 闹市区
12. 寺庙
重要的建筑物
铁路
主火车站
混合公交系统
公共交通

无题

Kim, Kyung Hwan（韩国）

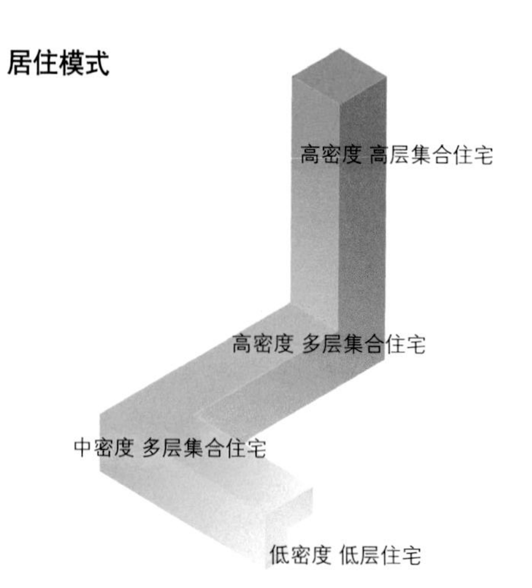

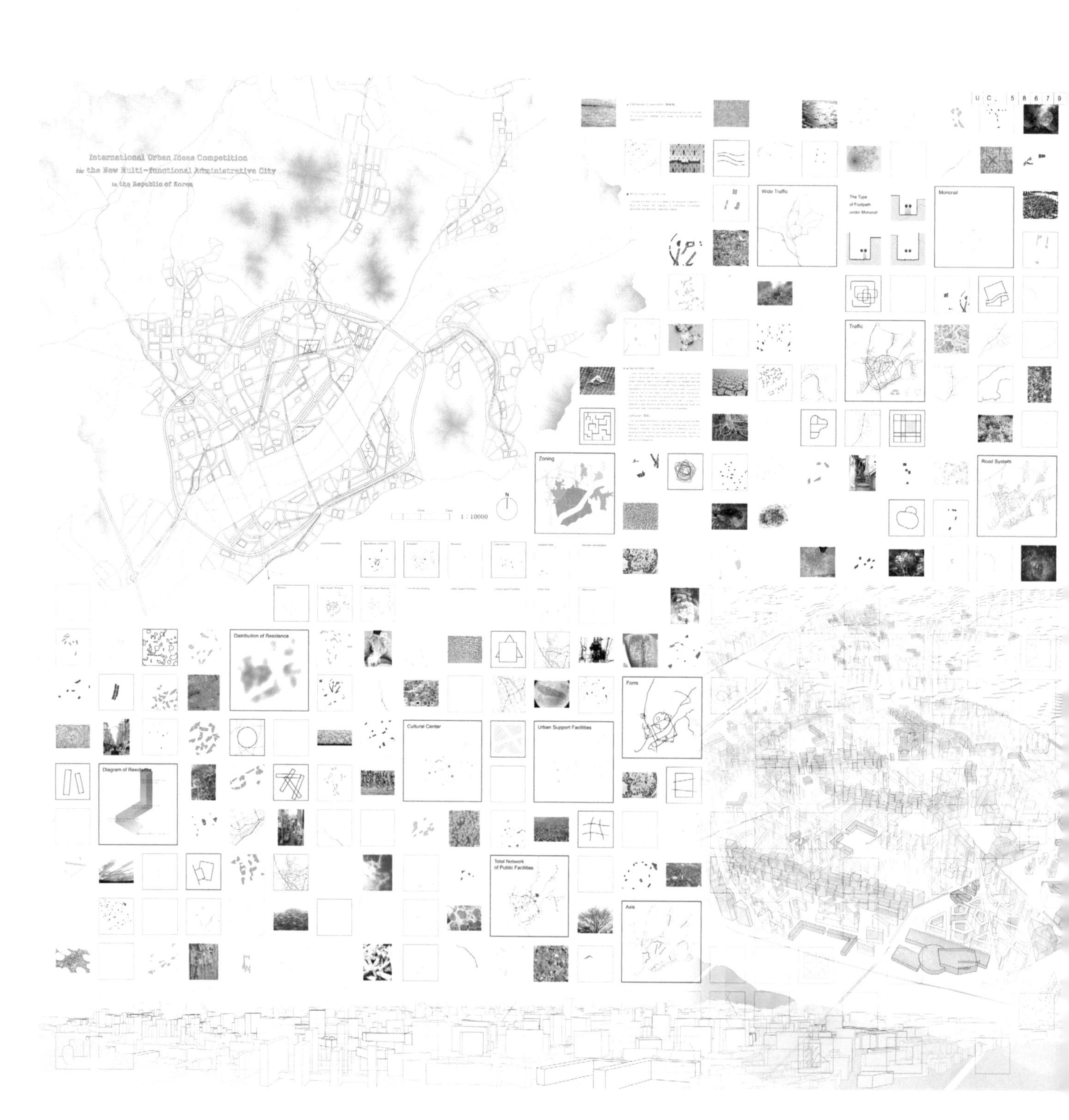

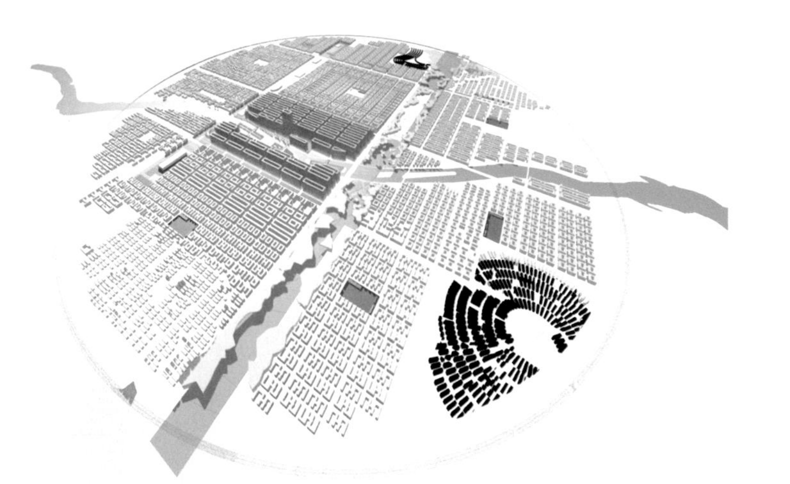

韩国金城

O.S.A Ochs. Schmidhuber. Architects（德国）

Heiner,Luz

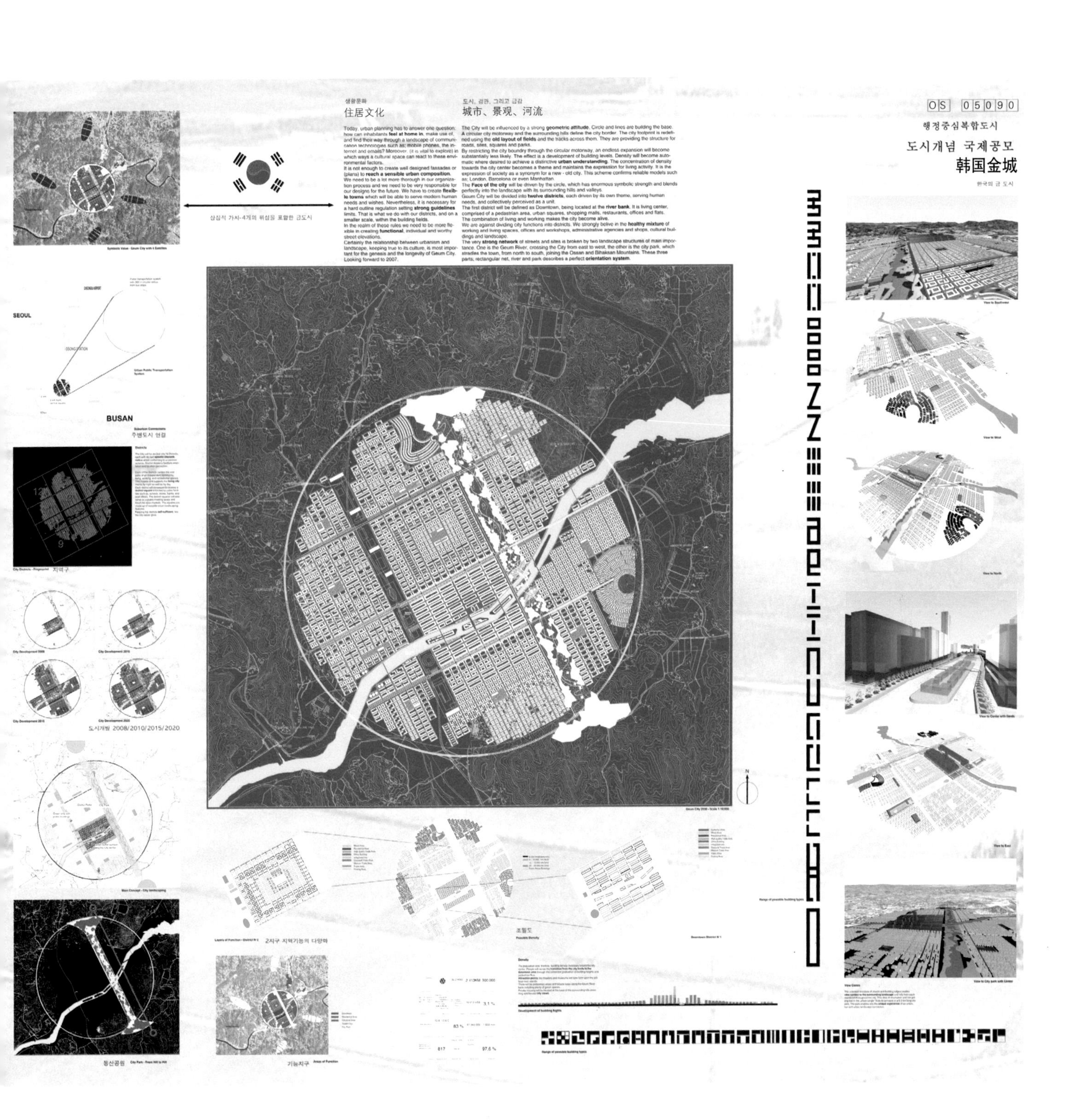

绿色通道

Chang, Chung Kul (中国香港)
Yamane, Cintia / Hsieh, Alice /Kim, Sang Ok

Spatial Synapse

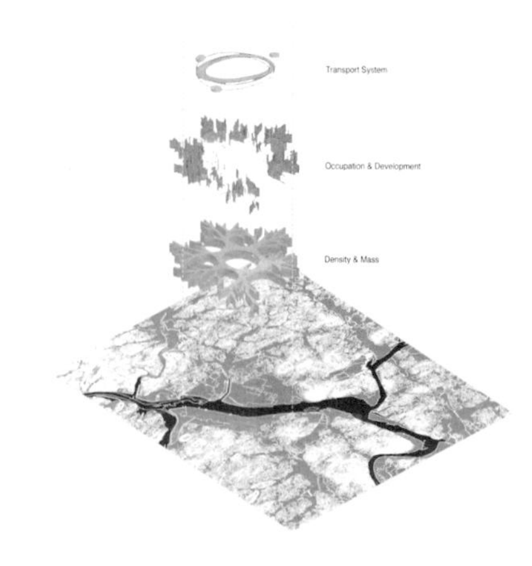

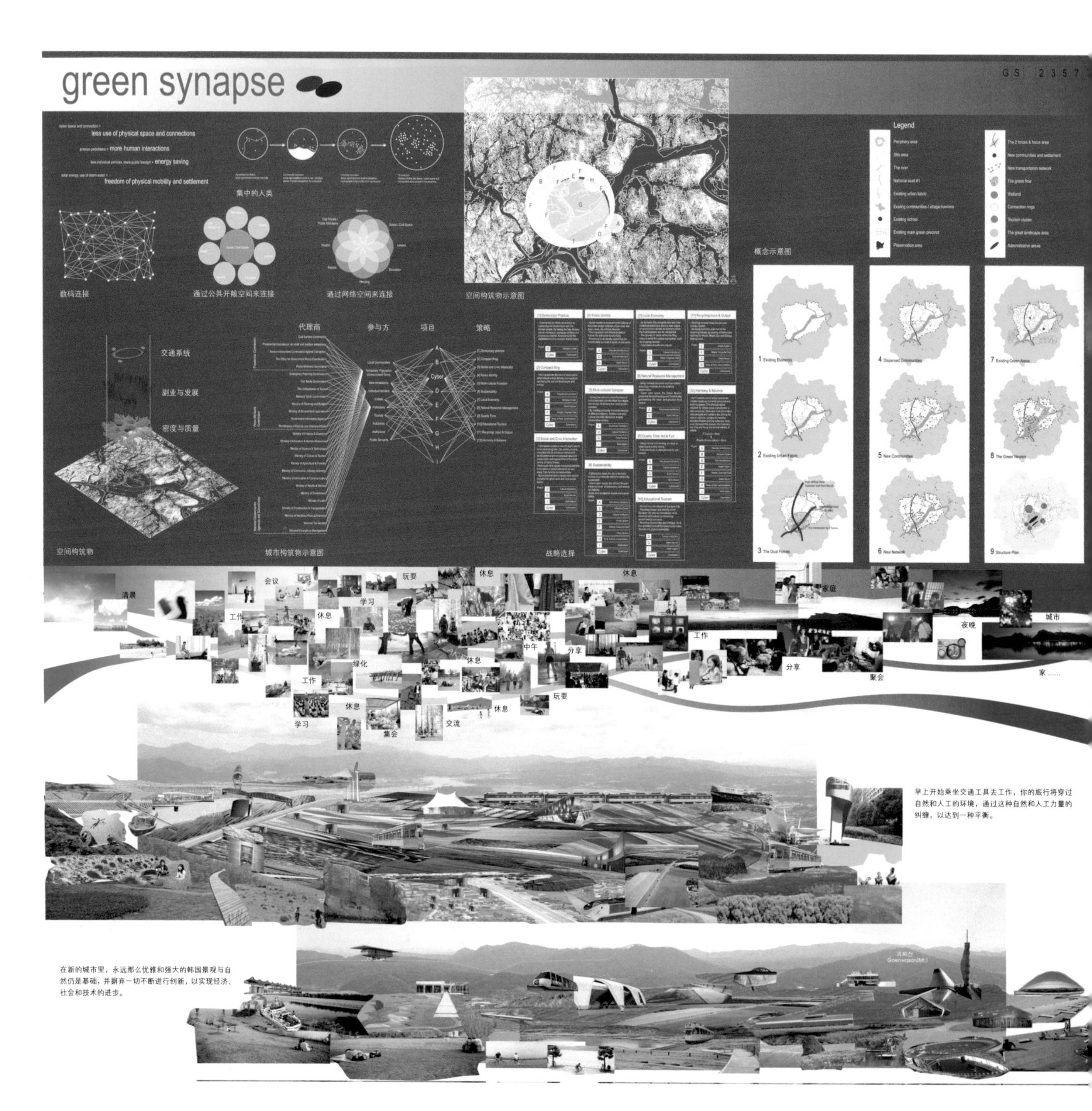

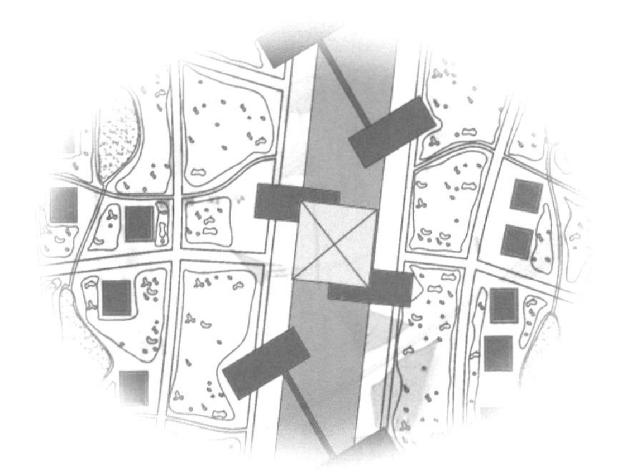

人类的绿色之城

Cho, Sung Eun（韩国）
Kim, Jeong Min / Back, Seung Mok

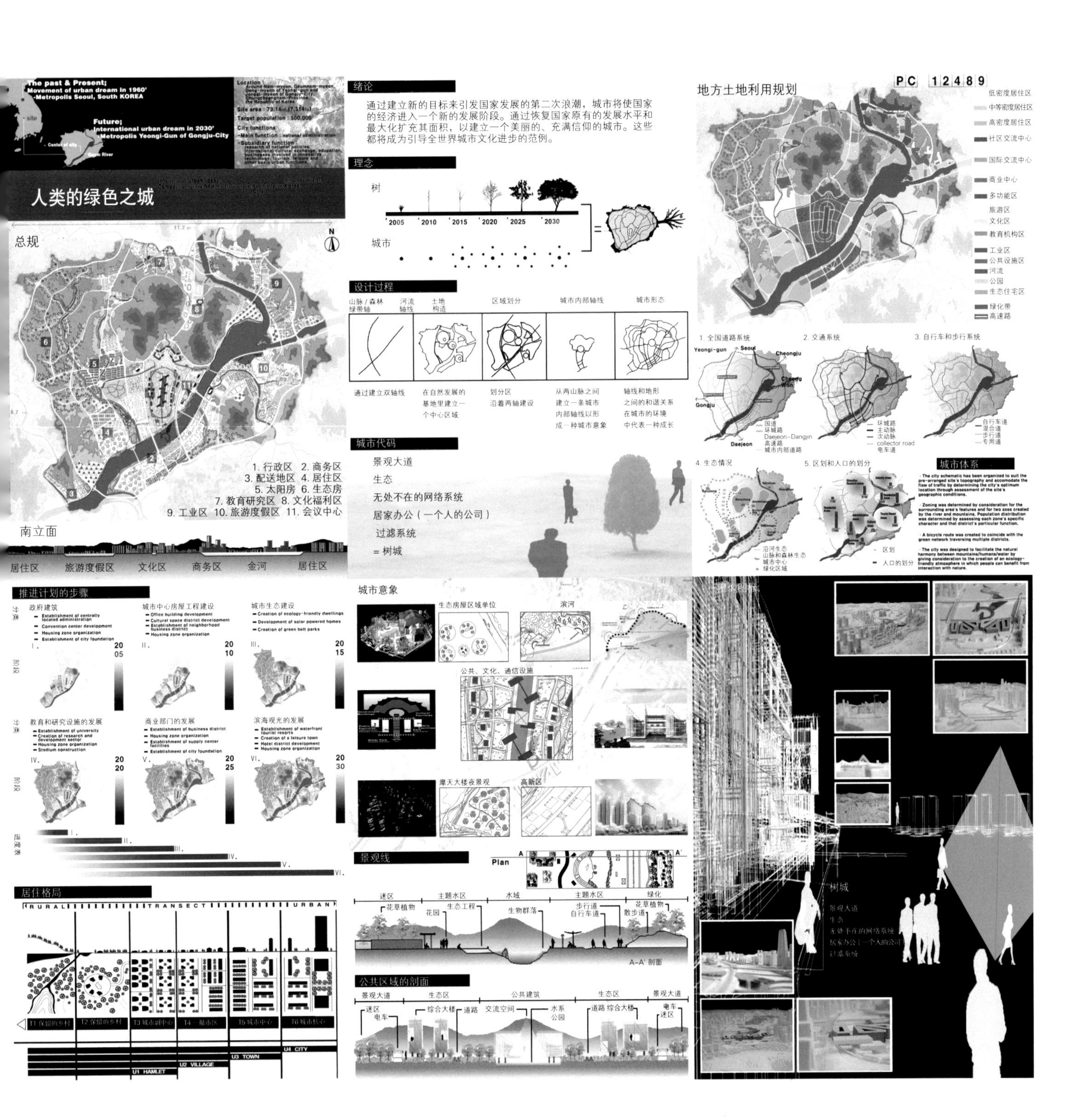

无题

Milaca, Bajic Brkoric(塞尔维亚和黑山共和国)
Archi 5 team

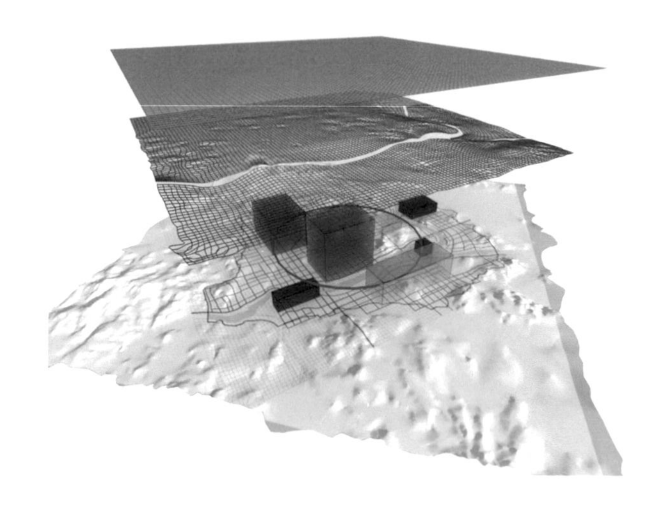

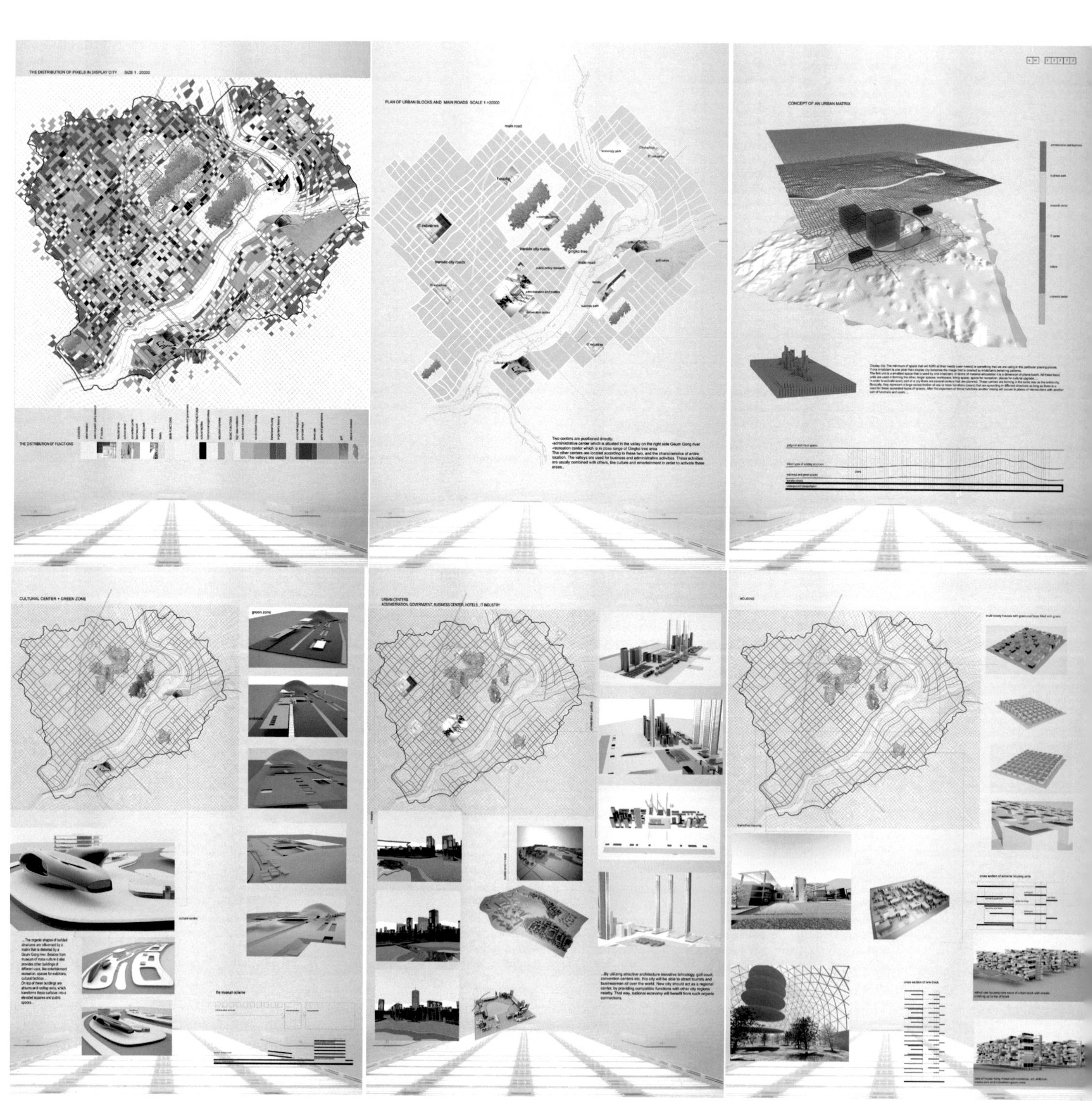

忠清新城

Jumsai, Sumet(泰国)
Michael, Sorkin / SJA 3D Company Limited
Architects 49 Limited / Buro, Happold

新都市主义

Lee, Young Chun（韩国）

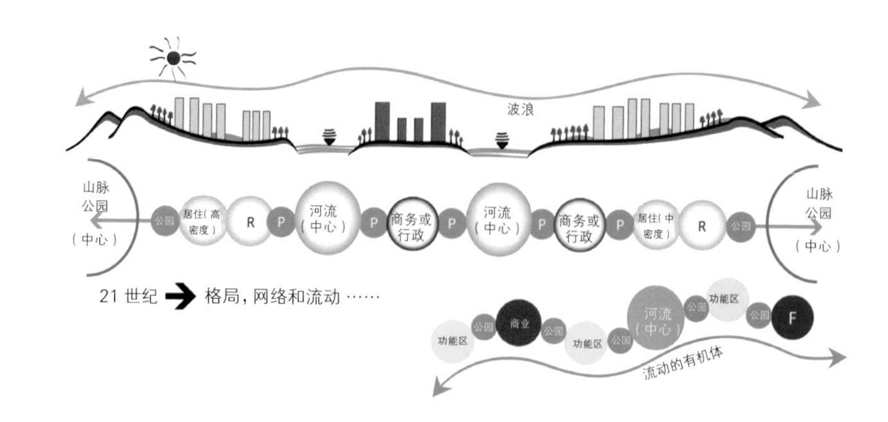

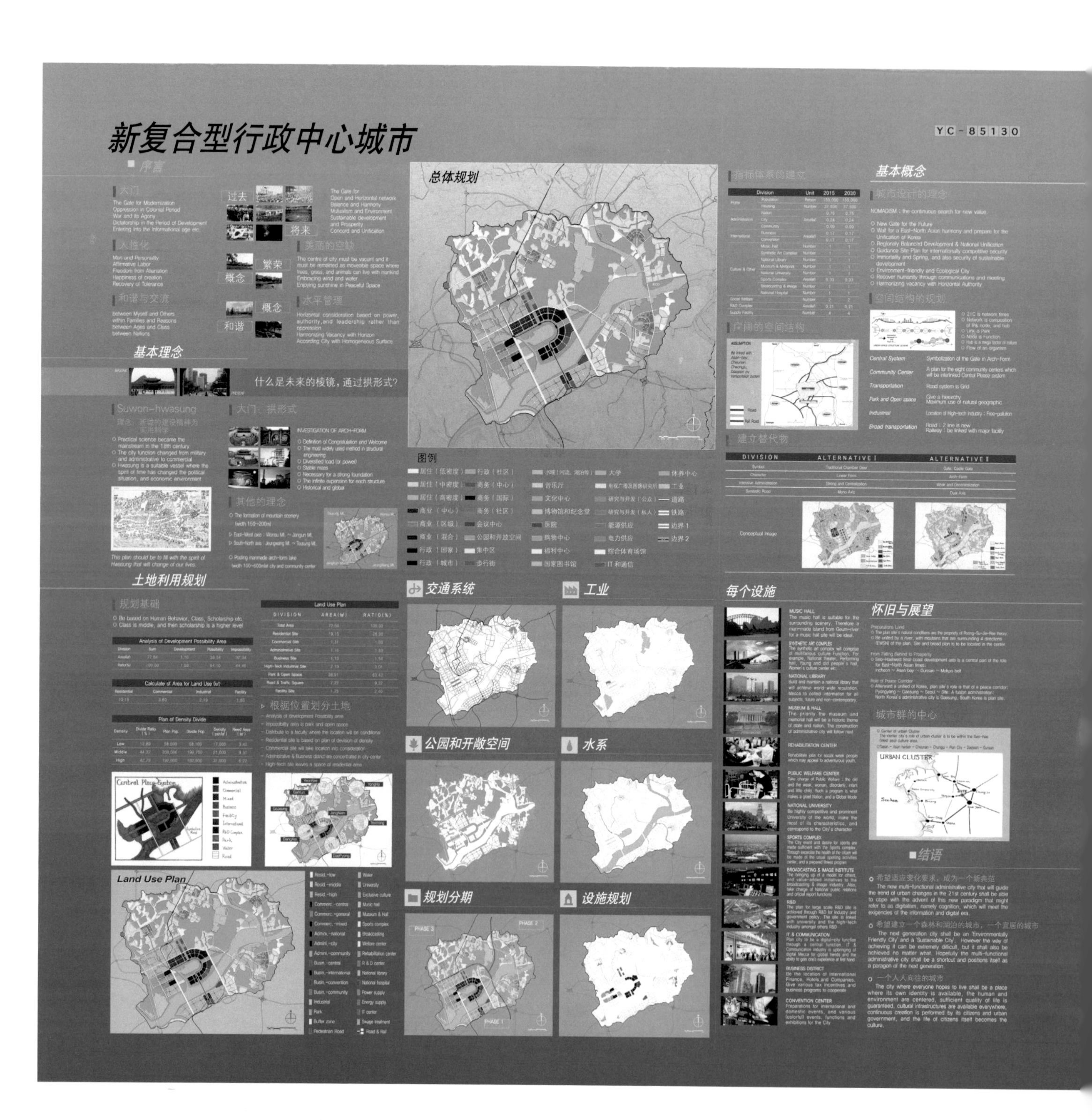

RID

Park, Won Woo（韩国）
Yoon, Jae Woong

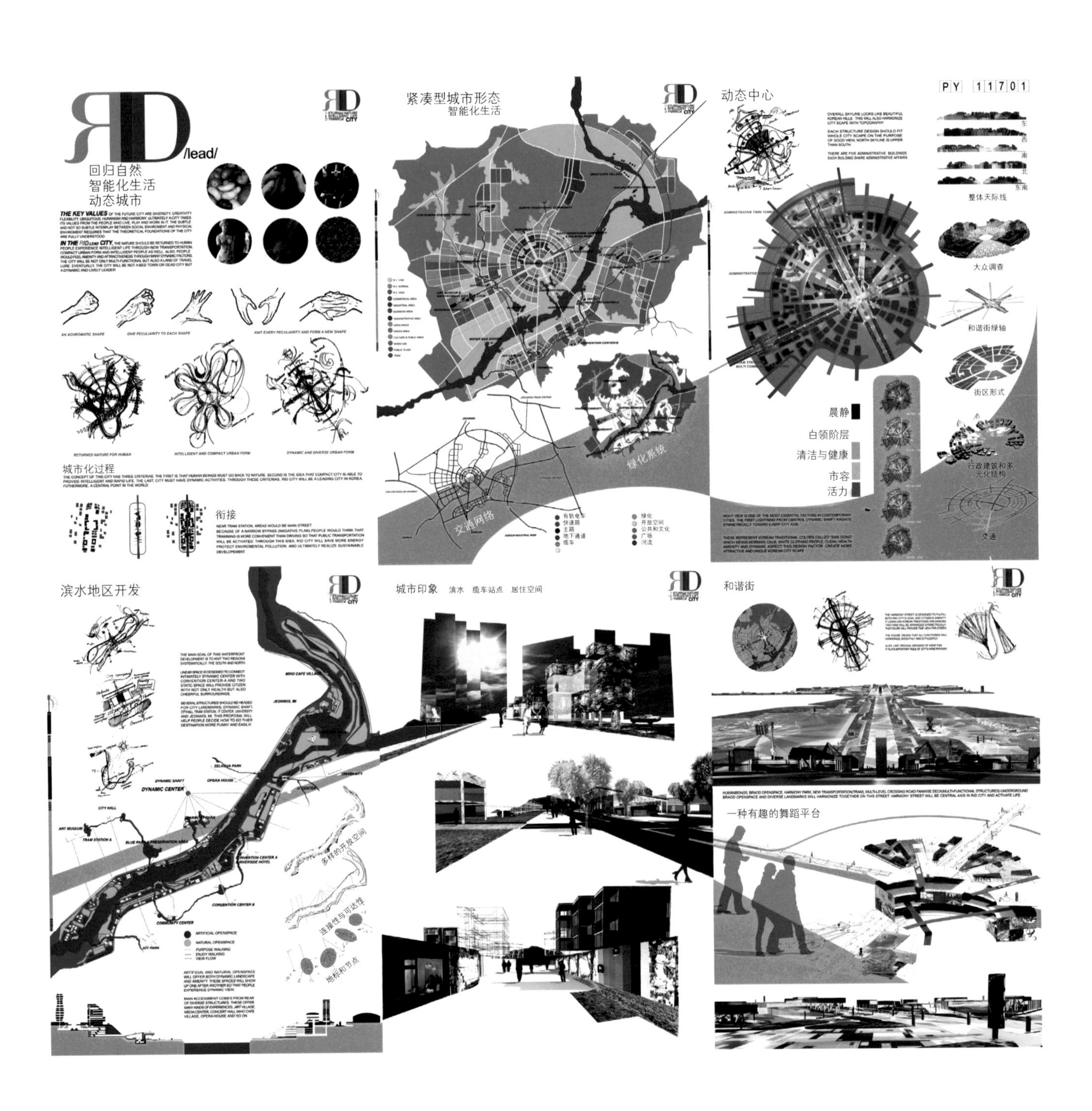

无题

Kohki Hiranuma (日本)

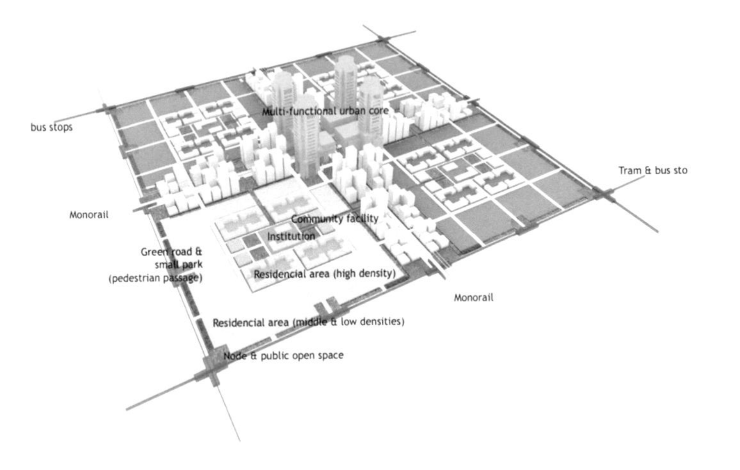

混合单元体

Min, Bum Kee(韩国)
Sir, Min+D&A Associates INC. / Kim, Sei Yong / Lee, Jae June / Park, Byung Hun

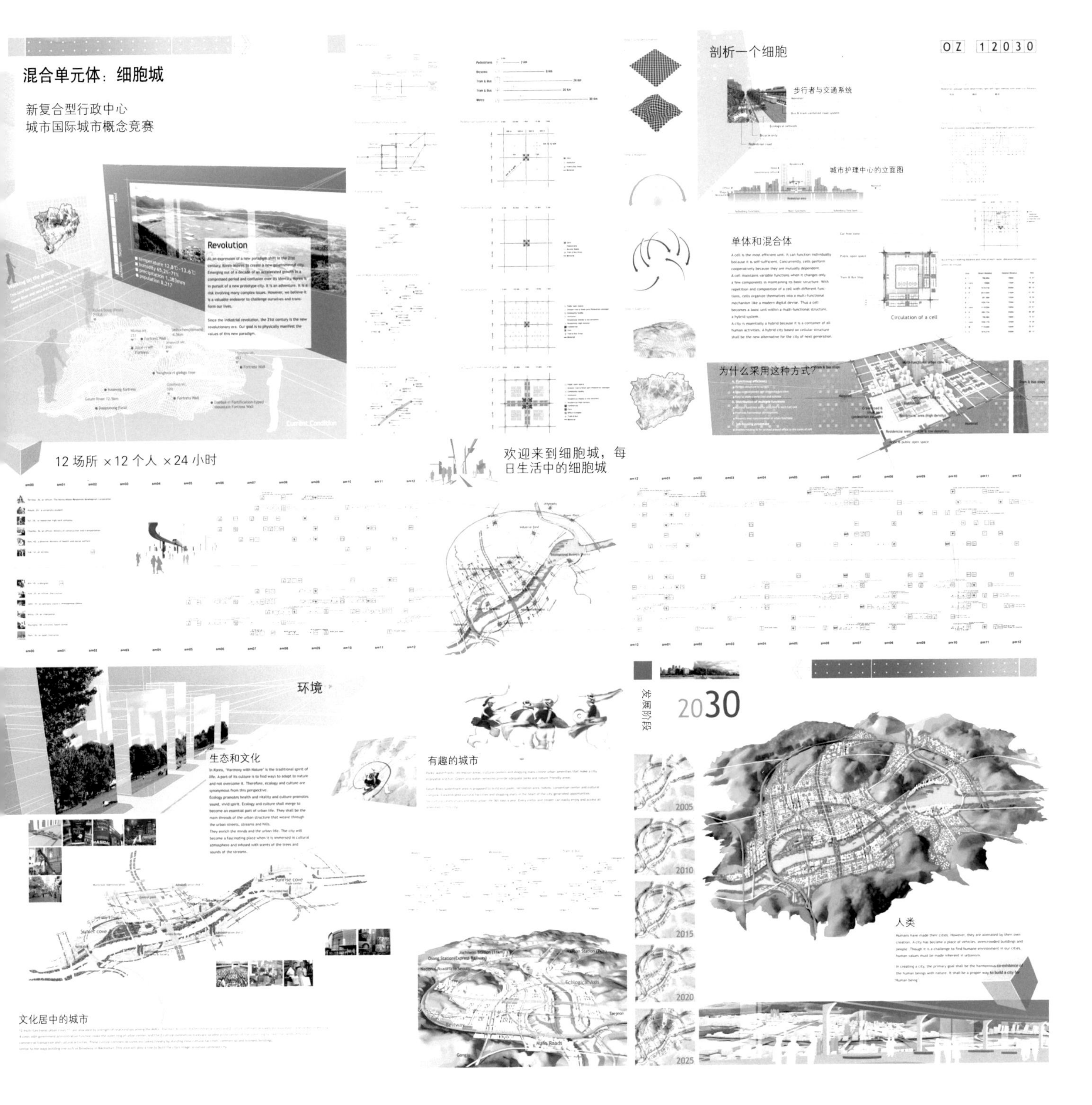

齿状城市

Nam, Soo Hyoun(韩国)
Leem, Jea Eun / Choe, Sun Mi

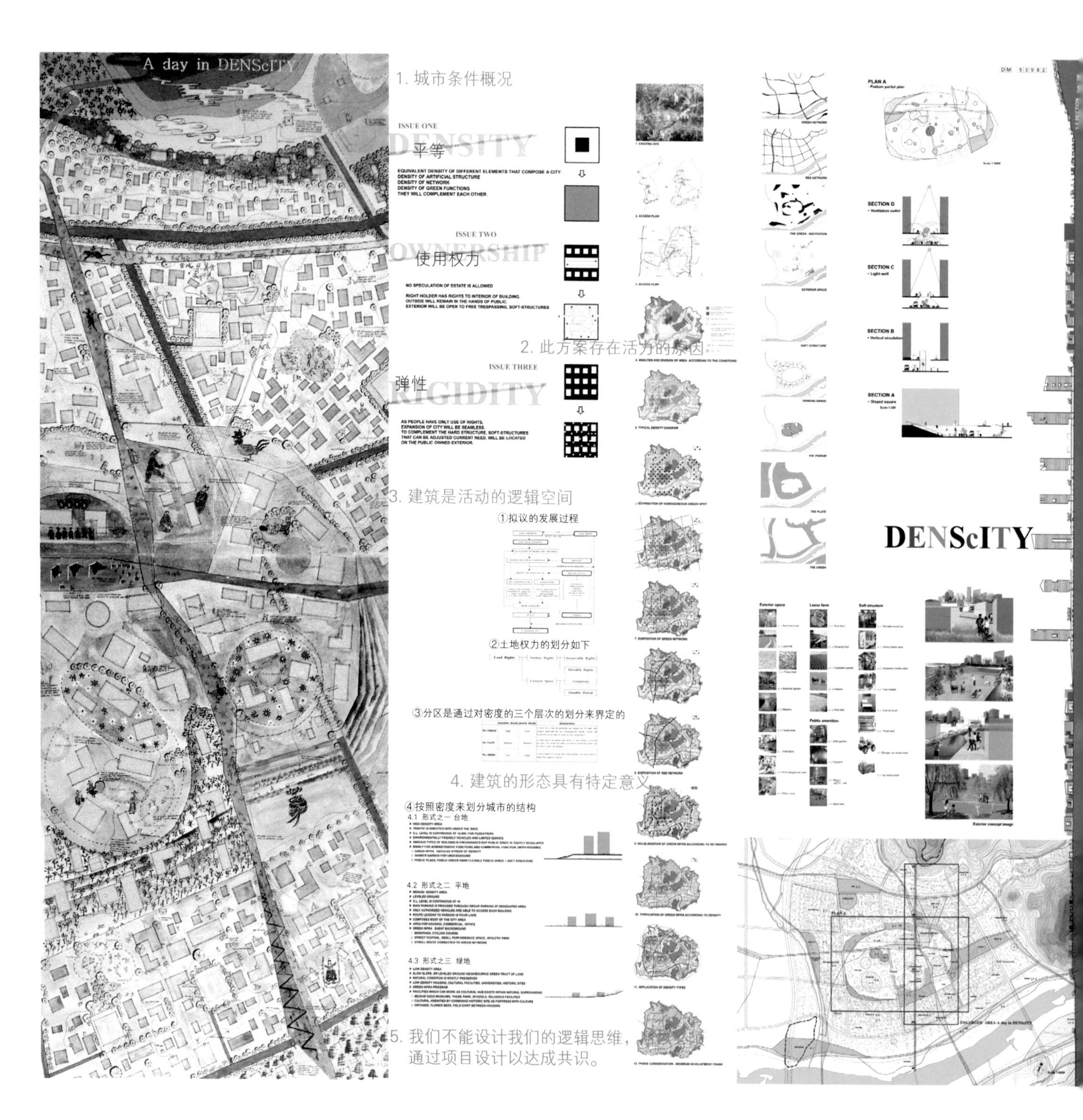

无题

Yoon, Woong Won(韩国)
Kim, Jeong Joo / Yoon, Seung Hyun

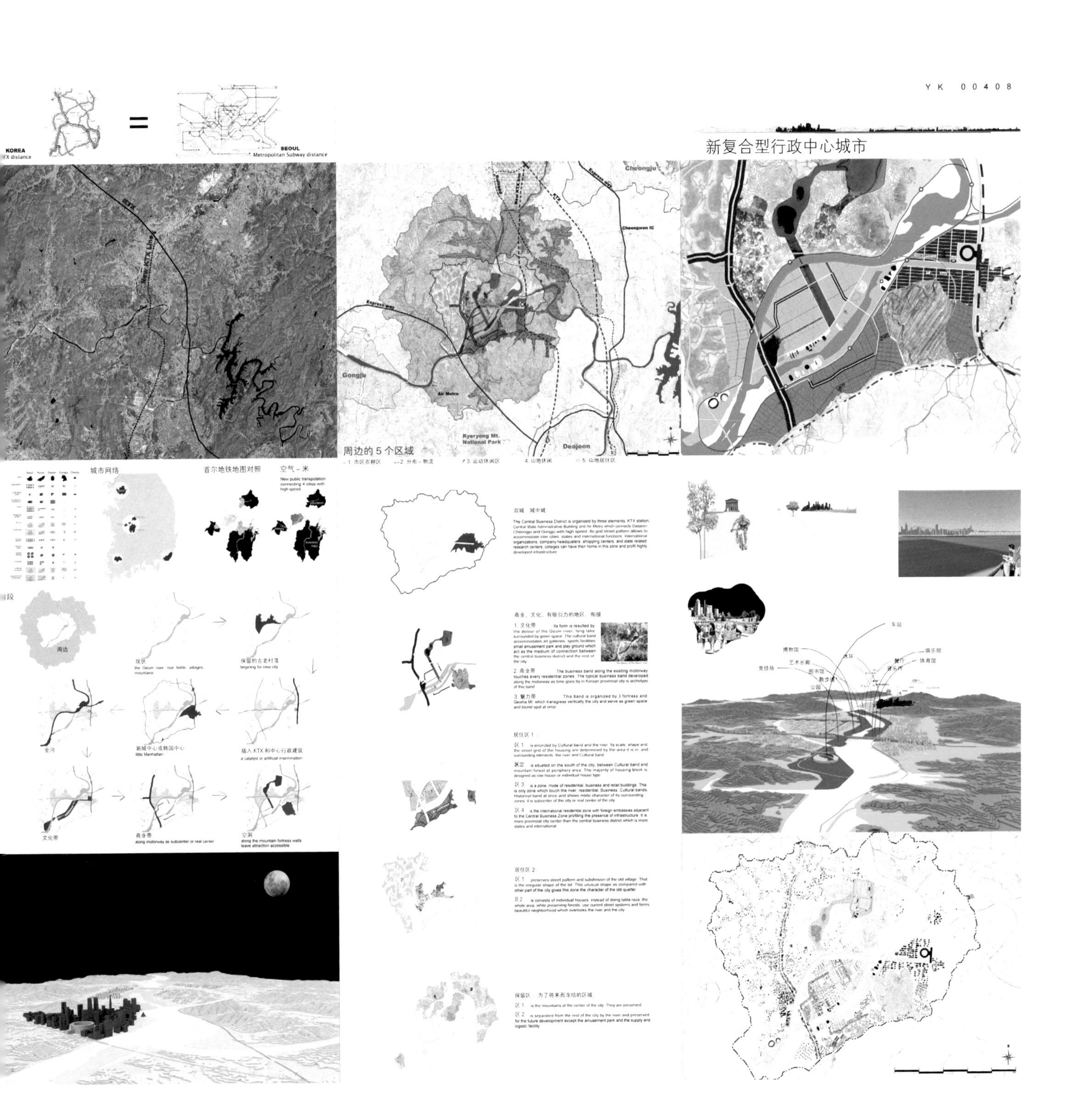

幸福之城

Lew, Deok Hyun（韩国）
Yeom, Yoon Sook

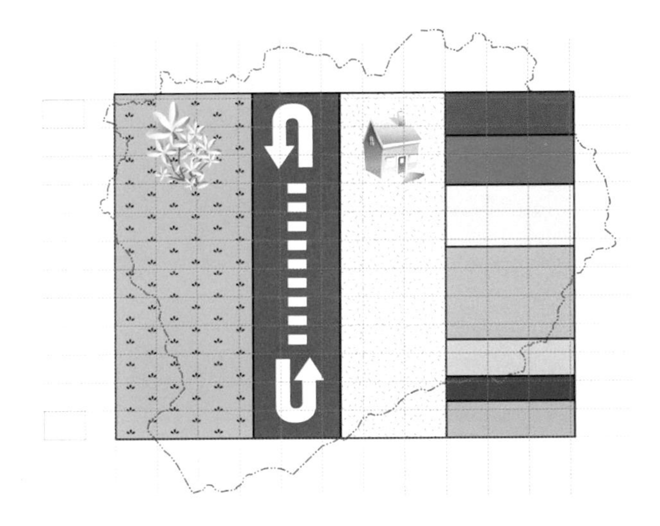

重叠之城

Lee. Dong Shin（韩国）
Lee. Sung Geun / Choi. Min Ah

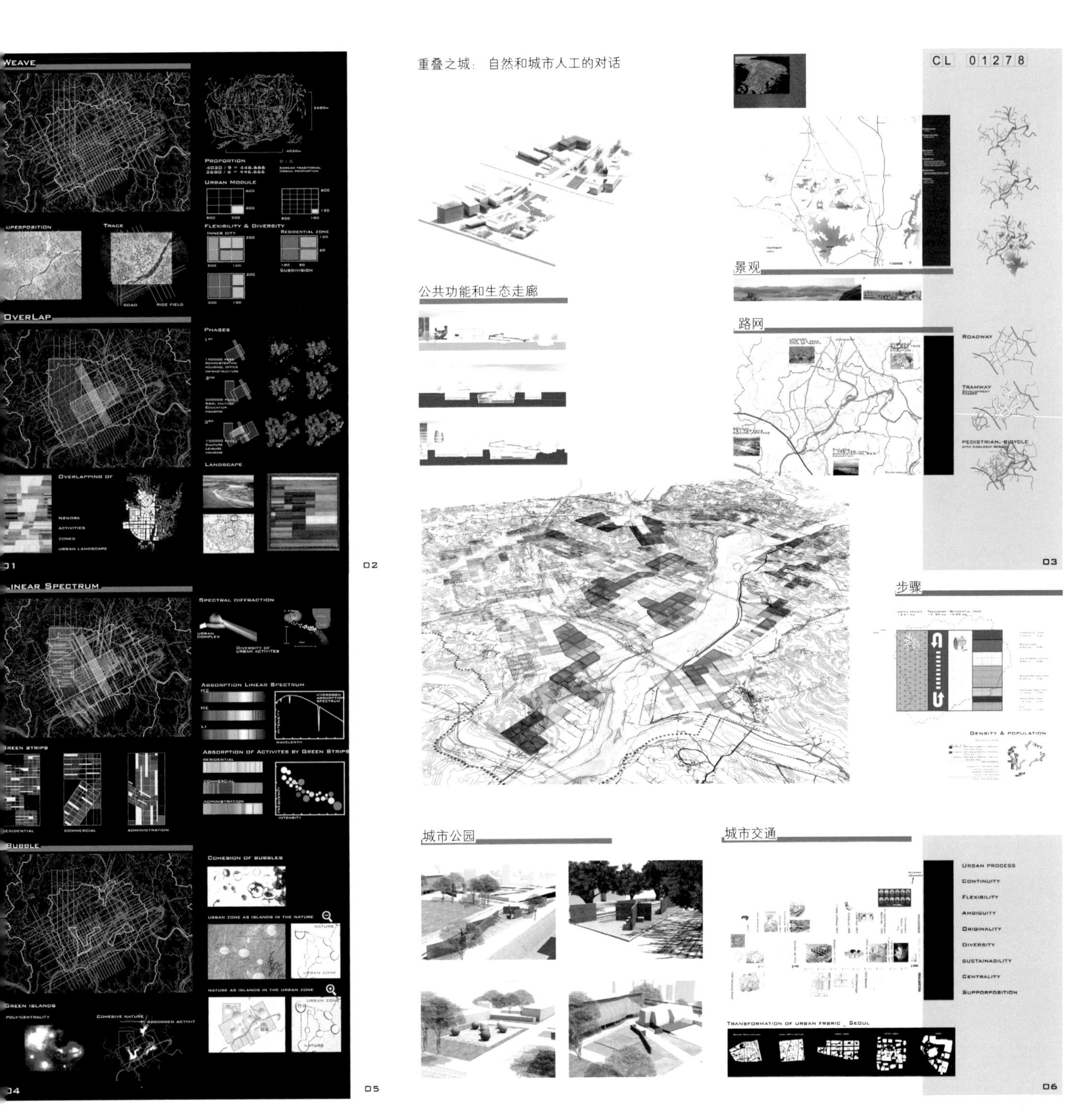

市民心中的城市

Lee, Seok Woo (韩国)
Kim, Woong Tae / Yoo, Young Mo /
Lee, Seung Bae / DongRim P&D, Co.

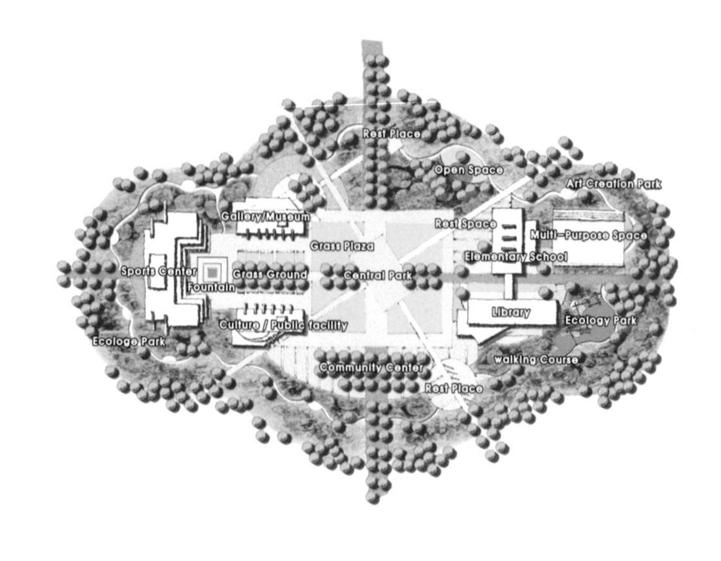

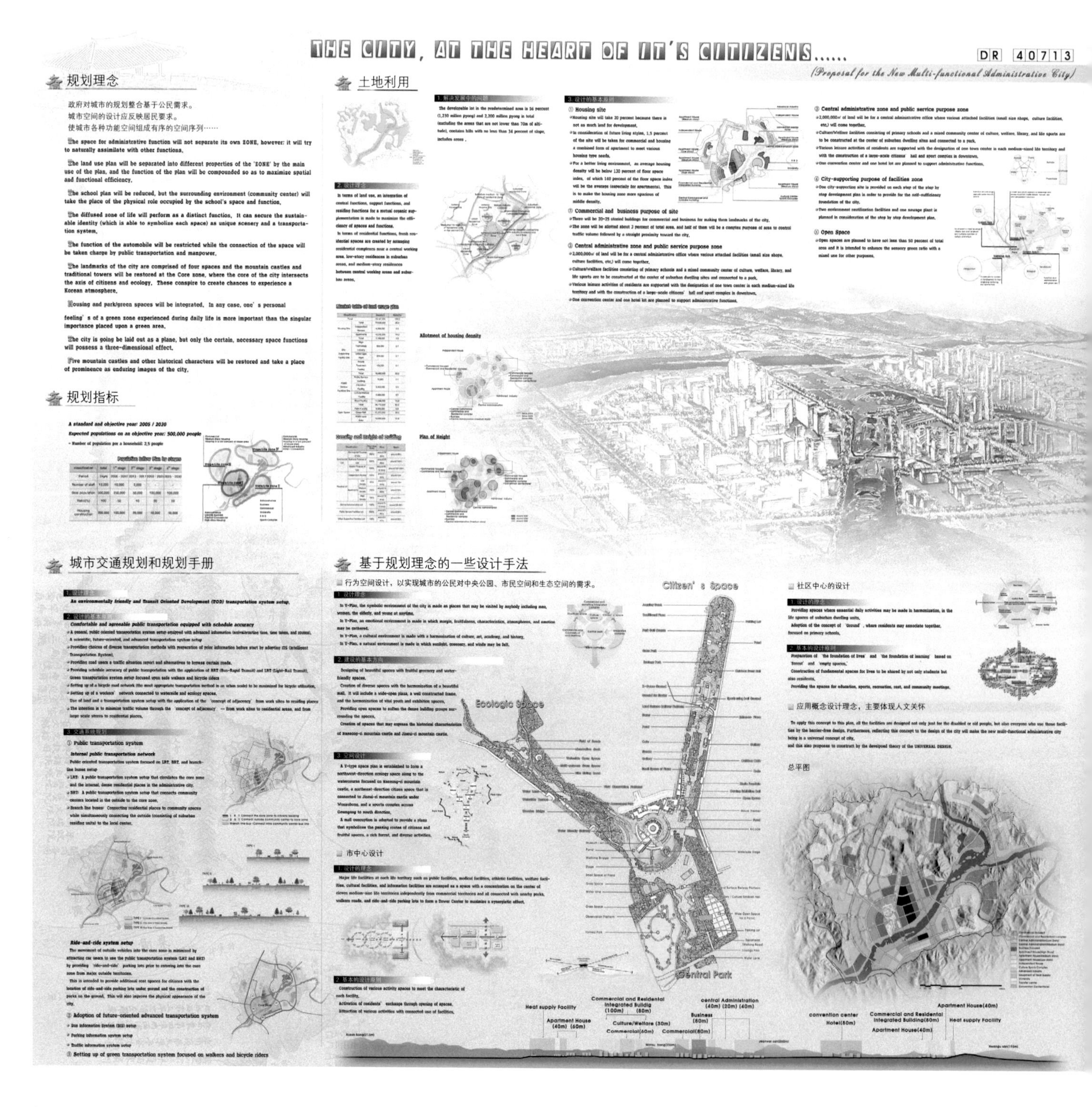

模糊之城

Park, Young Woo（韩国）
Kim, Yong Kyun / Ji, Hyun Ae /
Kim, Sung Kyum / Kim, Yun Jung

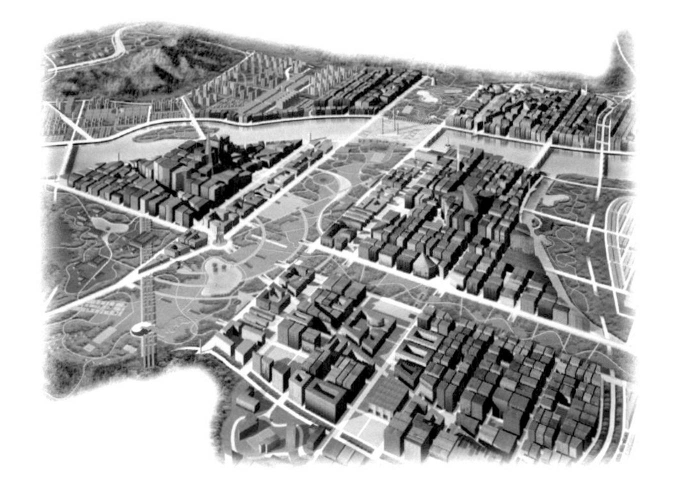

无题

Lee, Pill Soo (韩国)
Lee, Jong Ho

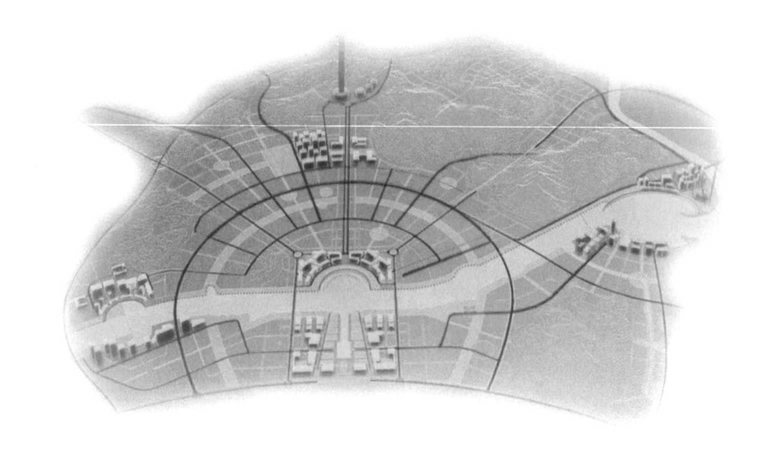

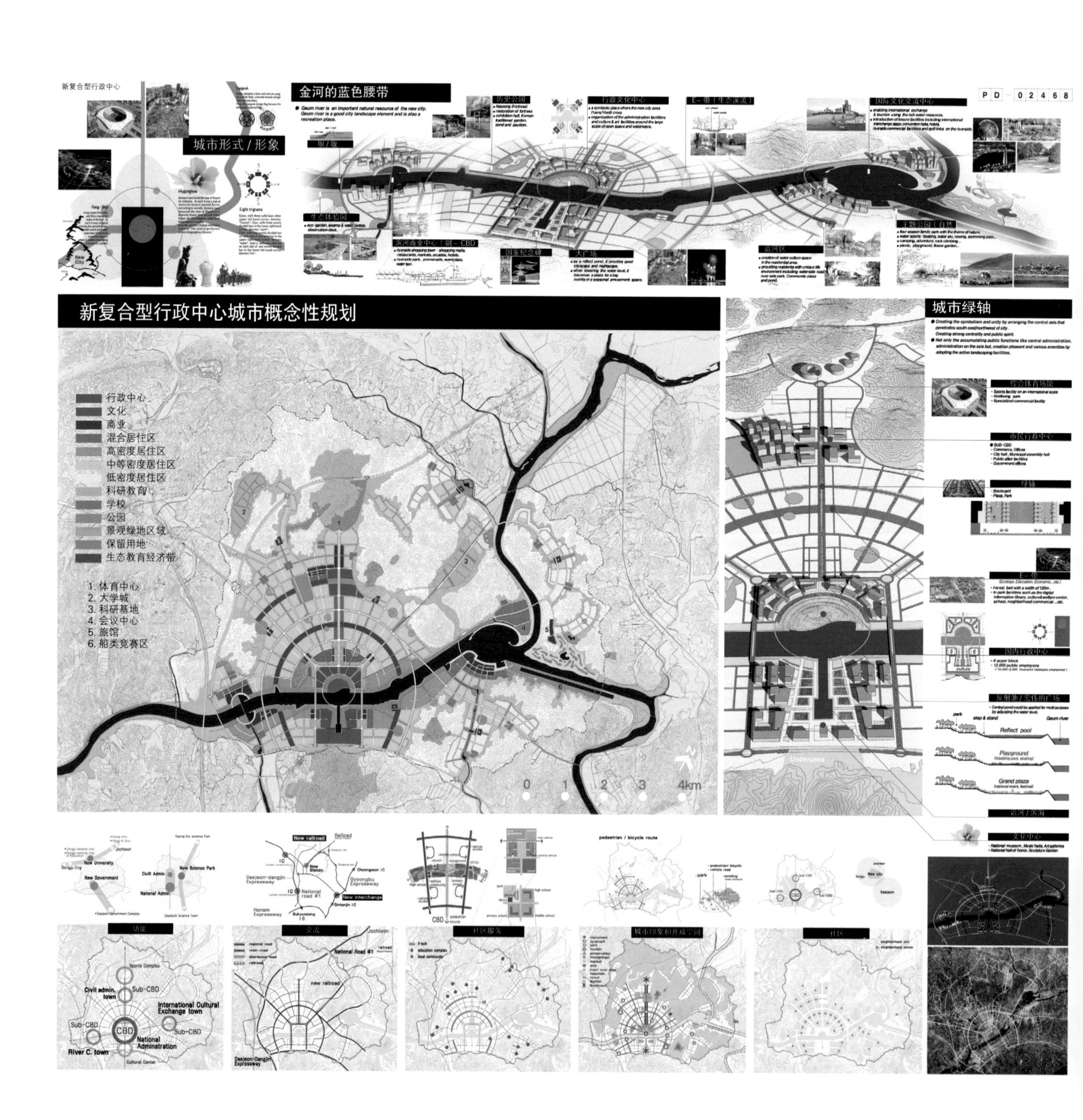

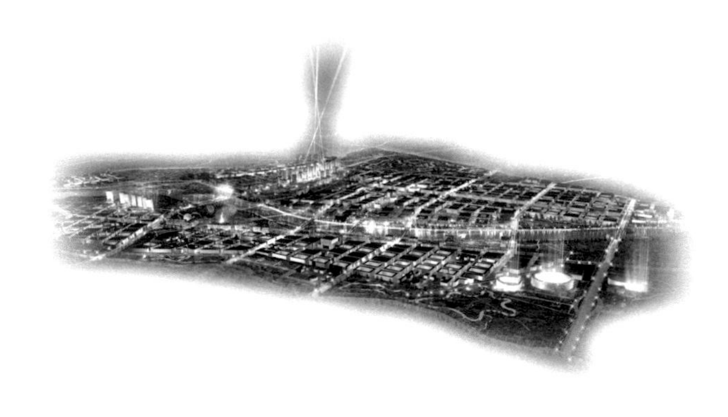

山水画

Jeong, Young Kyoon（韩国）
Kim, Don Yun / Chung, Jae Yong / Han, Gwang Ya /
Heerim Architects&Planners Co., Ltd.

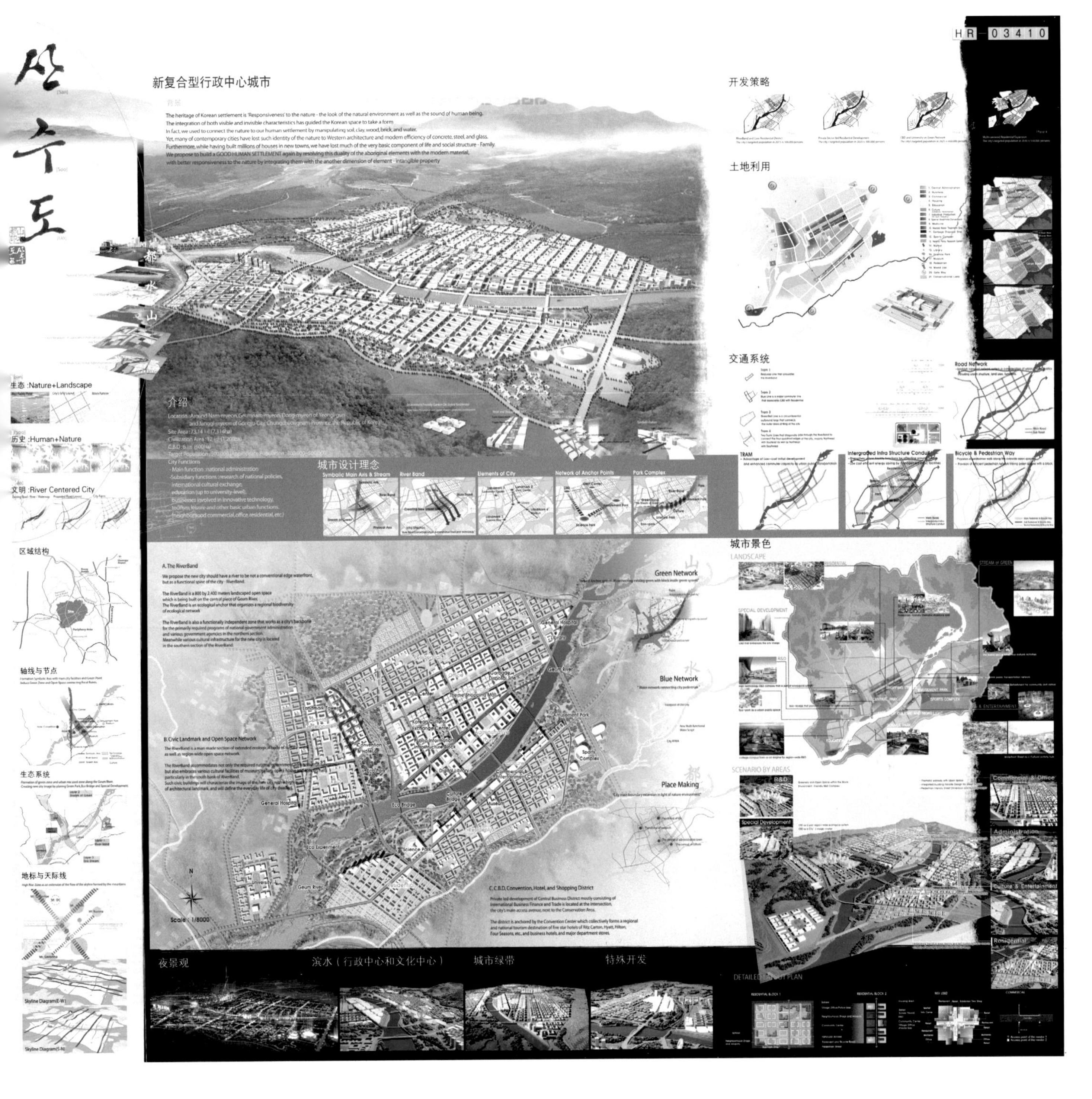

生态型轻巧城市

Chung, Gu Yon（韩国）
Sanin, Francisco / Magerand, Jean / Mortamais, Elizabeth

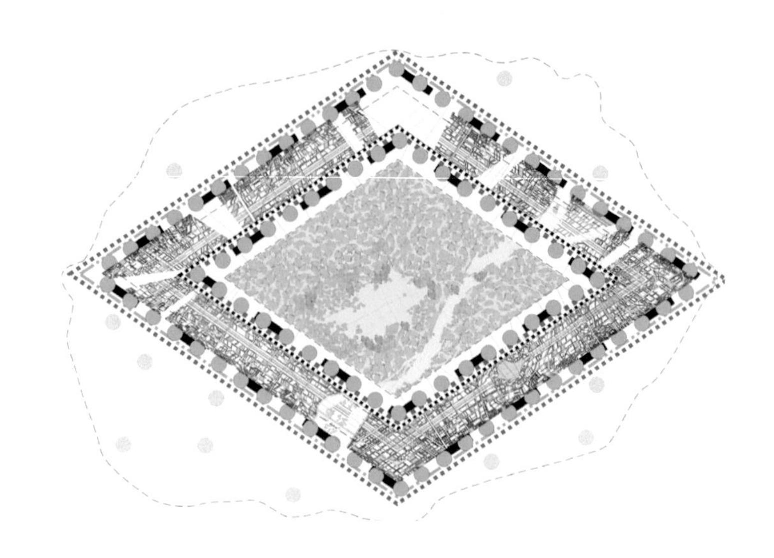

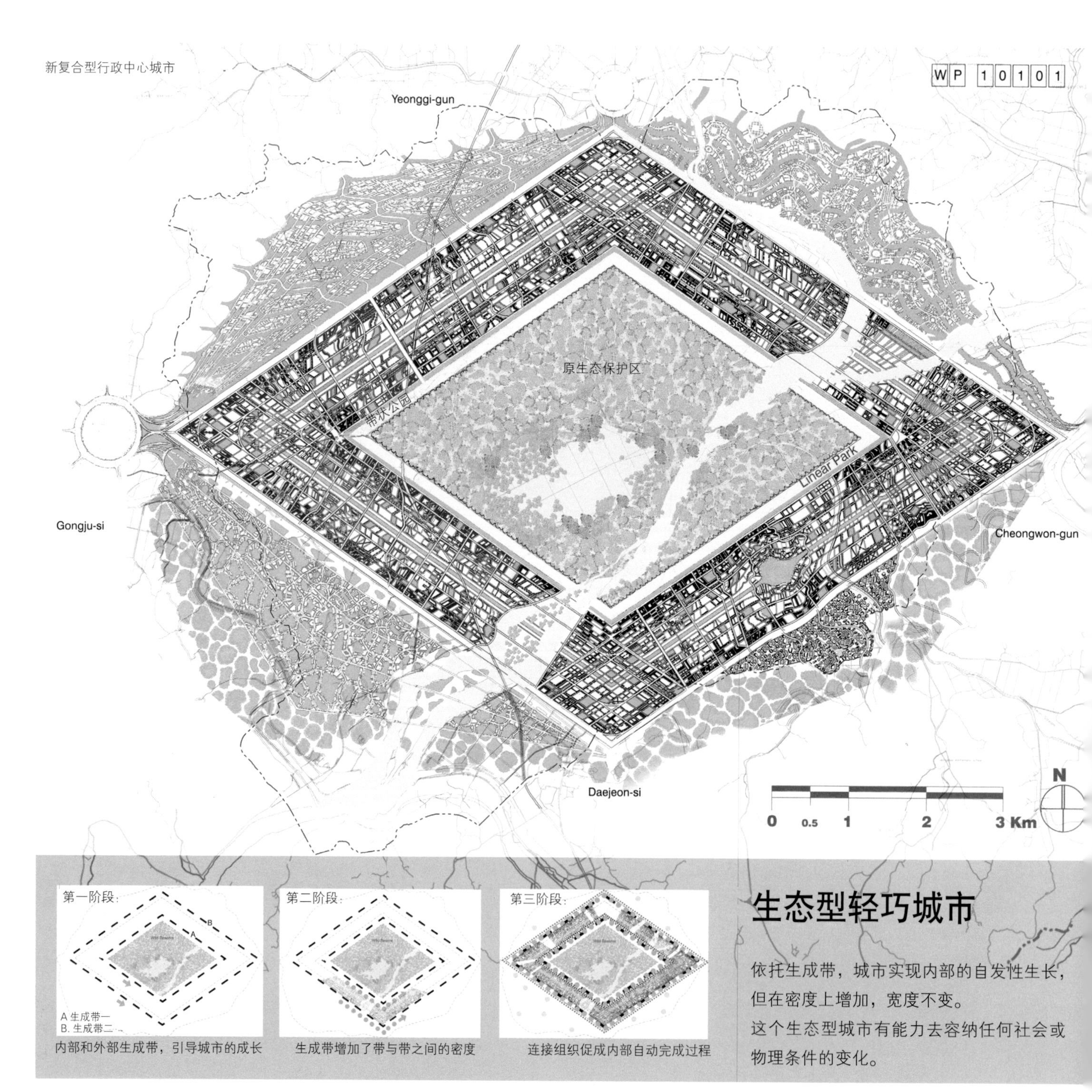

生态型轻巧城市

依托生成带，城市实现内部的自发性生长，但在密度上增加，宽度不变。
这个生态型城市有能力去容纳任何社会或物理条件的变化。

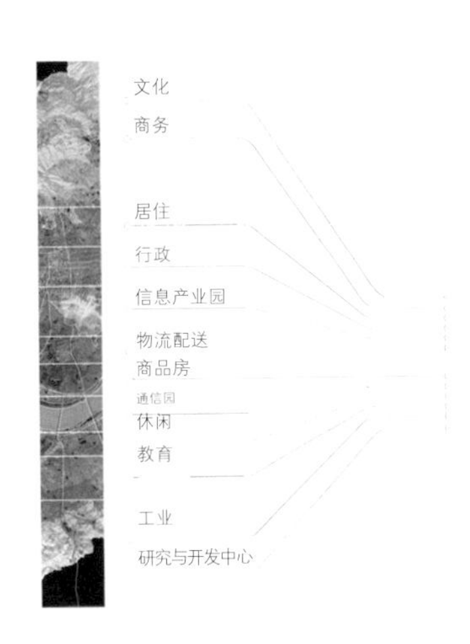

多形态城市

Lee, Hee Chung（韩国）
Kim, Yong Sung / Nam, Seung Kyun / Ko, Kwang Sung /
Myoung In Architects & Engineers

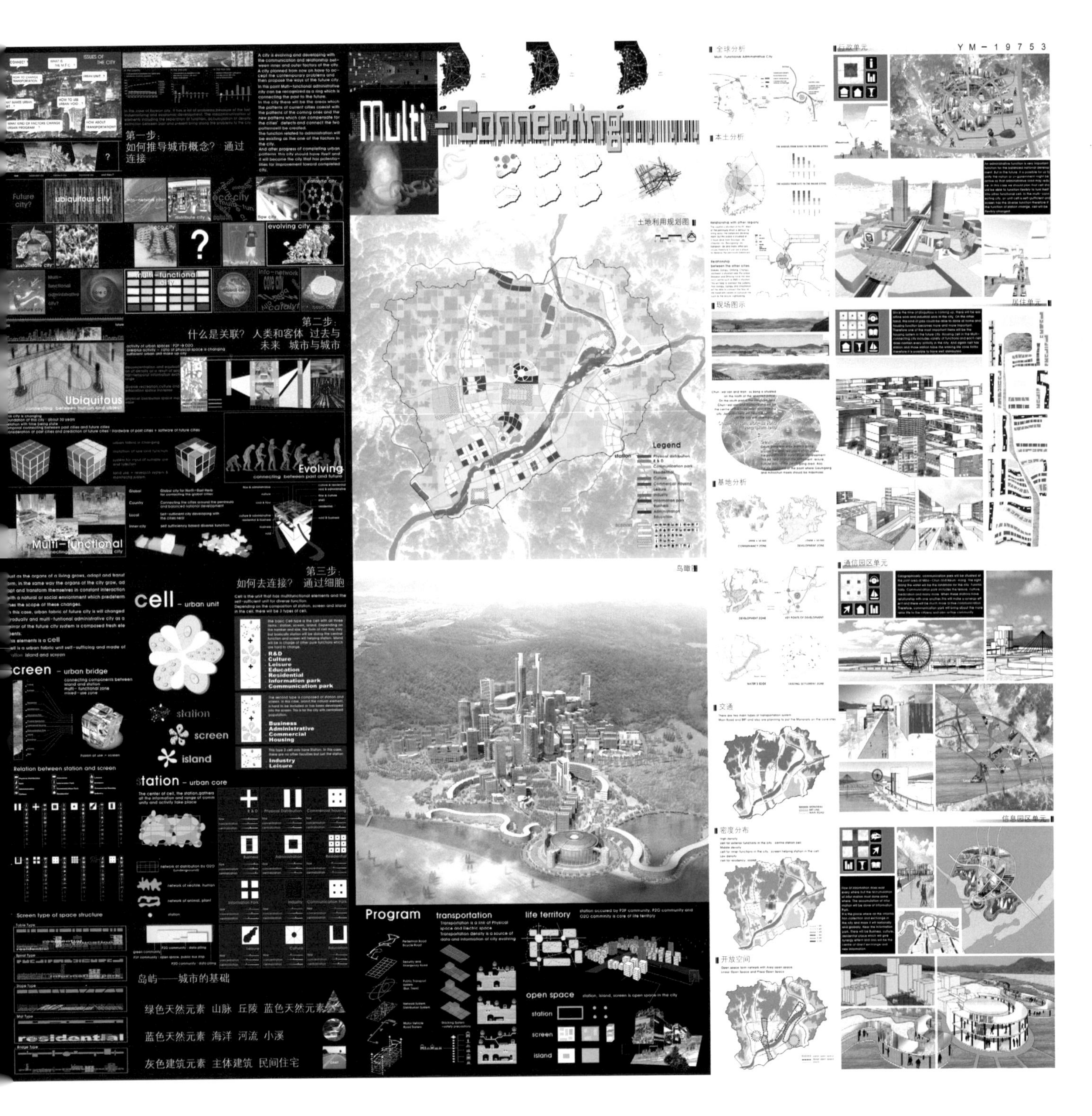

新世纪的不朽之城

Park, Hyun Jin (韩国)
Toshiken Korea co., Ltd

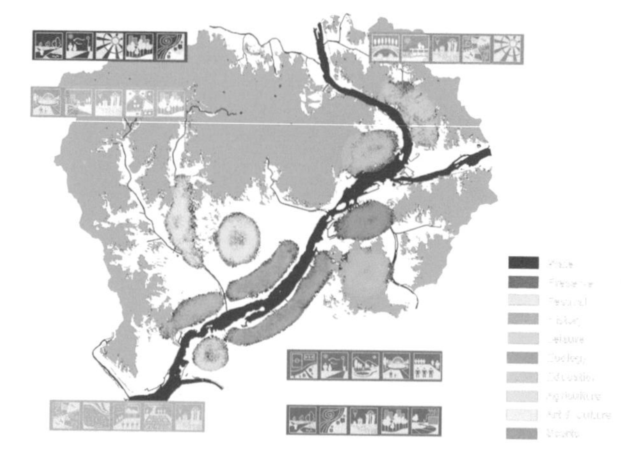

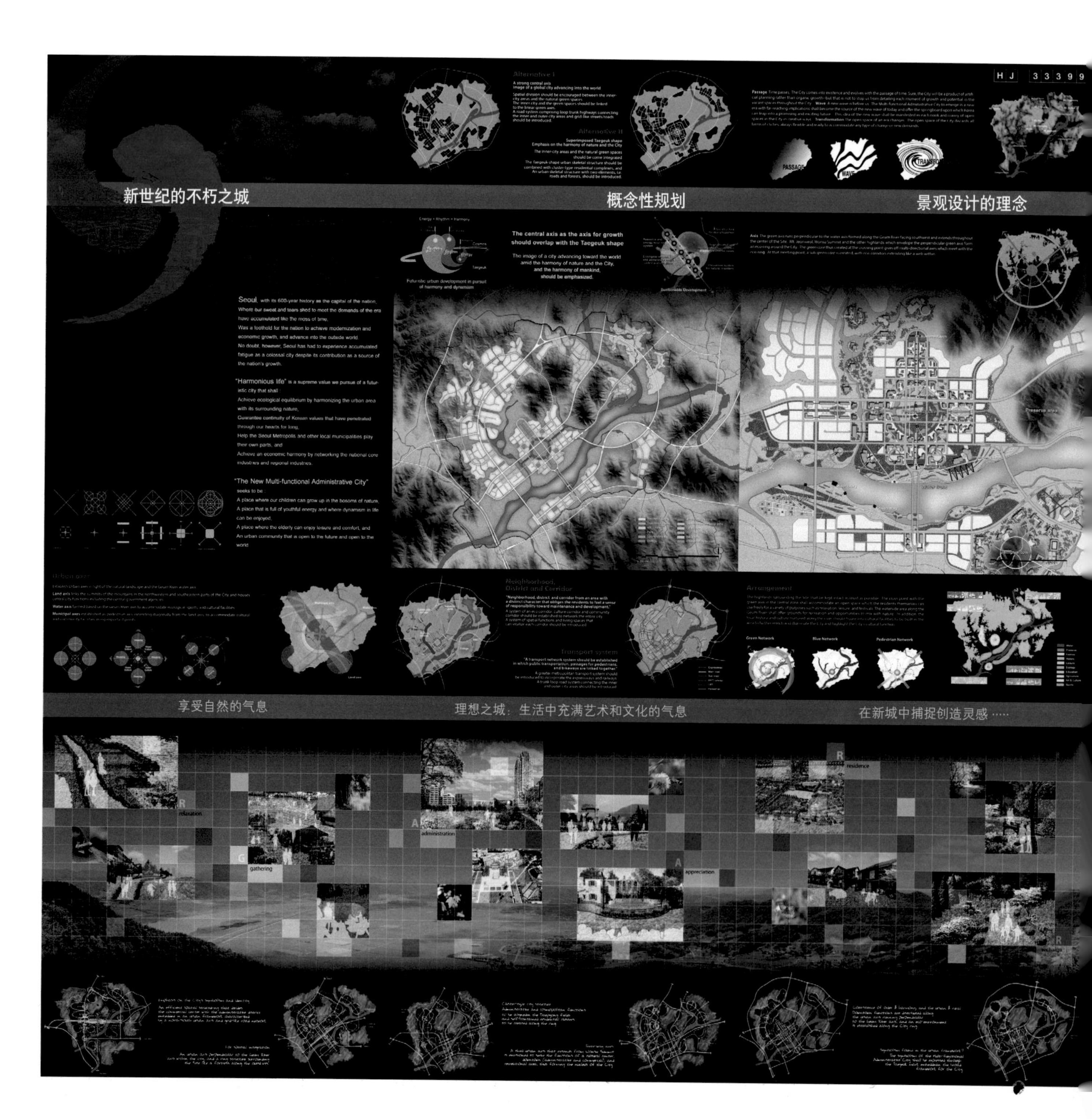

芙蓉城

Lee, Byung Jin（韩国）
Yu, Jung Hyuk / Hu, Sung Kyun /
Cho, Sung Gil / Cho, Young Gyu

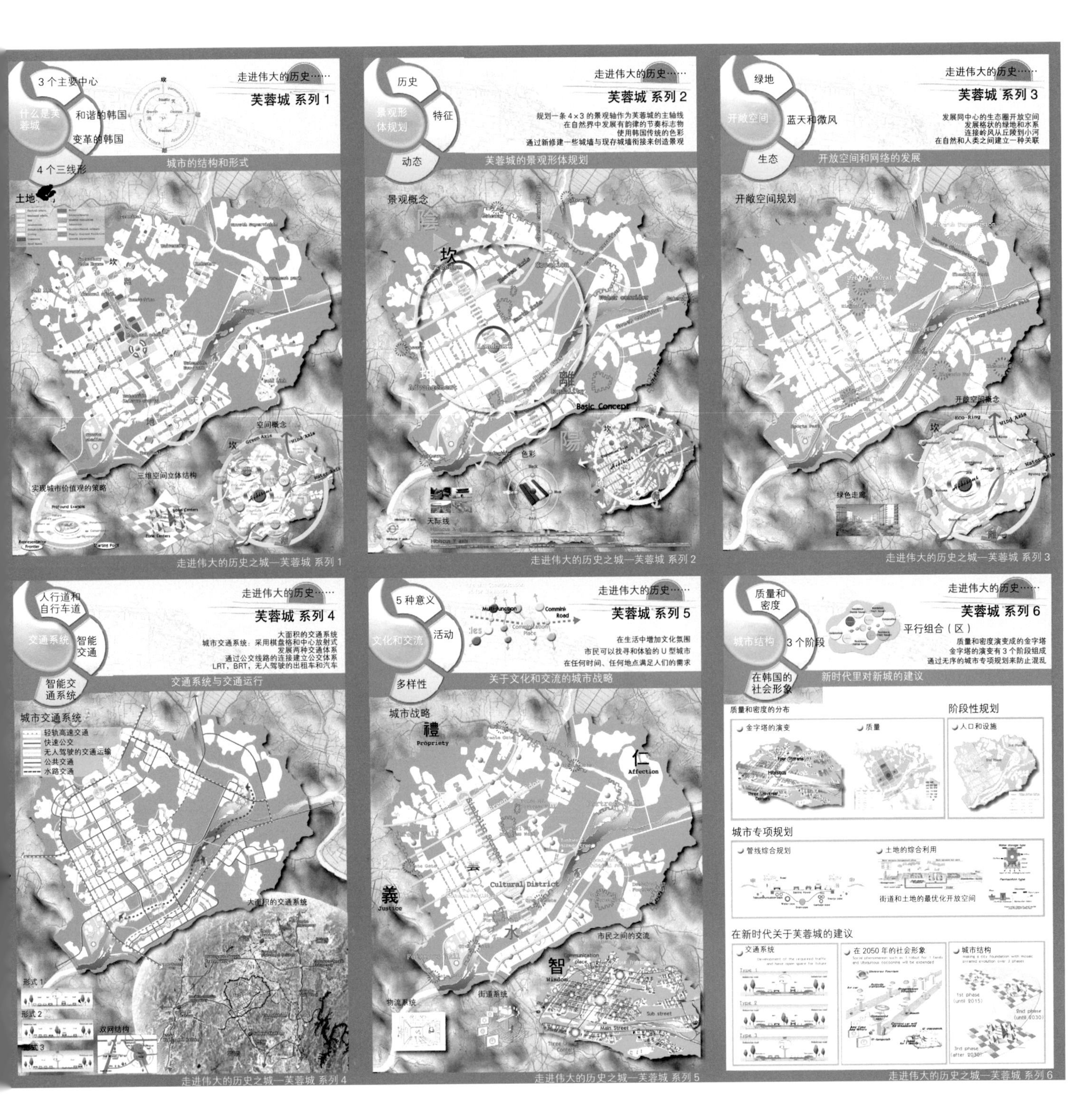

活力之城

Ahn, Young Su（韩国）
Shin, Min Joong / Joo, Sung Hak / Kim, Hyeon A

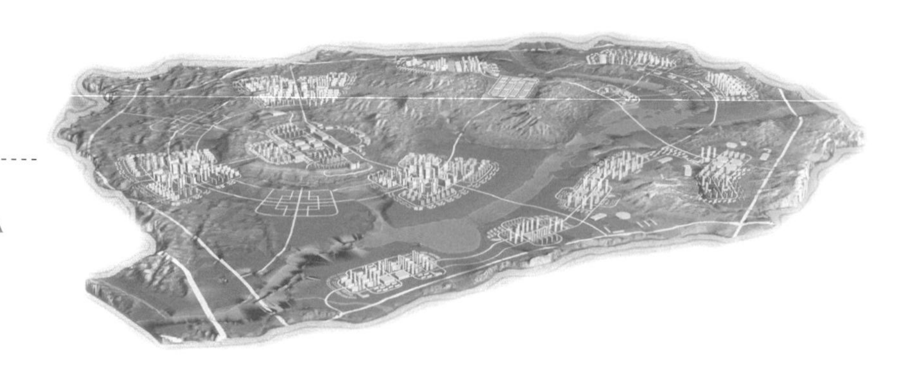

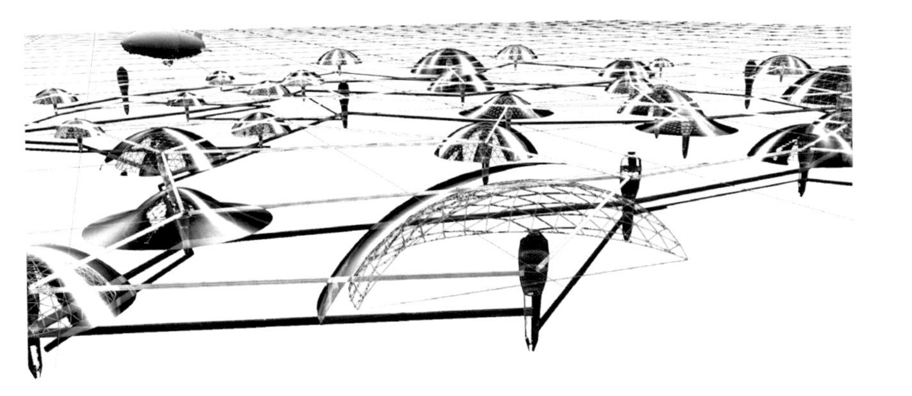

融入自然

Kim, Uk（韩国）
Kang, Chul Hee / Lee, Sang Ho /
Kang, Jun Mo / Bahn, Sang Chul

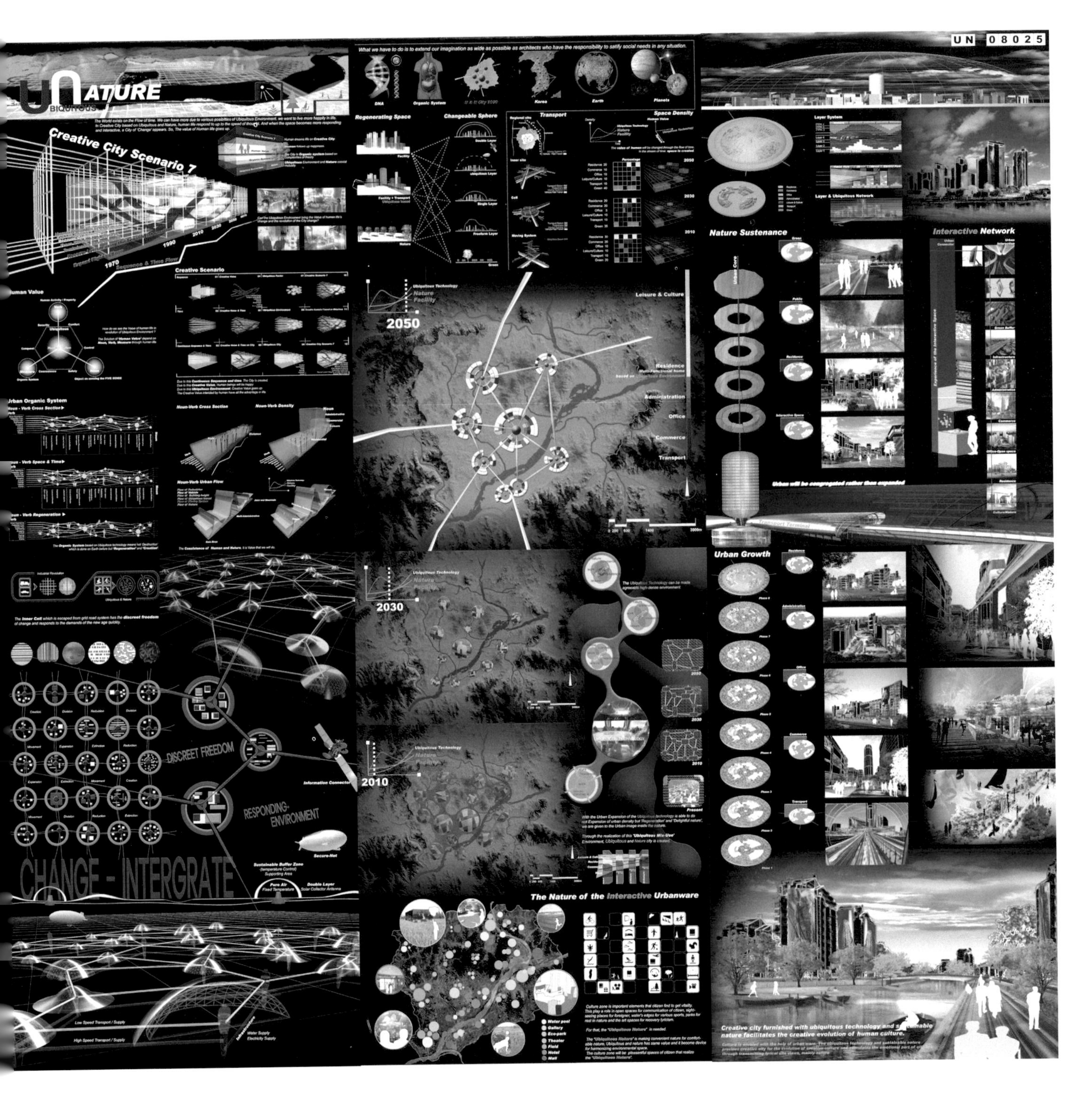

林中筑巢

Kim, Young Sub（韩国）
Waro Kishi+K. Associates Architects

进化中的都市体系

Kim, Woo Sung（韩国）
Lim, Jong Ah / Archiplan INC

梦幻之城

Shin, Min Jeong（韩国）
Park, Sung Nam / Kim, Hee Sung

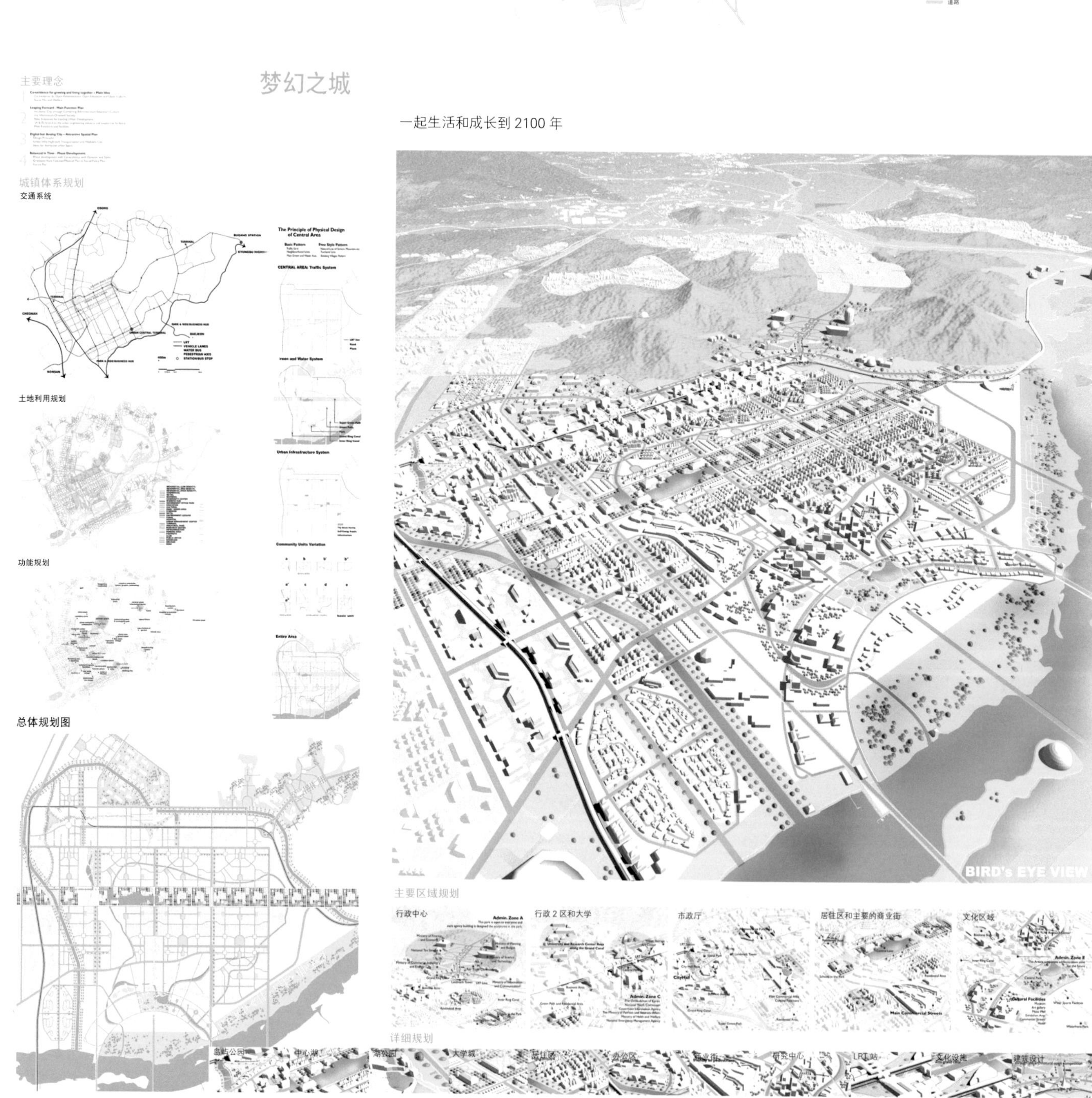

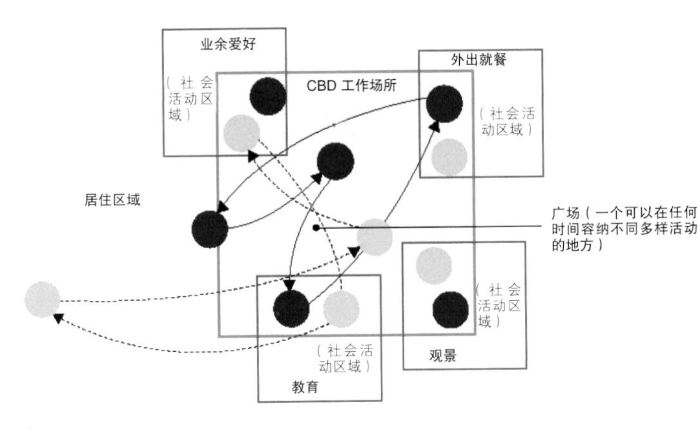

人类的活动与城市

Kim, Nam Jung（韩国）
Kang, Joo Yeon / Kim, Jin Hee

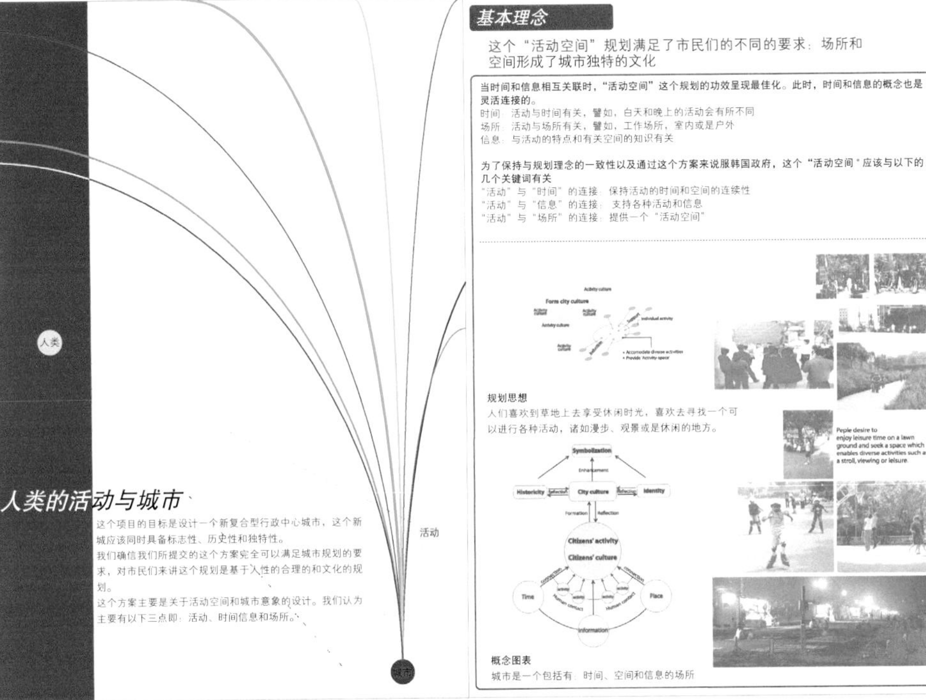

人类
人类的活动与城市
这个项目的目标是设计一个新复合型行政中心城市，这个新城应该同时具备标志性、历史性和独特性。
我们确信我们所提交的这个方案完全可以满足城市规划的要求，对市民们来讲这个规划是基于人性的合理的和文化的规划。
这个方案主要是关于活动空间和城市意象的设计。我们认为主要有以下三点即：活动、时间信息和场所。
活动
城市
基本理念
这个“活动空间”规划满足了市民们的不同的要求：场所和空间形成了城市独特的文化
当时间和信息相互关联时，“活动空间”这个规划的功效呈现最佳化。此时，时间和信息的概念也是灵活连接的。
时间：活动与时间有关，譬如，白天和晚上的活动会有所不同
场所：活动与场所有关，譬如，工作场所，室内或是户外
信息：与活动的特点和有关空间的知识有关
为了保持与规划理念的一致性以及通过这个方案来说服韩国政府，这个“活动空间”应该与以下的几个关键词有关
“活动”与“时间”的连接：保持活动的时间和空间的连续性
“活动”与“信息”的连接：支持各种活动和信息
“活动”与“场所”的连接：提供一个“活动空间”
Activity culture
Form city culture
Individual activity
规划思想
人们喜欢到草地上去享受休闲时光，喜欢去寻找一个可以进行各种活动，诸如漫步、观景或是休闲的地方。
Symbolization
Historicity
City culture
Identity
Citizens' activity
Citizens' culture
Time
Place
Information
People desire to enjoy leisure time on a lawn ground and seek a space which enables diverse activities such as a stroll, viewing or leisure.
概念图表
城市是一个包括有：时间、空间和信息的场所

富有活力和吸引力的场所

KJ 26334

这个提议集中在韩国文化中关于“水”的特殊含义上。
滨水规划的理念对于建设一个理想的城市是必要的，同时这也是韩国政府寻求的一种新的城市理念。
我们期望这个滨水规划可以起到催化剂的作用，可以增加城市的魅力，来弥补“可爱城市”和“只值一游的城市”的缺憾。

概念

滨水规划可以提高城市的舒适度，同同给居民提供有吸引力的活动空间并且有利于人与人之间的交往

通过自然和人类的平衡和共存来追求协调

允许实体有变化的可能性
保护好被金河分开的两个区域之间的联系
滨水区不仅是一个有水的地方，还可以作为一个真正有吸引力的场所。在那里：自然、人类以及人工建筑共存，平衡和发展这些因素都包含在内。通过实施“和谐”这个关键理念，这些要素才可以得到共存。

Provide with attractiveness
Nature (Waterfront) Water-green land
Linkage Coexistence Balance Evolution
Human (Activity) Dynamic-static activity
Space Divided zone Nature-Human-Structure
HARMONIZATION

The waterfront is not simply places with water. To be a real charming space, such important concepts as the nature, human, coexistence of artificial structures, balance, and evolution have to be incorporated. These concepts can be implemented by the key planning concept of 'harmonization'.

Plan of idea

计划

1. **和谐：设施之间的连接**
   通过自然和人工建筑的和谐来构建友好城市的环境形象
   在自然和人工建筑，动静活动之间来寻找一个切入点
   在步行空间集中的地方，可以安排主要的活动空间，包括开敞空间、公园和文化设施
   构建城市空间的中心轴，主要集中在步行区或步行桥周围

2. **共存、平衡、发展：设施管理**
   设施管理包含了波状图像
   提供空间感来匹配海滨的景色：保护活动的一致性
   通过提供风路与自然协调来管理设施
   赋予滨水区延伸和发展的可能性

   将广场、公园和文化设施以及滨水区的活动空间相连接
   在滨水中心区域规划一个开敞空间
   从开敞空间到滨水区来寻找一个切入点

3. **吸引点：设施的利用**
   步行桥象征着开始和结束
   作为城市景观的中心轴来促进区域的发展
   滨水区动态的视角

   通过夜景景观规划来增加城市的吸引力
   通过时间段来组织滨水区的景观
   通过灯光和水的和谐来创造一种特别的景观效果

   通过自然环境和人类活动来形成一种文化的核心区域

Axis of city view
View panorama
Cultural facility
Linkage space
Pedestrian only bridge
Green space
Core area of culture
Cultural space
Axis of waterfront view
Pedestrian only bridge (Linkage between Zone A and B)
Linkage(pedestrian path)
建筑设计

流动的水，就像河流拥有线条，河岸拥有韵律一般。这些都应该在土地利用规划中有所反映，以此来实现真正的“和谐”。

在地面上看到的夜景观

在空中看到的夜景观

活动、信息和场所

该规划条款是从这个方案中的主要概念中引申出来的：
建筑、交通和各种各样的人类活动的交叉点是地区的核心
广场有各种各样的功能

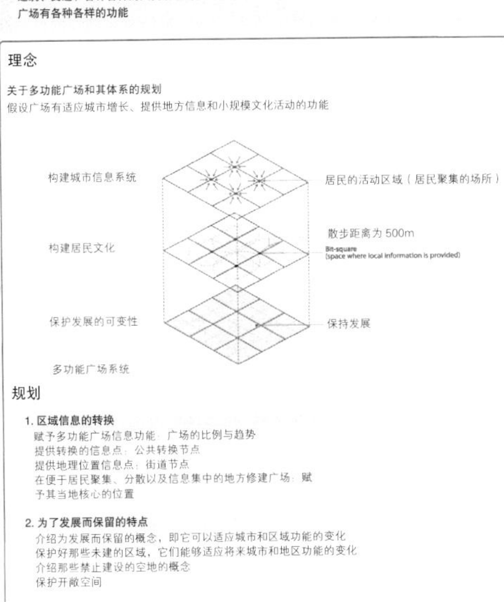

理念
关于多功能广场和其体系的规划
假设广场有适应城市增长、提供地方信息和小规模文化活动的功能
构建城市信息系统
居民的活动区域（居民聚集的场所）
构建居民文化
散步距离为 500m
Bit-square (space where local information is provided)
保护发展的可变性
保持发展
多功能广场系统
规划
1. 区域信息的转换
赋予多功能广场信息功能：广场的比例与趋势
提供转换的信息点：公共转换节点
提供地理位置信息点：街道节点
在便于居民聚集、分散以及信息集中的地方修建广场：赋予其当地核心的位置
2. 为了发展而保留的特点
介绍为发展而保留的概念，即它可以适应城市和区域功能的变化
保护好那些未建的区域，它们能够适应将来城市和地区功能的变化
介绍那些禁止建设的空地的概念
保护开敞空间
3. 构建多功能广场体系
通过信息的基本单位，即多功能广场网络信息化来构建城市的信息系统。
通过行走来获得一个行走距离的界定。即一个大约为 500m 的行走距离（5~10 分钟的路程）

活动与时间

这个提议的目的是为了空间的规划，它可以改善 CBD 的利用效率。以时间划分活动的种类，在综合规划中寻求解决办法。
这就意味着应该有机地组织白天和夜间的活动，并将相关的地方与工作地方在空间上相连接。

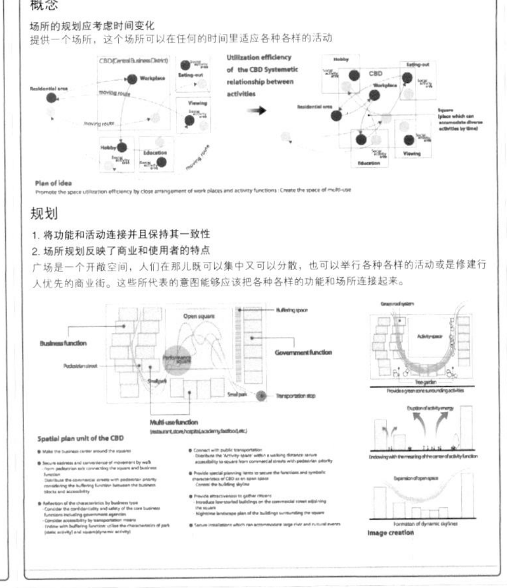

概念
场所的规划应考虑时间变化
提供一个场所，这个场所可以在任何的时间里适应各种各样的活动
Utilization efficiency of the CBD Systematic relationship between activities
Residential area
Workplace
Viewing
Education
Plan of idea
规划
1. 将功能和活动连接并且保持其一致性
2. 场所规划反映了商业和使用者的特点
广场是一个开敞空间，人们在那儿既可以集中又可以分散，也可以举行各种各样的活动或是修建行人优先的商业街。这些所代表的意图能够应该把各种各样的功能和场所连接起来。
Business function
Open square
Government function
Multi-use function
Spatial plan unit of the CBD
Image creation

未来之城

Lee, Jong Ho（韩国）
Kim, Jong Sik

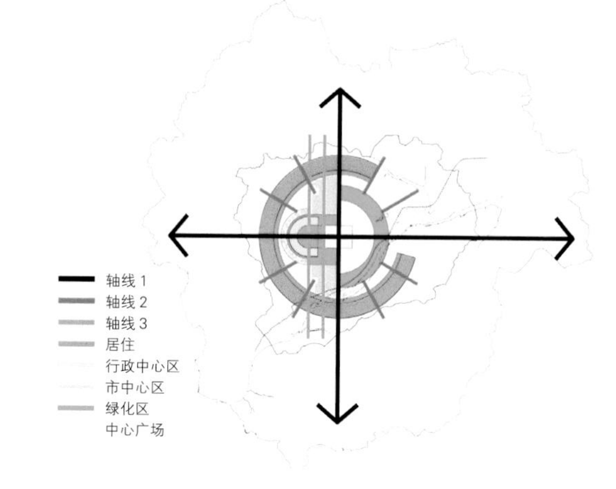

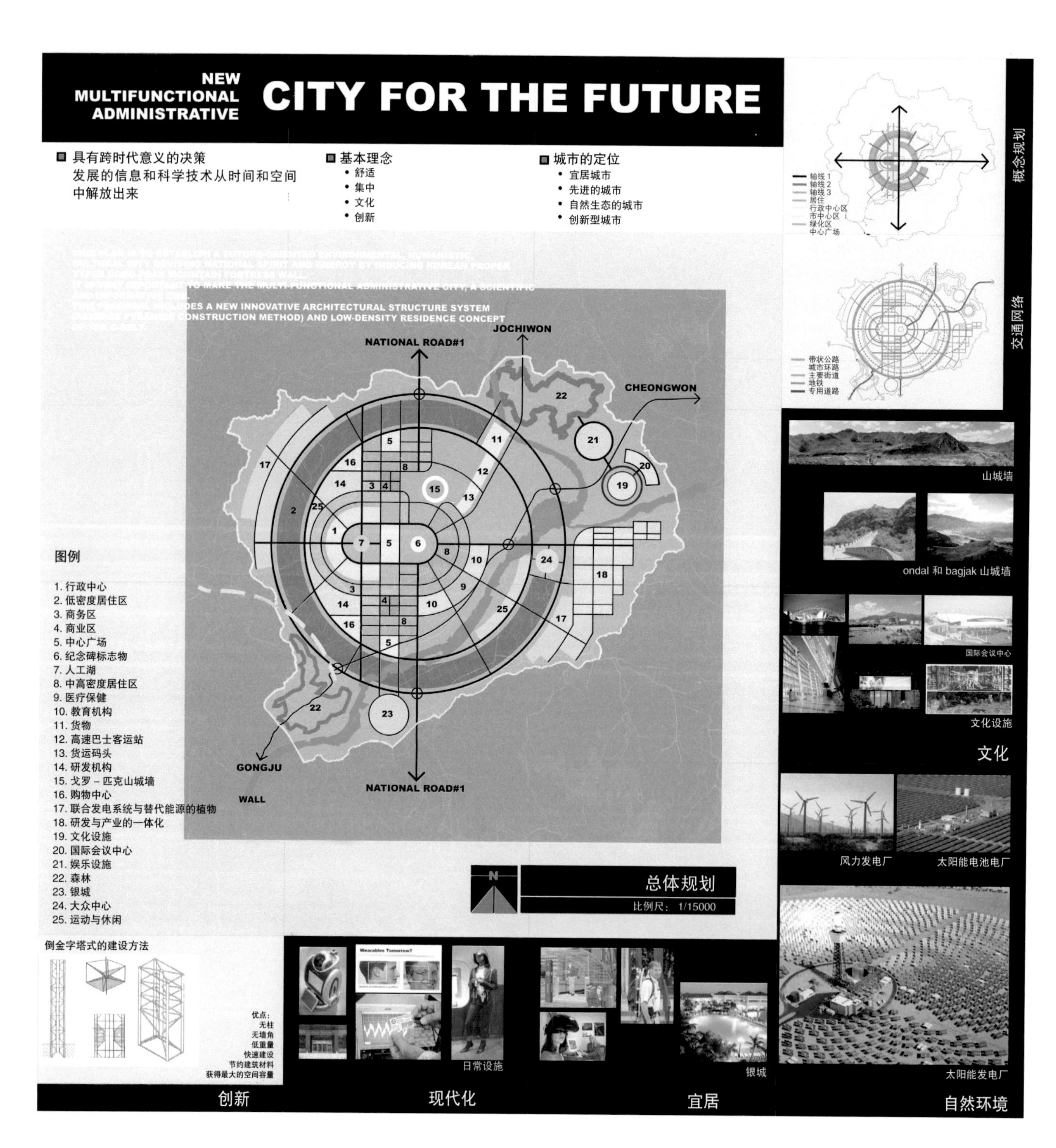

永恒之城

Lee, Jae Won（韩国）
Jang, Seong Il / Shin, Sang Hyun /
Choi, Yun Jung / Kwag, Ji Hee

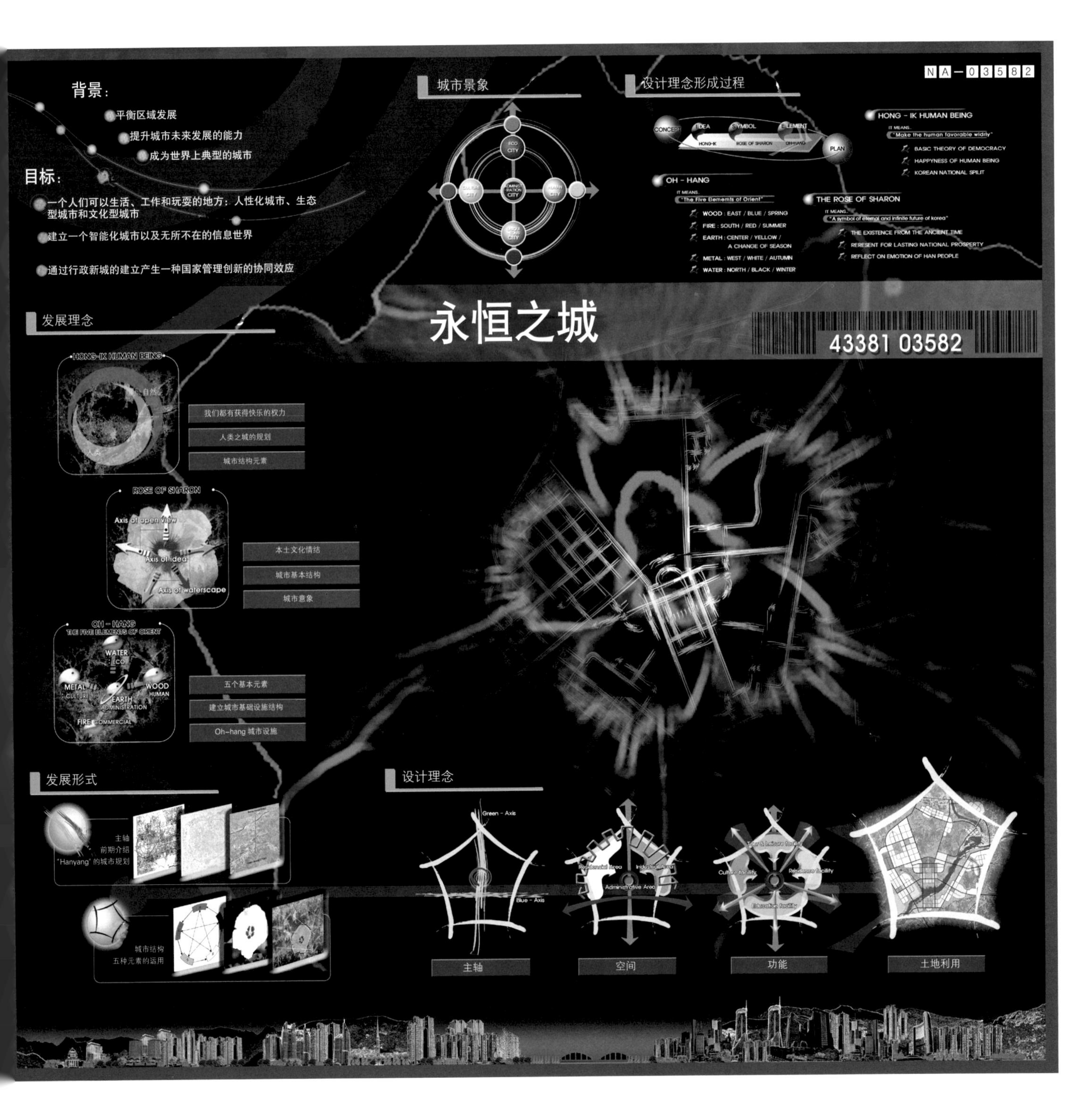

未来的主要模式

Choi, Won Seok（韩国）
Lee, Jun Hyung / Lee, Jee Young / Kim, Min Ho

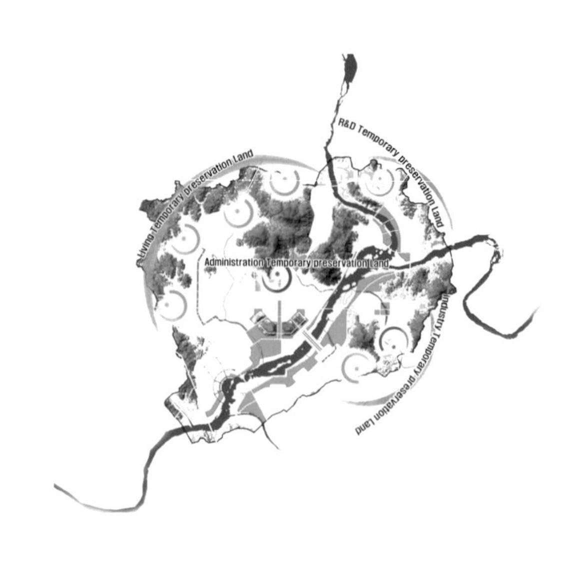

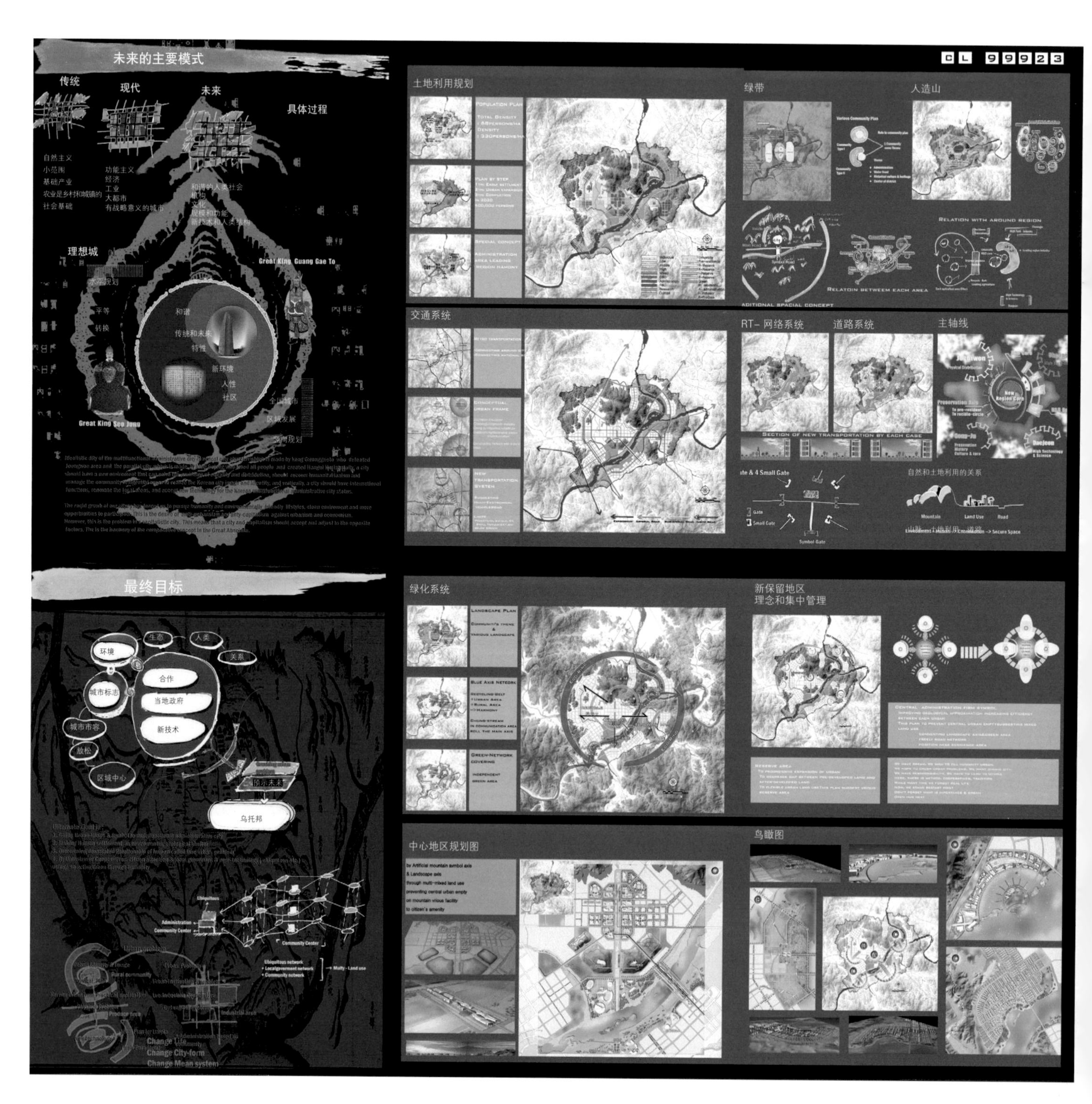

无题

Lee, Ho Joon（韩国）
Sung, In Jung

网络基础城市单元

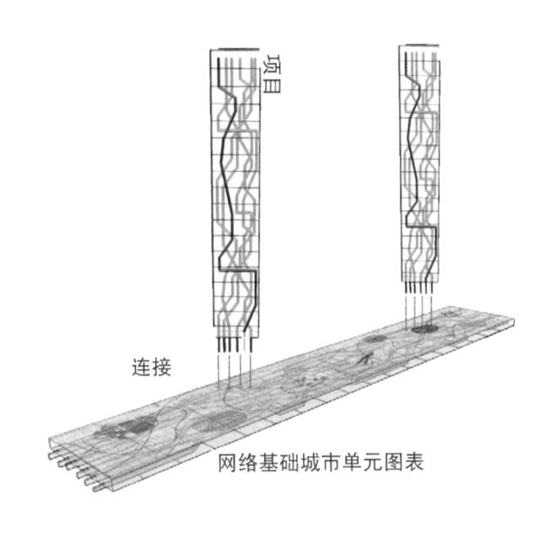

网络基础城市单元图表

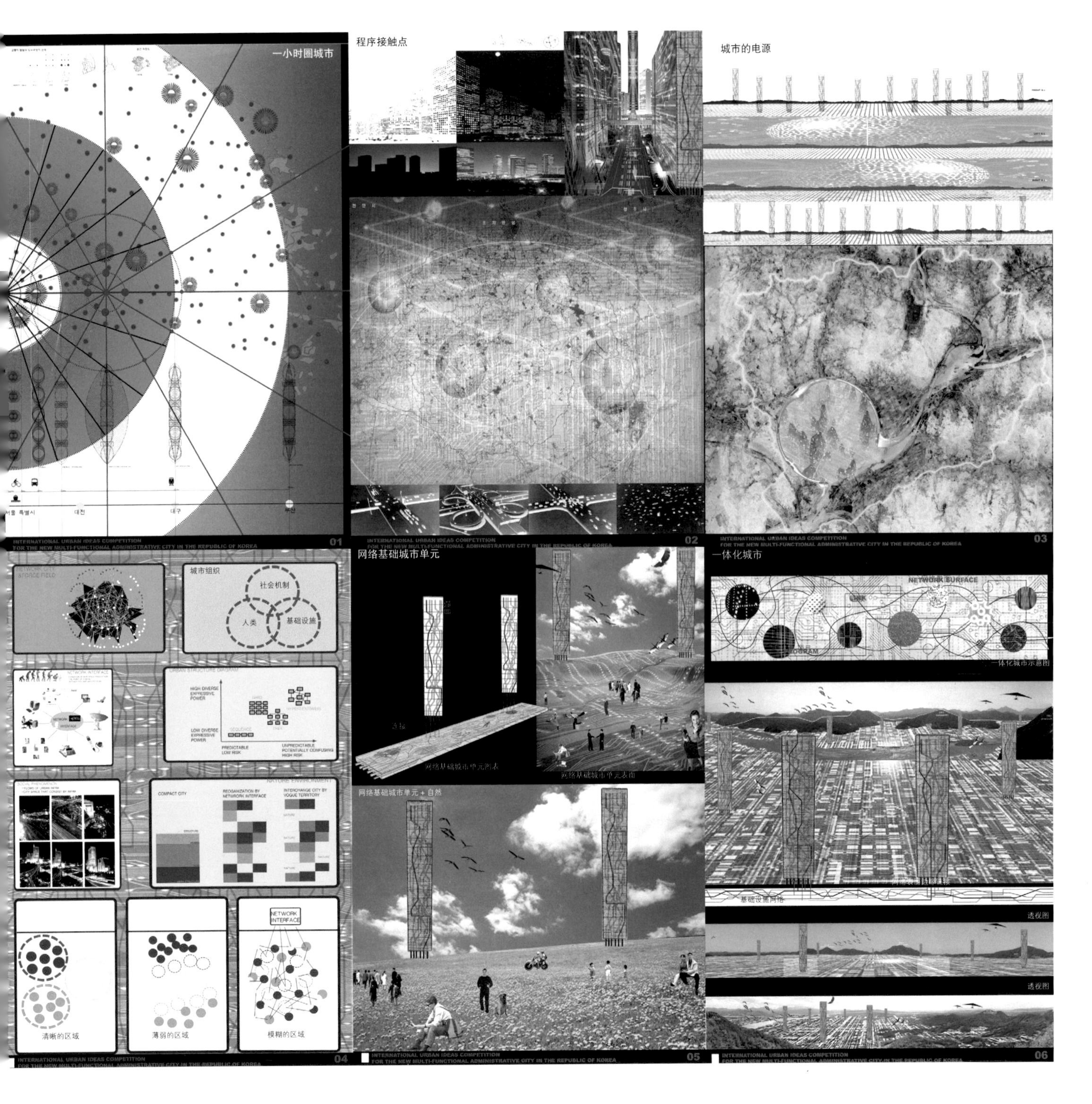

无题

Han, Ju Hyung(韩国)

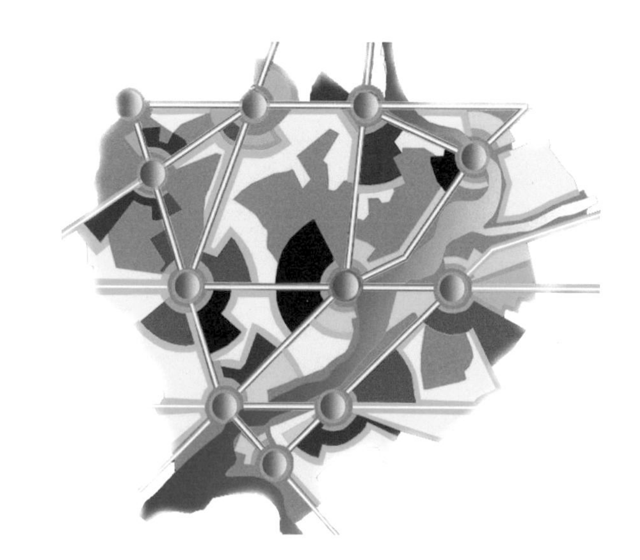

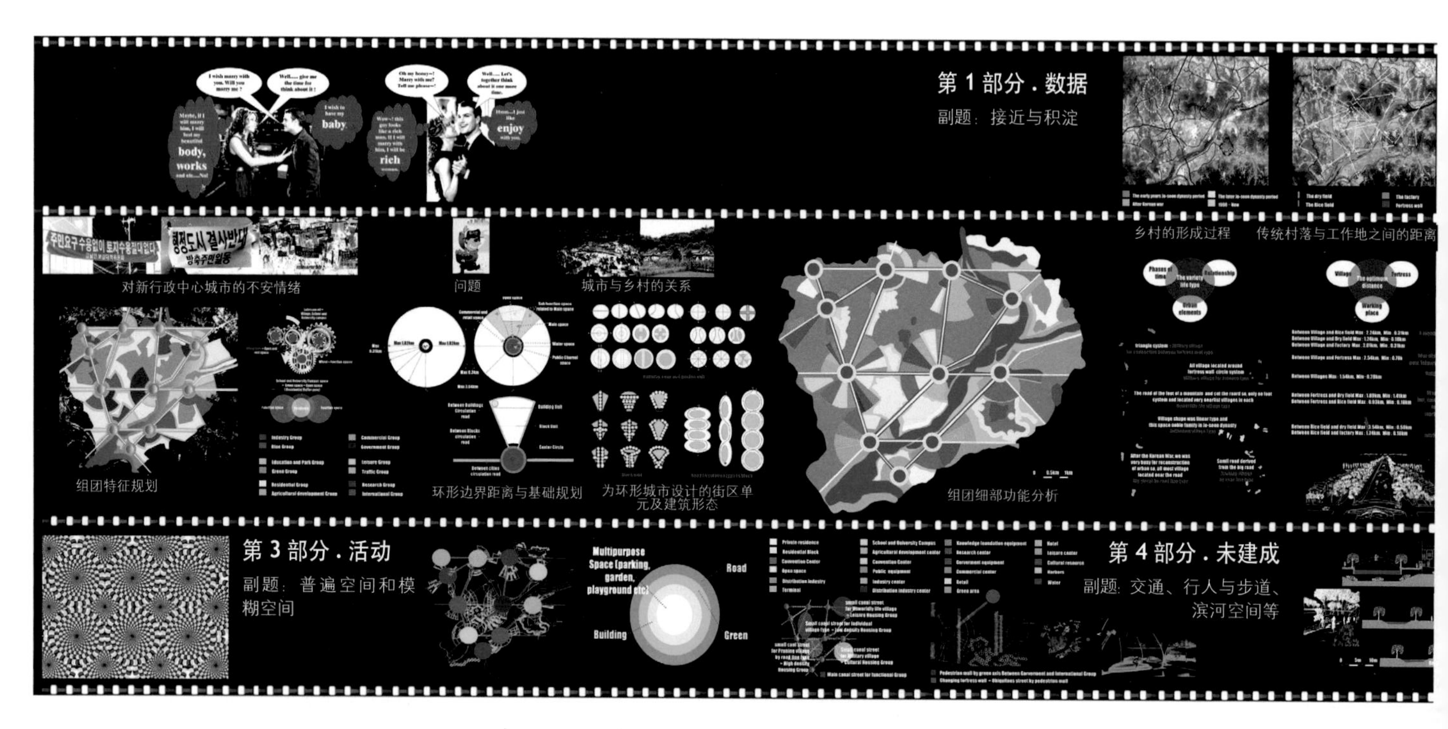

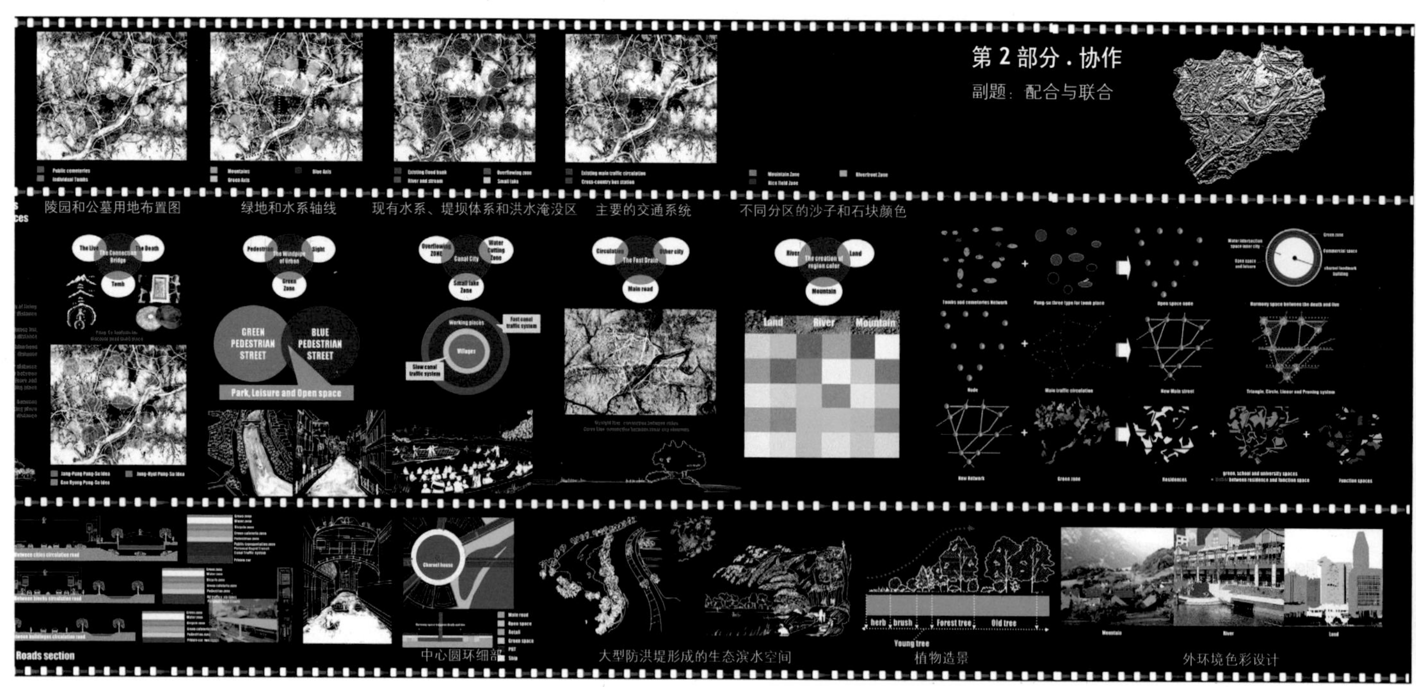

无题

Sim, Kil Je（韩国）
No, Tae Young

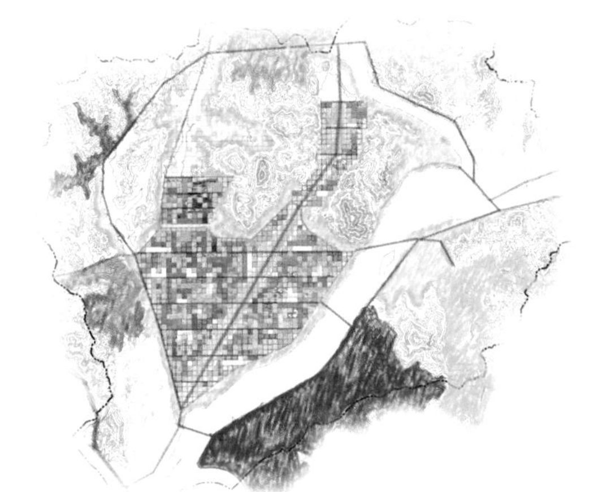

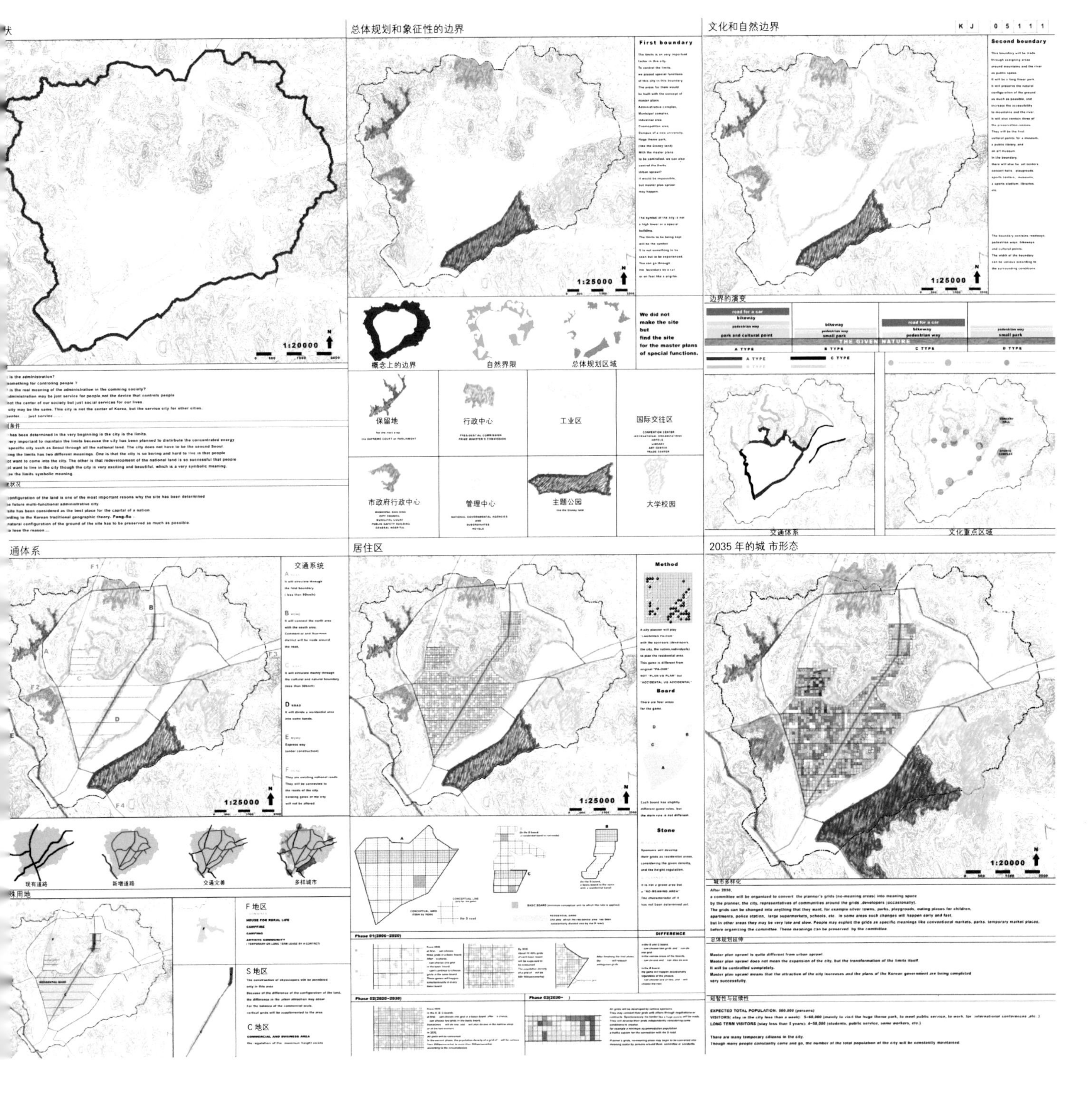

无题

Eom, Su Ryu（韩国）
Jung, Hee Joo / Lee, Eun Hee / Lee, Jae Woo

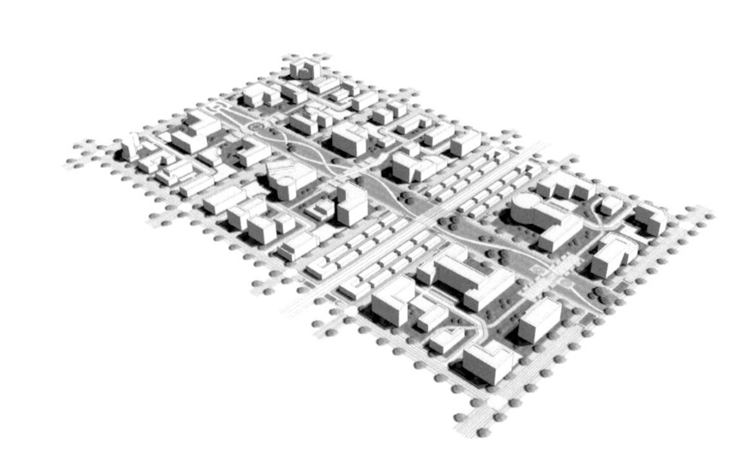

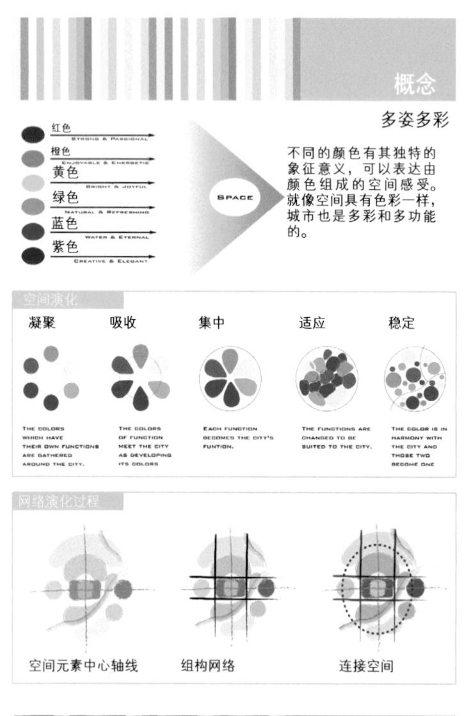

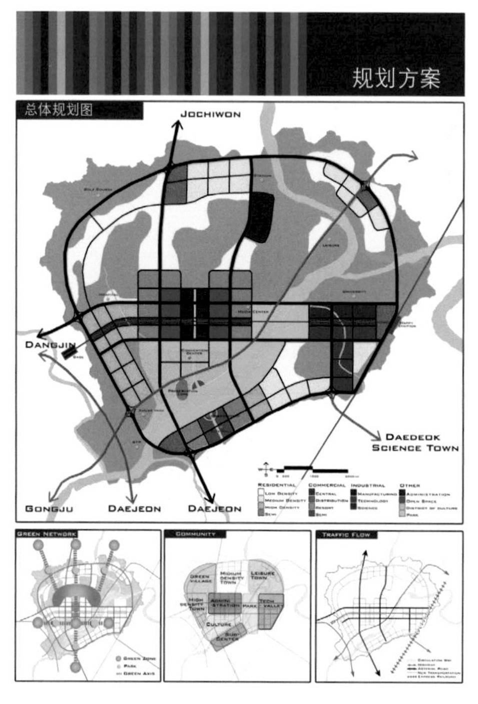

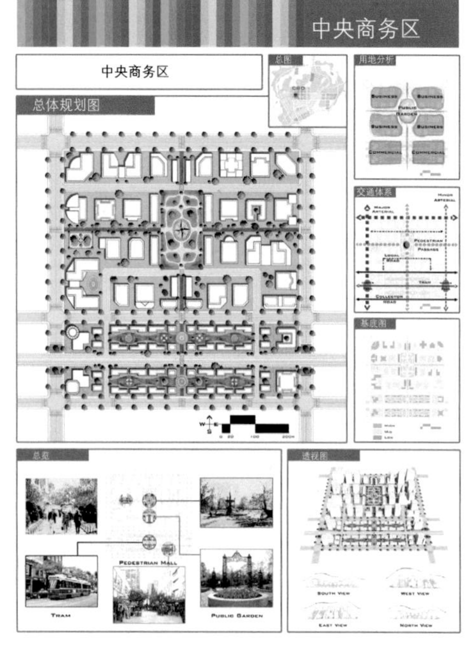

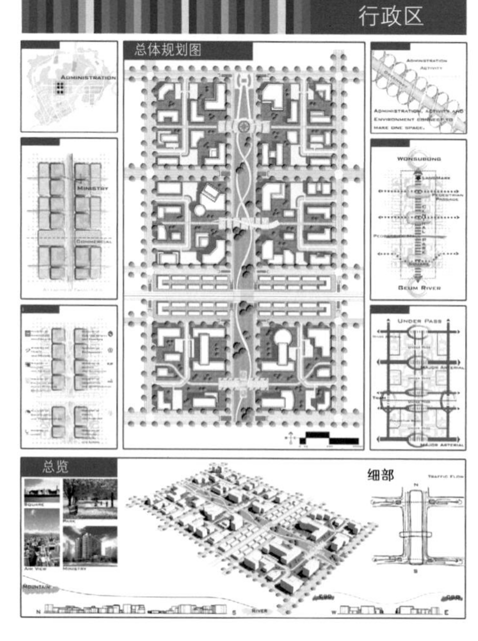

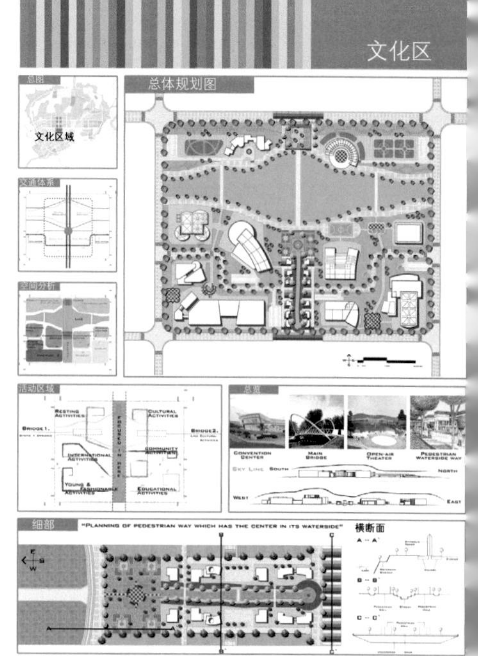

无题

Eun Jin Pyo（韩国）
Kim, Dong Young/Sung Hwan

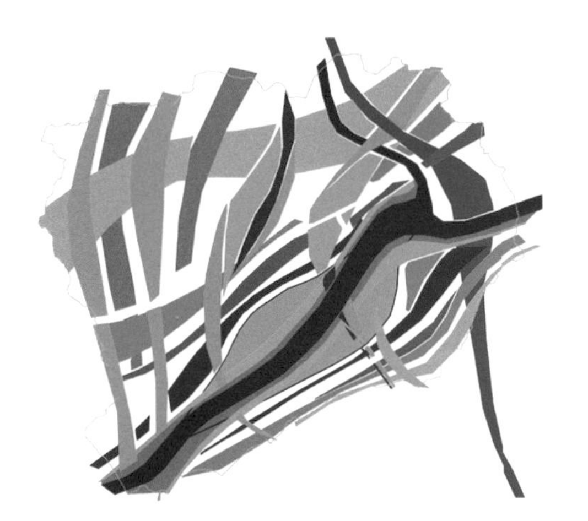

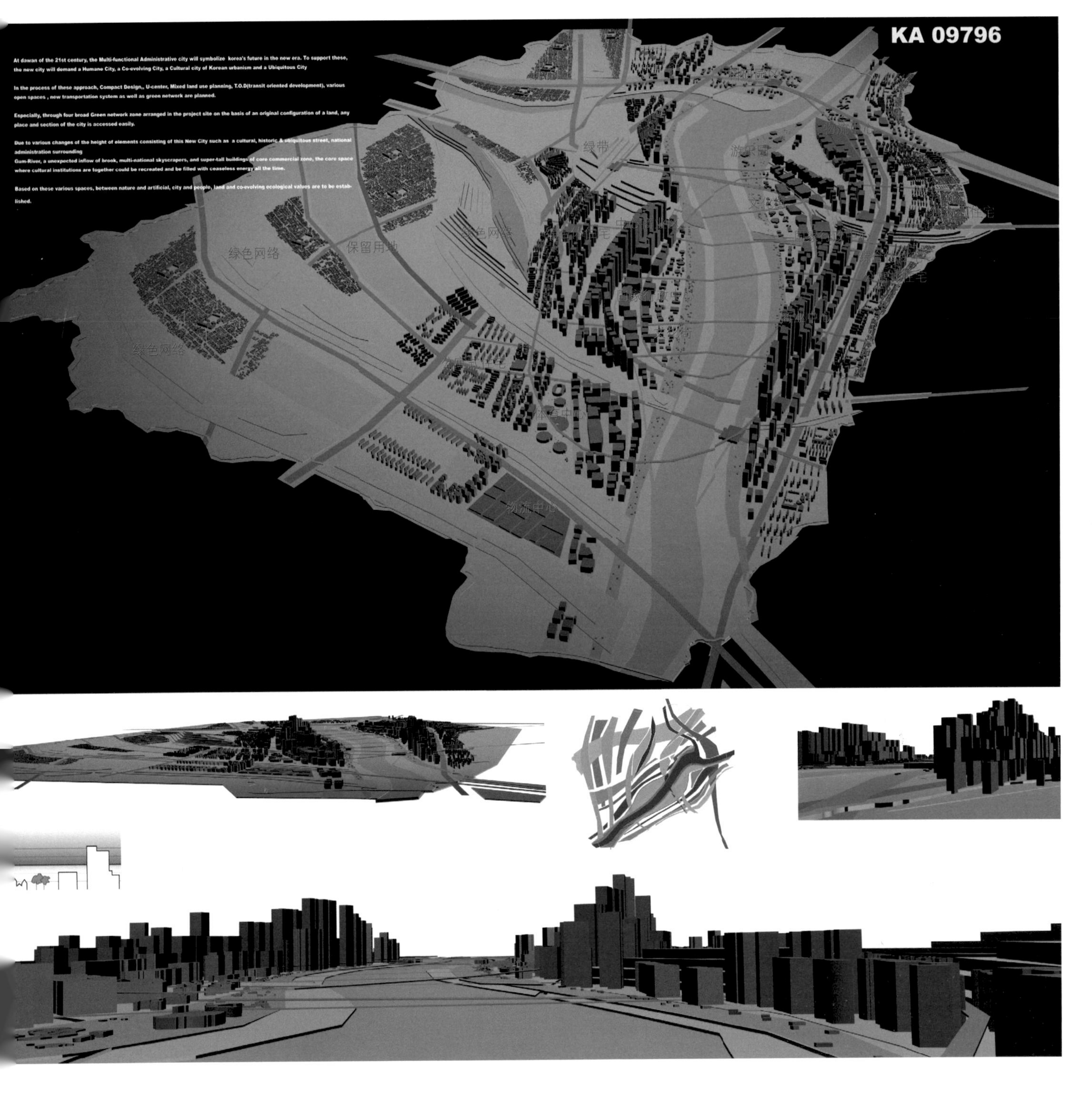

KA 09796
At dawan of the 21st century, the Multi-functional Administrative city will symbolize korea's future in the new era. To support these, the new city will demand a Humane City, a Co-evolving City, a Cultural city of Korean urbanism and a Ubiquitous City
In the process of these approach, Compact Design,, U-center, Mixed land use planning, T.O.D(transit oriented development), various open spaces , new transportation system as well as green network are planned.
Especially, through four broad Green network zone arranged in the project site on the basis of an original configuration of a land, any place and section of the city is accessed easily.
Due to various changes of the height of elements consisting of this New City such as a cultural, historic & ubiquitous street, national administration surrounding
Gum-River, a unexpected inflow of brook, multi-national skyscrapers, and super-tall buildings of core commercial zone, the core space where cultural institutions are together could be recreated and be filled with ceaseless energy all the time.
Based on these various spaces, between nature and artificial, city and people, land and co-evolving ecological values are to be estab-lished.
绿带
绿色网络
保留用地
绿色网络

幸福之城

Kim,Jin Young（韩国）

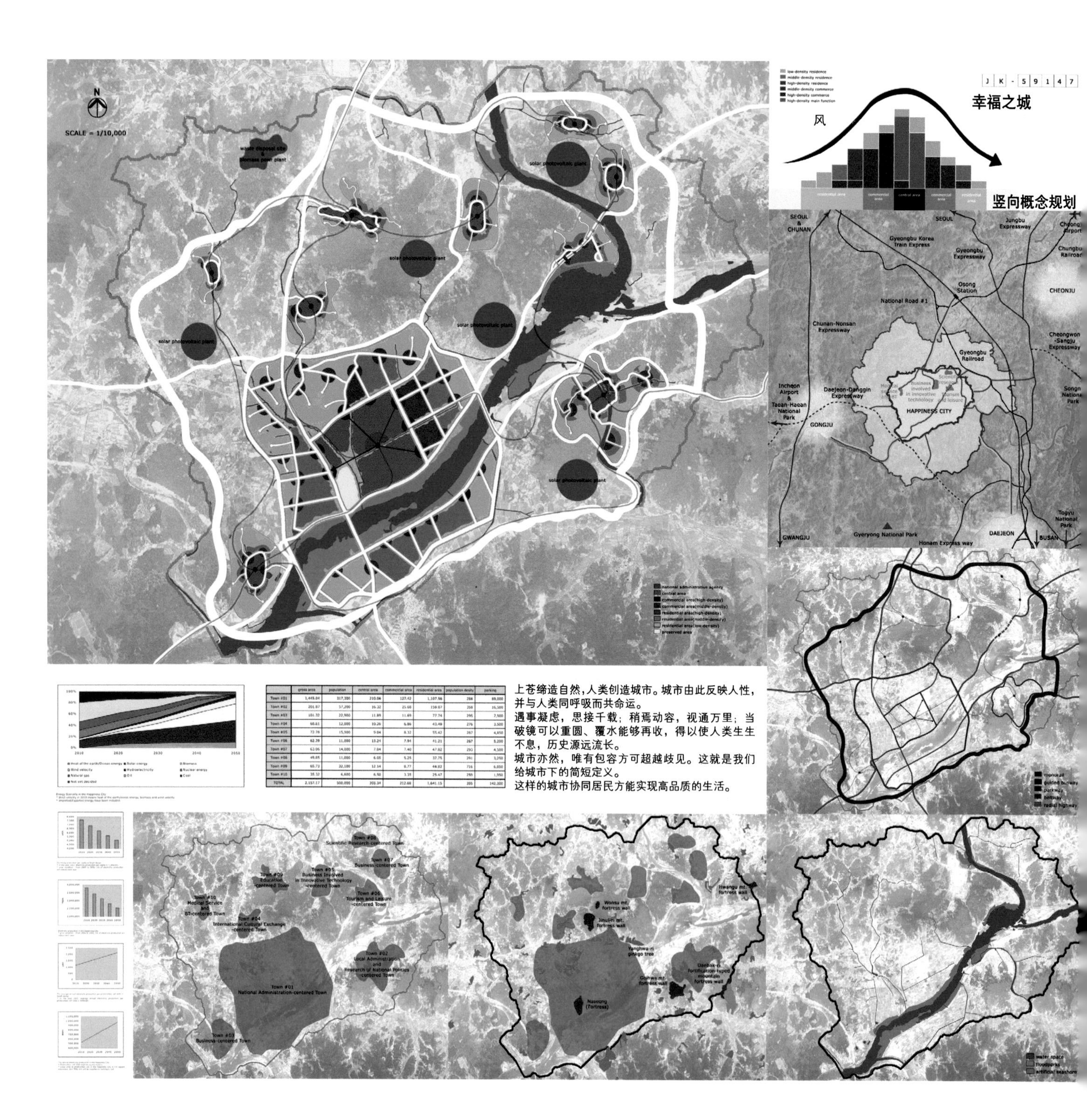

	gross area	population	central area	commercial area	residential area	population desity	parking
Town #01	1,445.04	317,300	210.06	127.42	1,107.56	288	89,000
Town #02	201.87	57,200	16.32	25.68	159.87	358	16,500
Town #03	101.32	22,900	11.89	11.69	77.74	295	7,500
Town #04	60.61	12,000	10.26	6.86	43.49	276	3,500
Town #05	72.78	15,900	9.04	8.32	55.42	287	4,850
Town #06	62.39	11,000	13.24	7.94	41.21	267	5,200
Town #07	62.06	14,000	7.04	7.40	47.82	293	4,500
Town #08	49.05	11,000	6.05	5.25	37.75	291	3,250
Town #09	65.73	32,100	12.14	8.77	44.82	716	6,050
Town #10	35.32	6,600	6.50	3.35	25.47	259	1,950
TOTAL	2,157.17	500,000	303.34	212.68	1,641.15	305	142,300

上苍缔造自然,人类创造城市。城市由此反映人性,并与人类同呼吸而共命运。
遇事凝虑，思接千载；稍焉动容，视通万里；当破镜可以重圆、覆水能够再收，得以使人类生生不息，历史源远流长。
城市亦然，唯有包容方可超越歧见。这就是我们给城市下的简短定义。
这样的城市协同居民方能实现高品质的生活。

无题

Nam,Young Hyun（韩国）
Jim,Young Sup/Lee, Myung Hee/Jin,Yoon

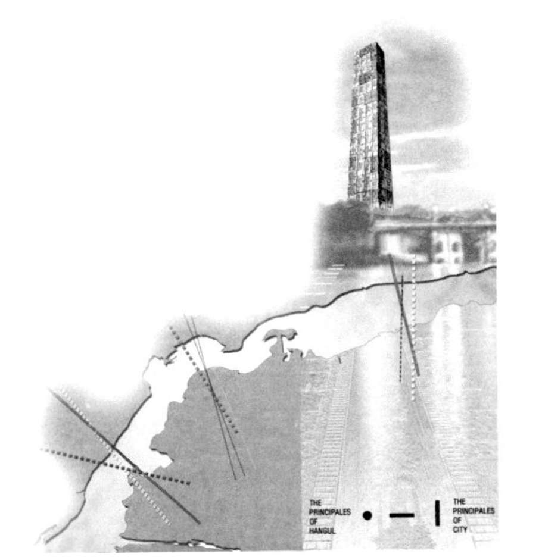

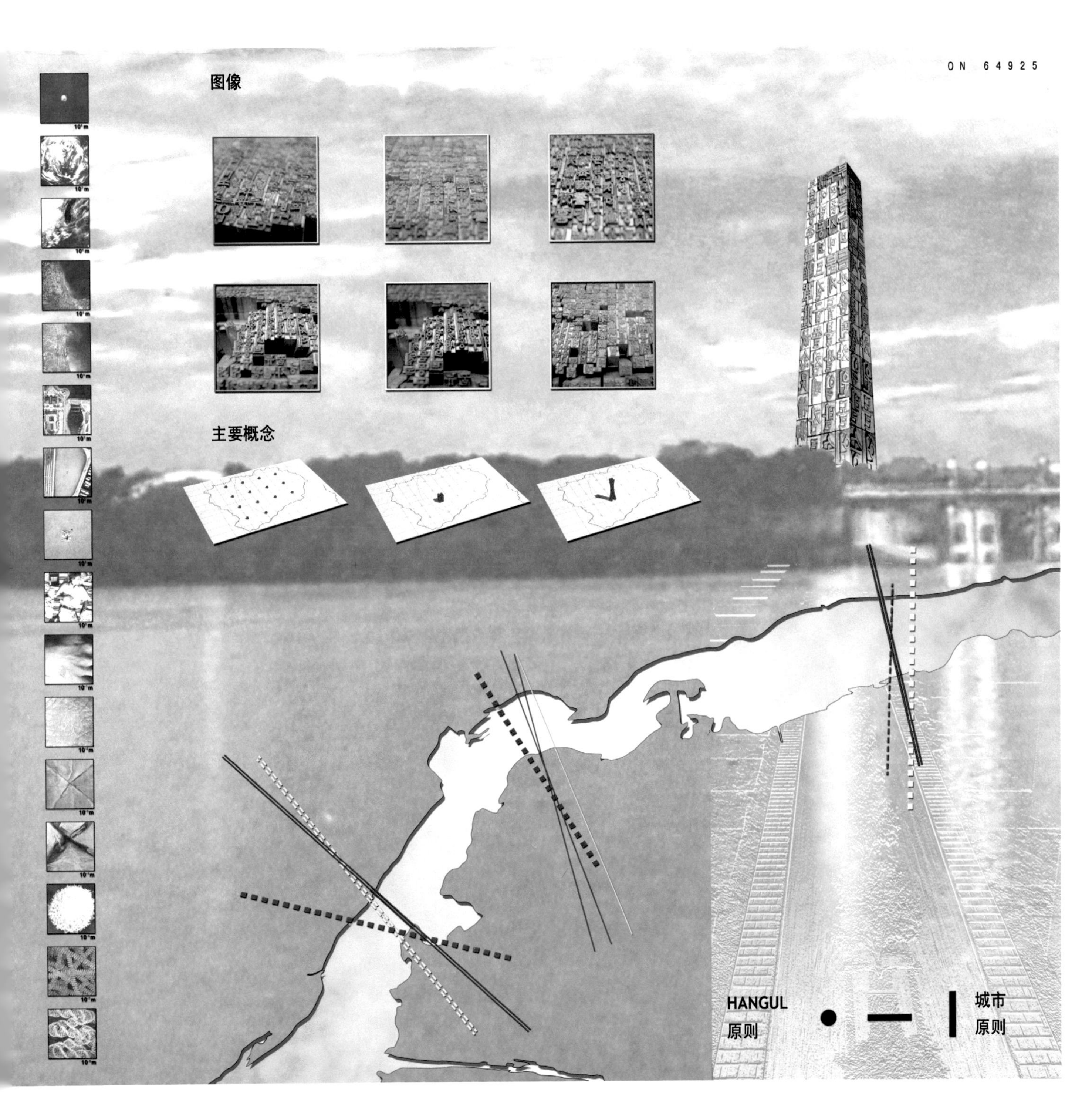

无题

Shin, Ye Kyeong(韩国 **)**
Lee, Jeong Houn

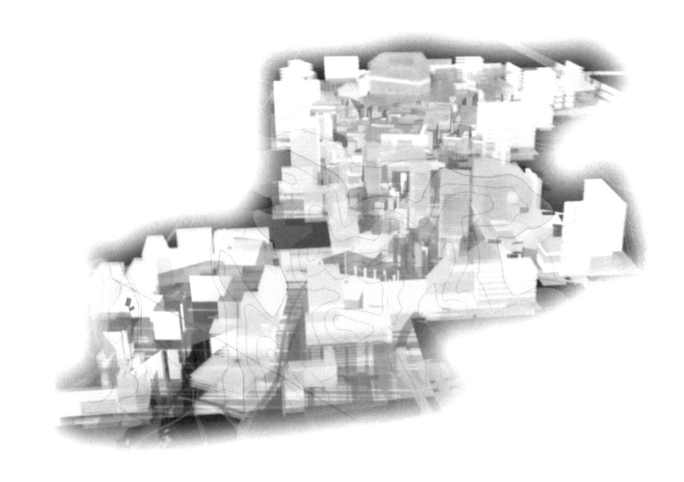

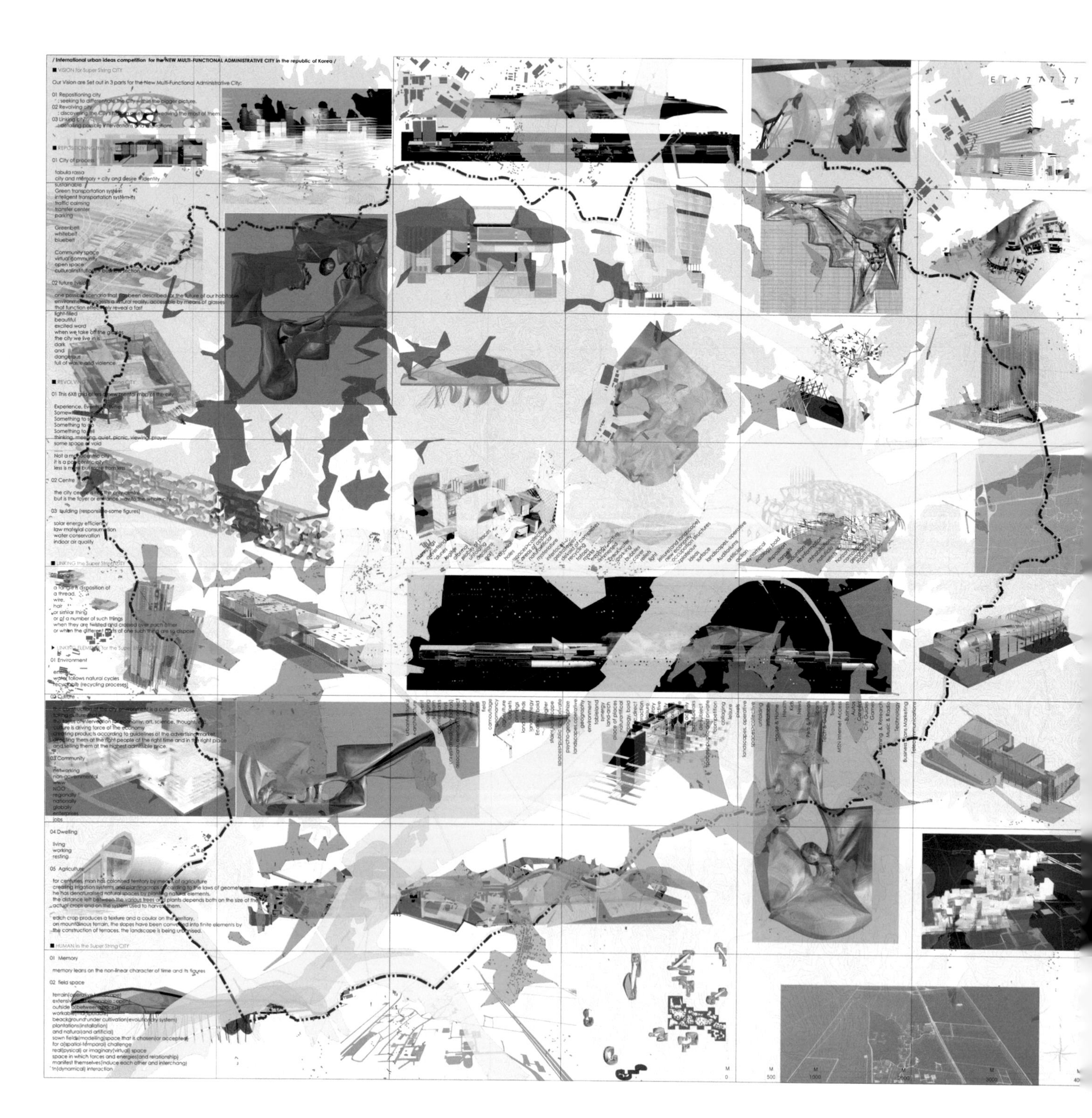

无题

Kim, Tai Young（韩国）
Seon Architects & Engineers Group /
School of Architecture and Engineering, Cheongju University

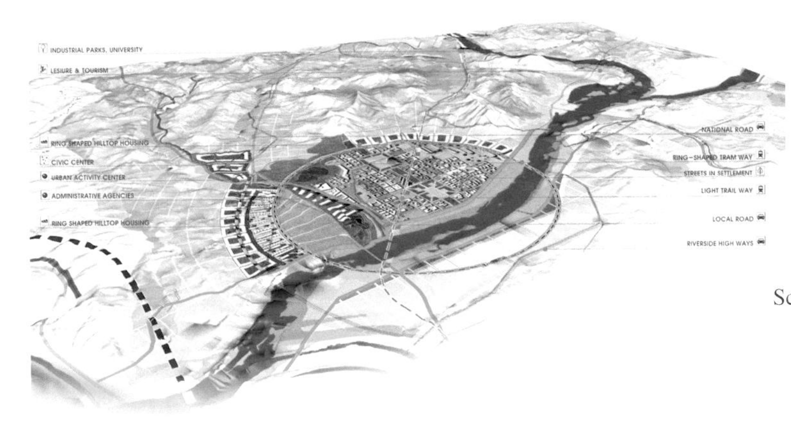

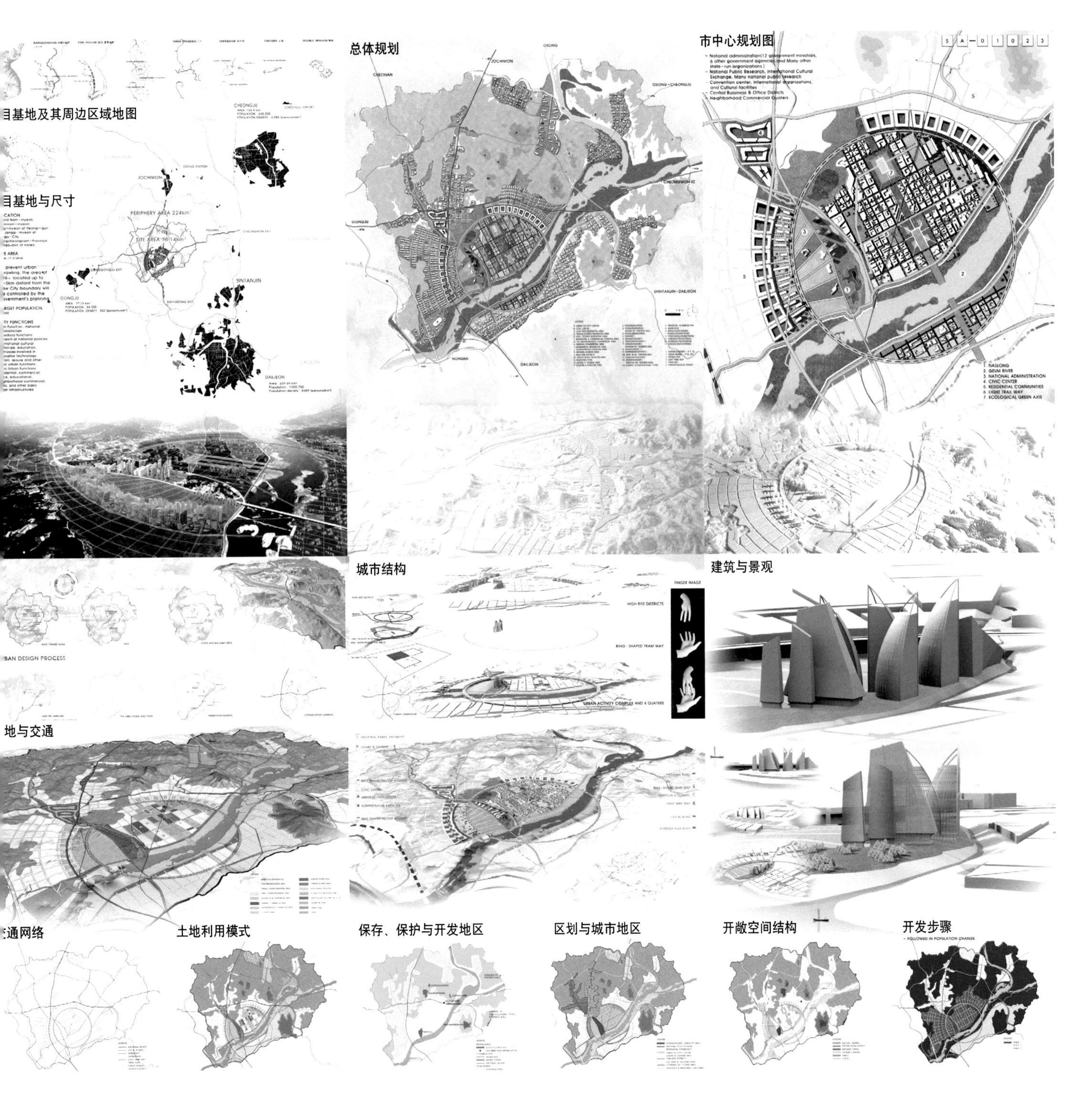

无题

Lee,Jang Gun(韩国)
Koh, Dong Wook/Jang, Yong Hun

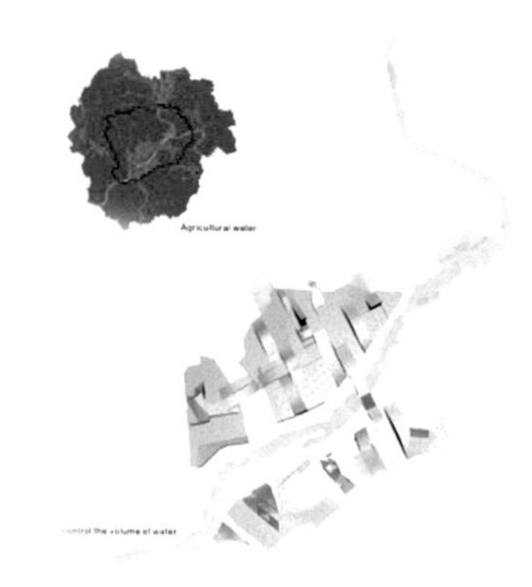

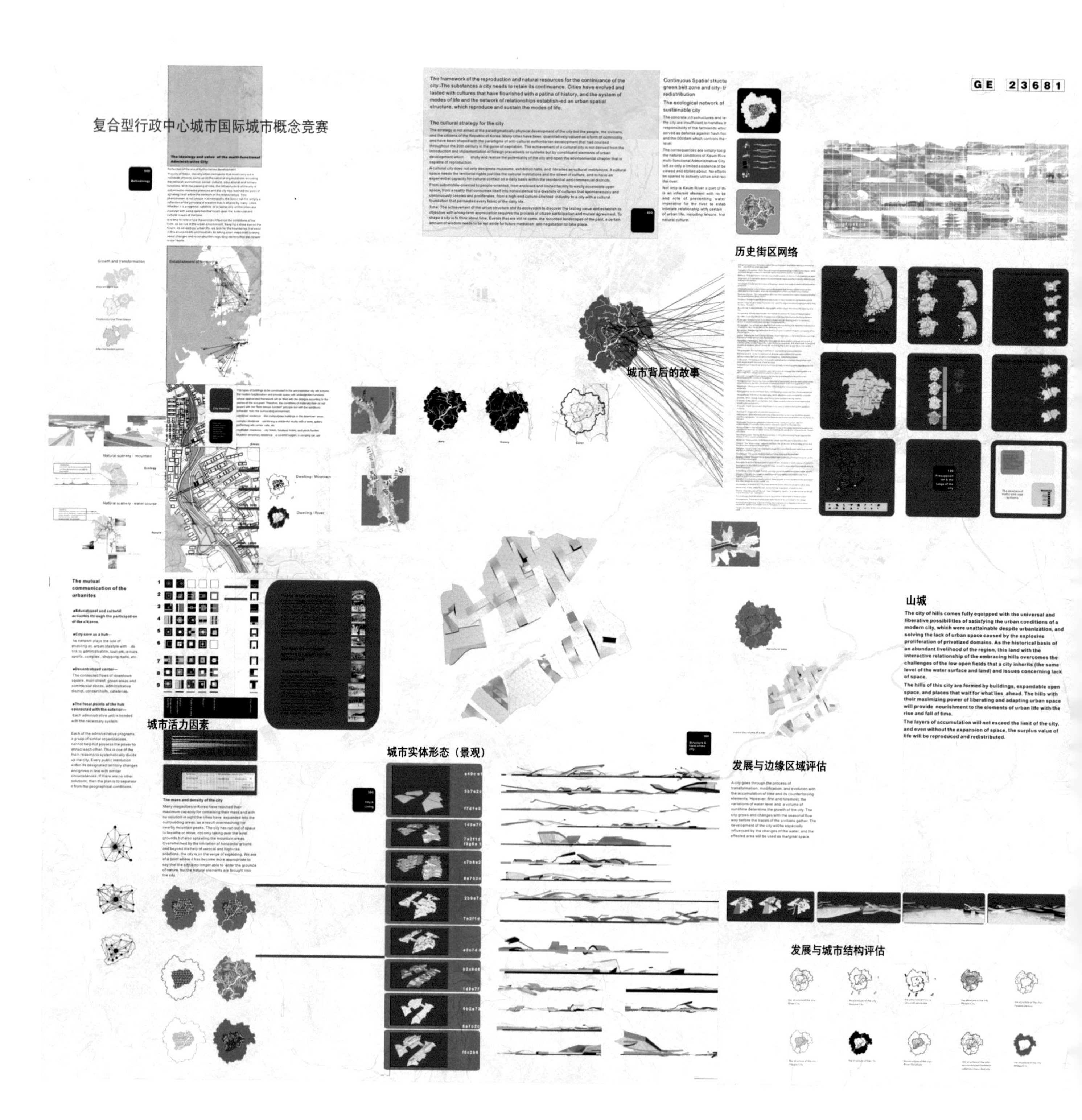

进化中的城市

Lee, Sang Hyun（韩国）

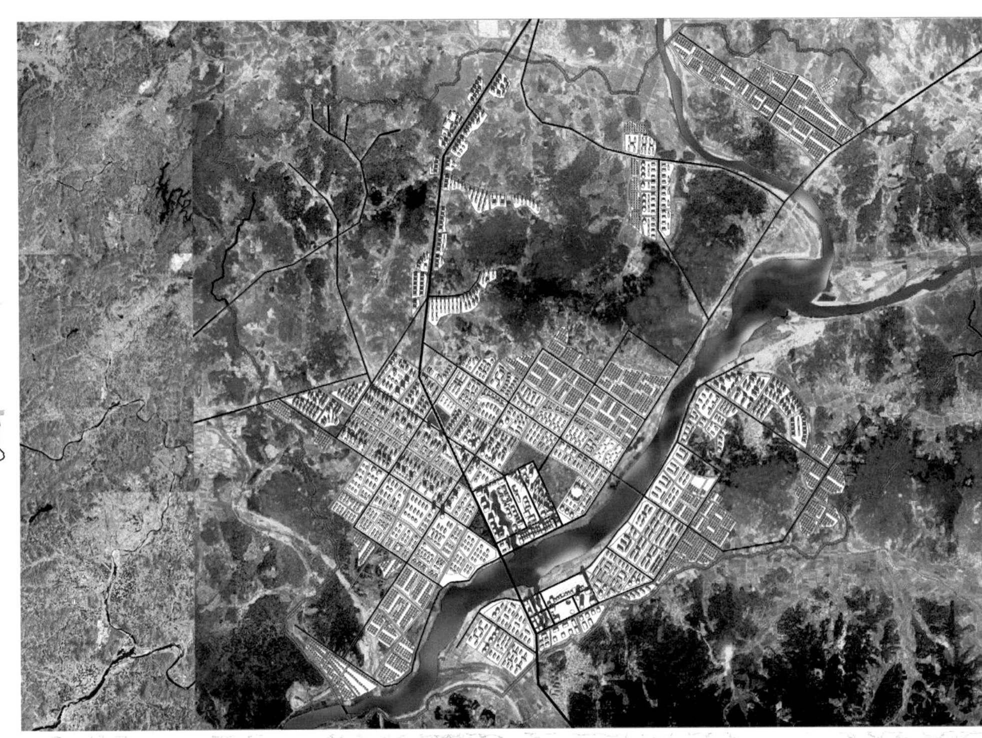

城市演进

城市从相互学习中不断完善：行政、商业、研究、大学和居住等。

城市绵延不绝的步行网络使我们可以任意穿行。

城市利用树木使我们辨明方位——如同迪卡尔网格里的特殊结构。

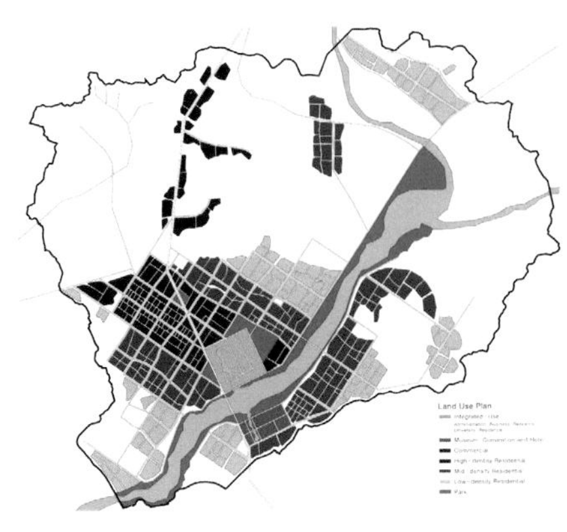

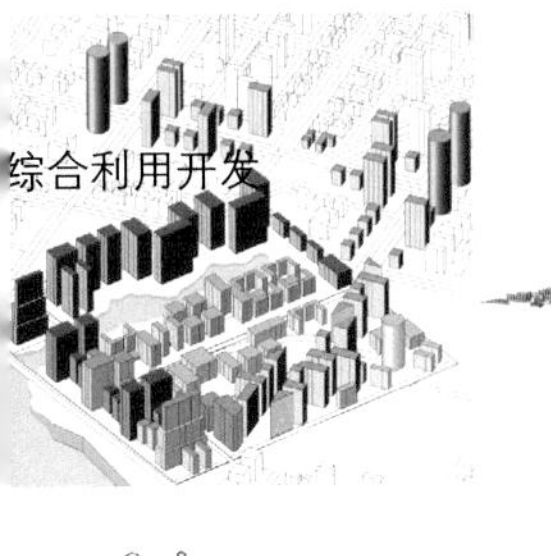

西侧视角

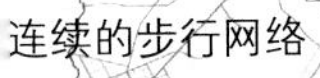

北侧视角

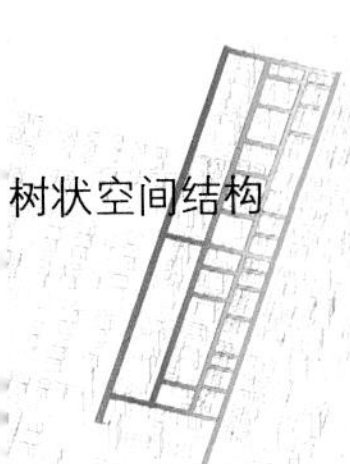

东侧视角

仿生态城市

Yun, Hee Jin(韩国)
Ji, Jang Hun/Han, Geang Nam/
Lee、Yu Mi/Braun, Adrea

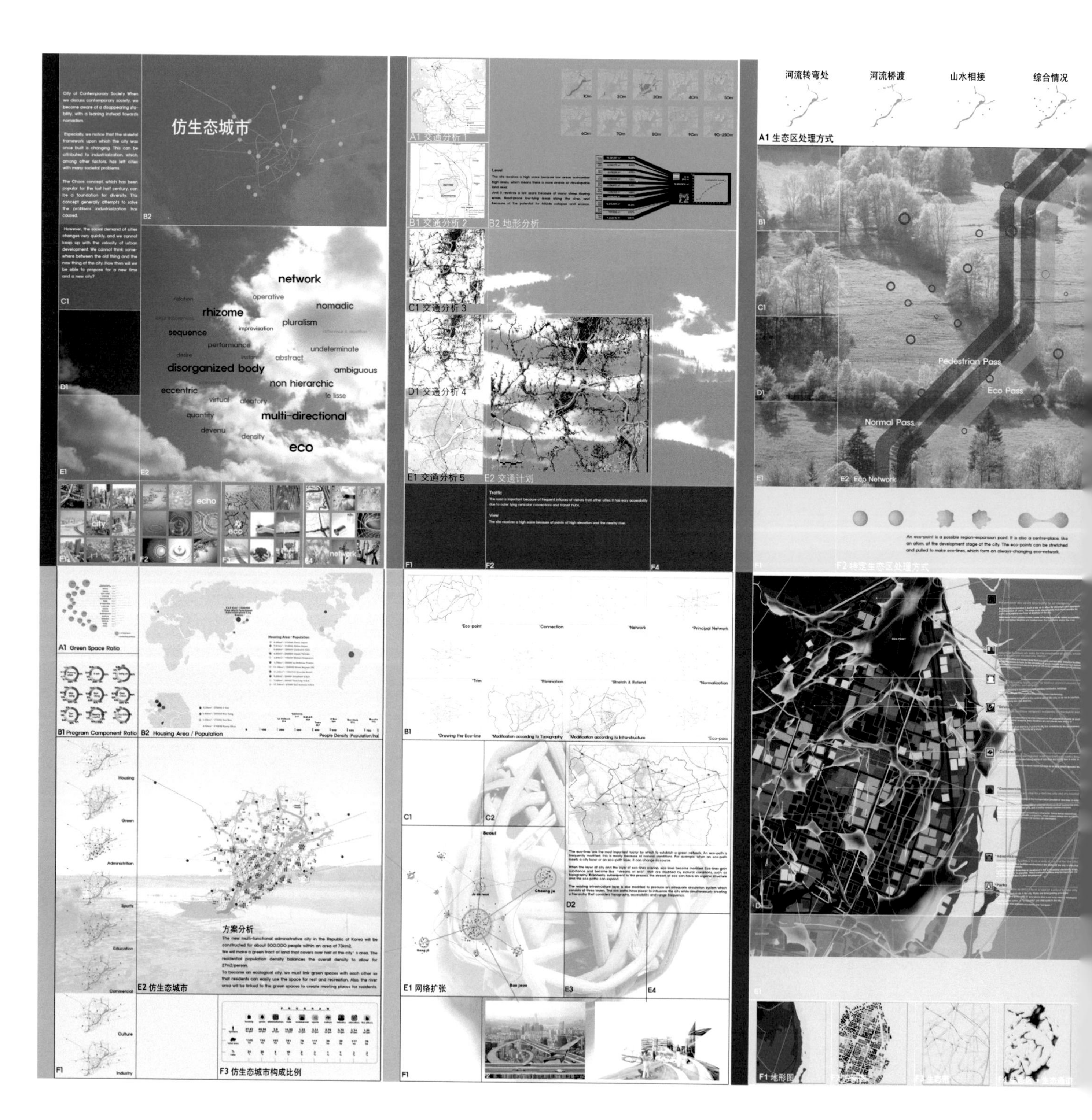

无题

Landstrom, Roger（澳大利亚）

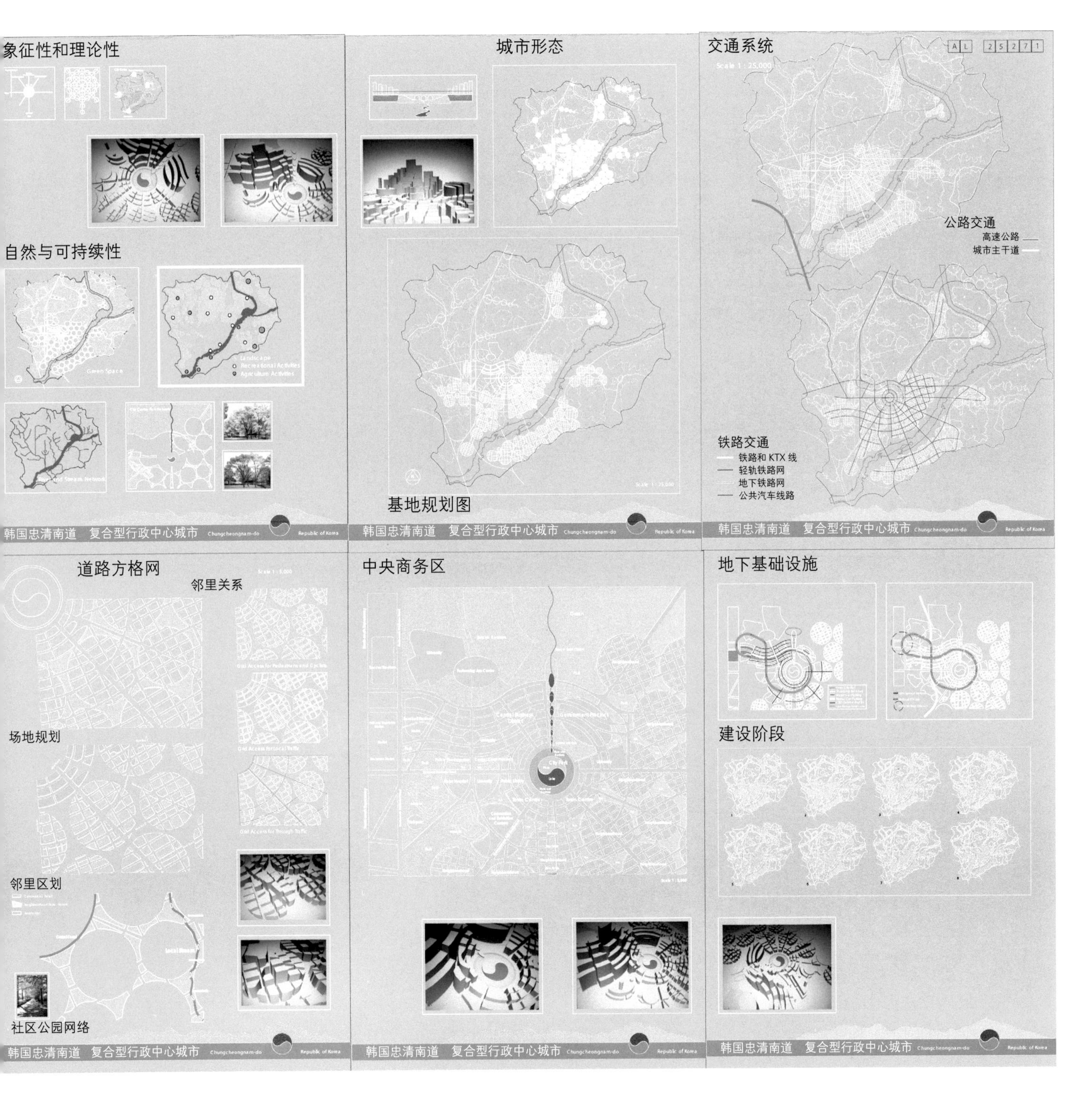

无题

Karakiewicz,Justyna（中国香港）
Kvan, Thoms/Hwang, Se Young Iris/
Zhai, Binging/Rotmeyer,Juliana

원형 물질 대사 円の新陳代謝
cirkel metabolisme
métabolisme circulaire
metabolismo circolare
круговой метаболизм
metabolismo circular
新陈代谢循环组织

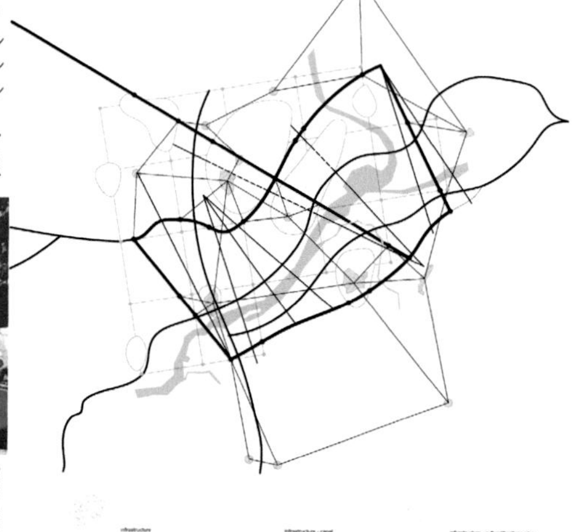

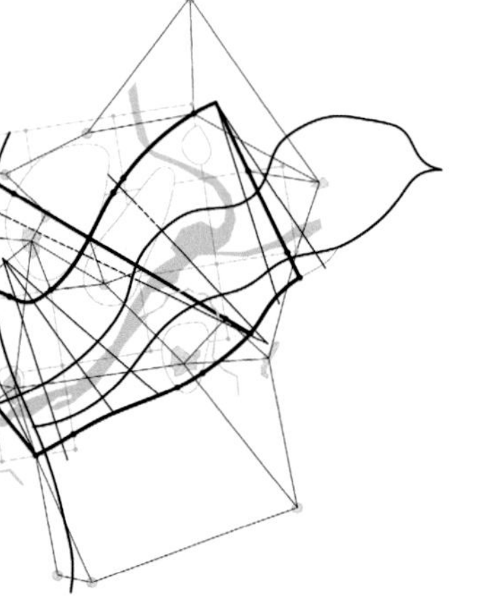

新陈代谢
创造一座有机的城市
一种与自然要素相结合的生长环境
可增长的细胞式社区
友好的步行系统
一座适于步行的城市
产生成效机制

基础设施
交通：铁路、电动渡船、宜人的步行空间、
车行道路
管道：制冷、清洁、排水、电力
水：给水、排水、中水、城市降水

协同发展
最大程度的将人工环境与景观营造结合起来
重新定义公共和私人空间并加以综合利用

BX 83800

InNovation
introduce a new dwelling type
communities that master plan themselves
schools within easy access to housing
an infrastructure defined by a circular metabolism
matrix of water and nutrients
maximizing urban growth with the natural environment
minimum disturbance to the environment itself
recycling the urban waste
benefiting from the elements provided by nature.

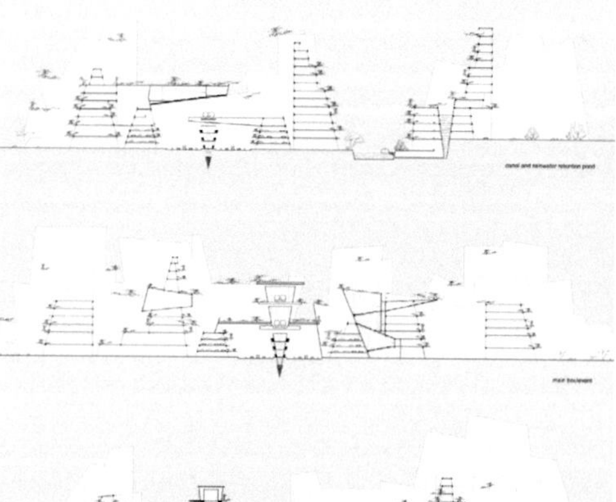

演变
创造一种全新的城市模式
面向未来的环境：神秘的、灵活的、繁荣的
强调动态和生长
环境改变

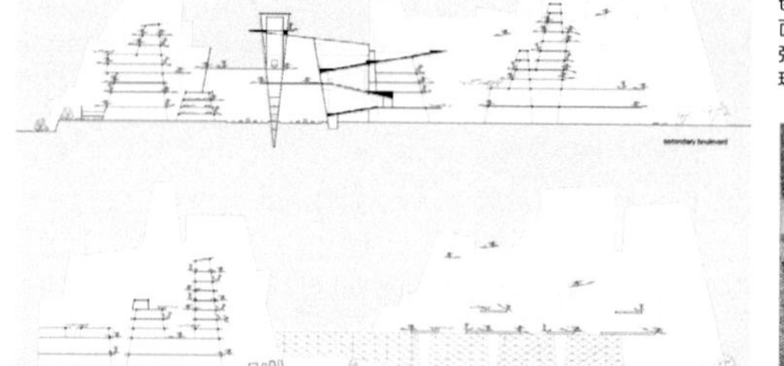

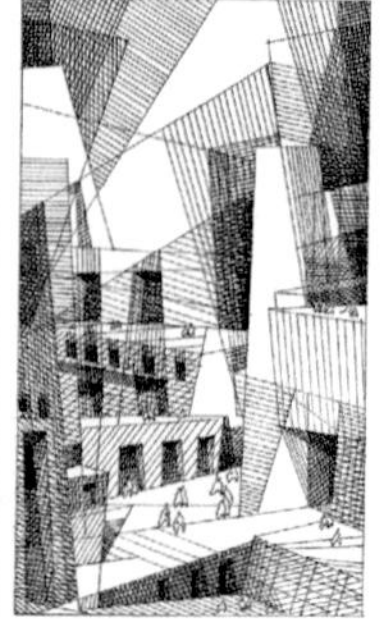

共生城市

Yu, Eric（澳大利亚）/Lee, Louise

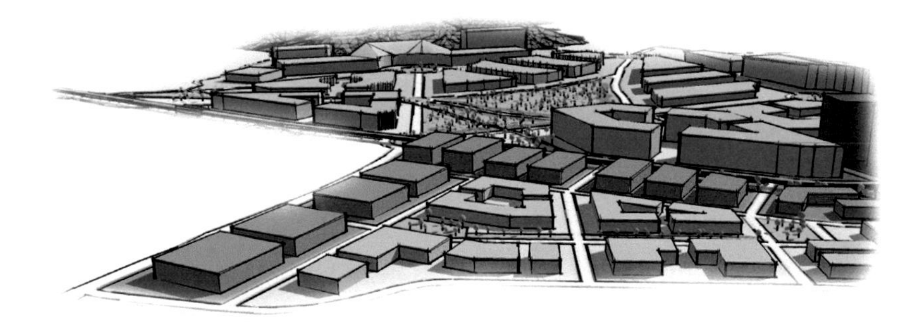

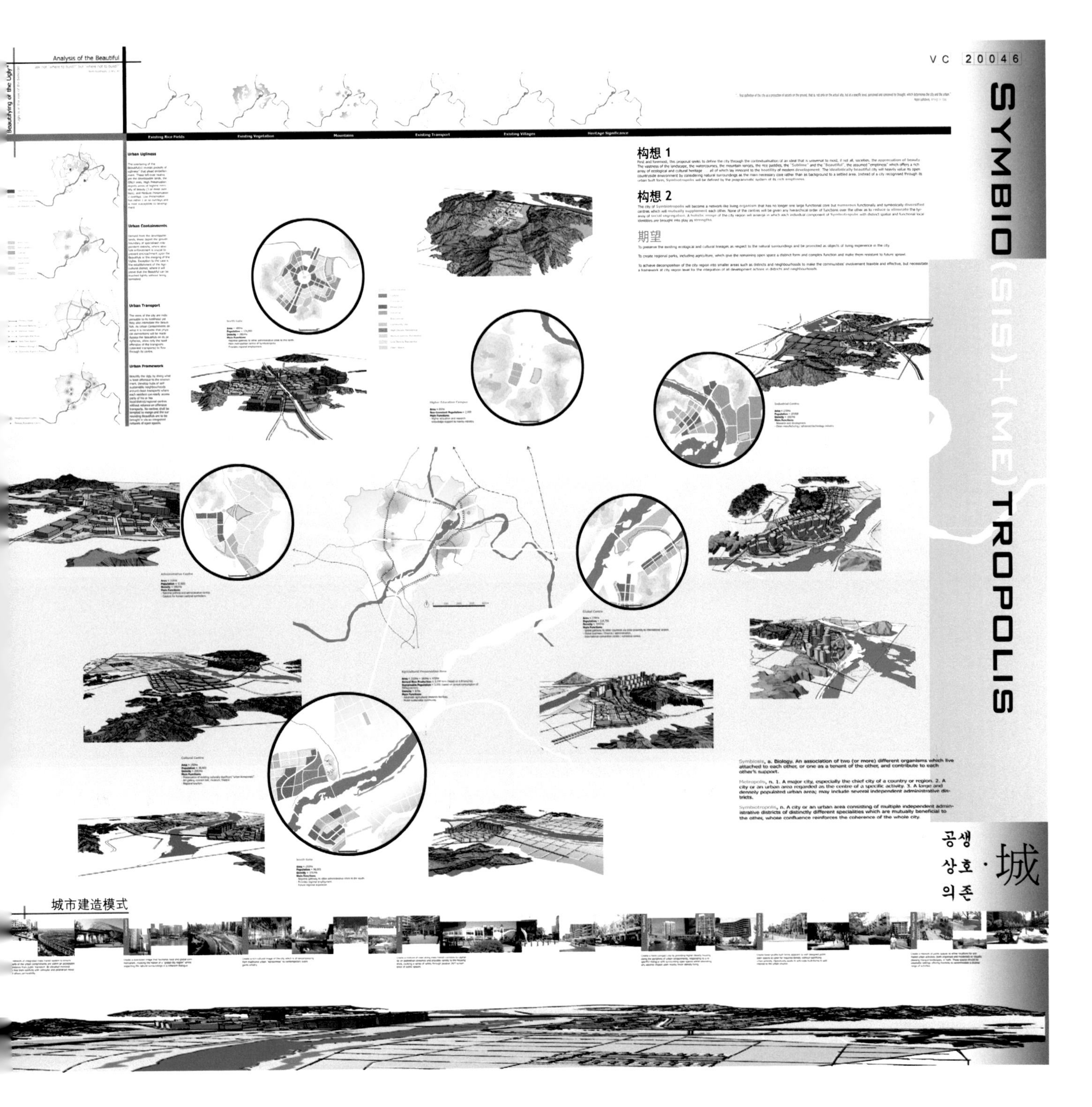

无题

Moshe, Salomon（以色列）

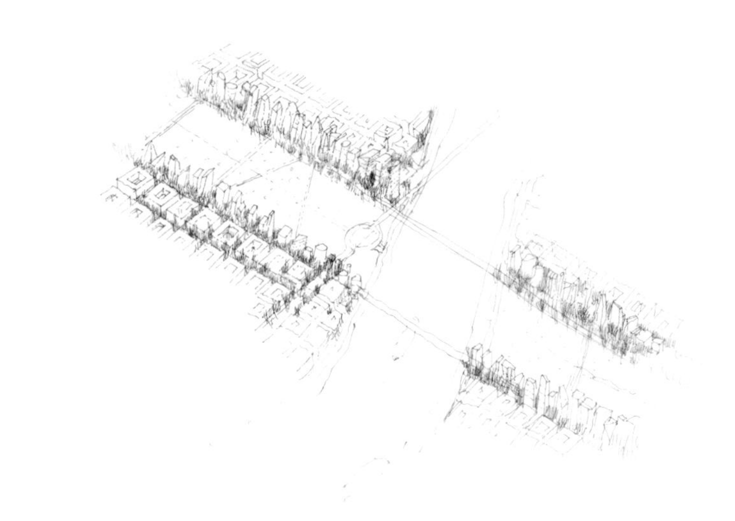

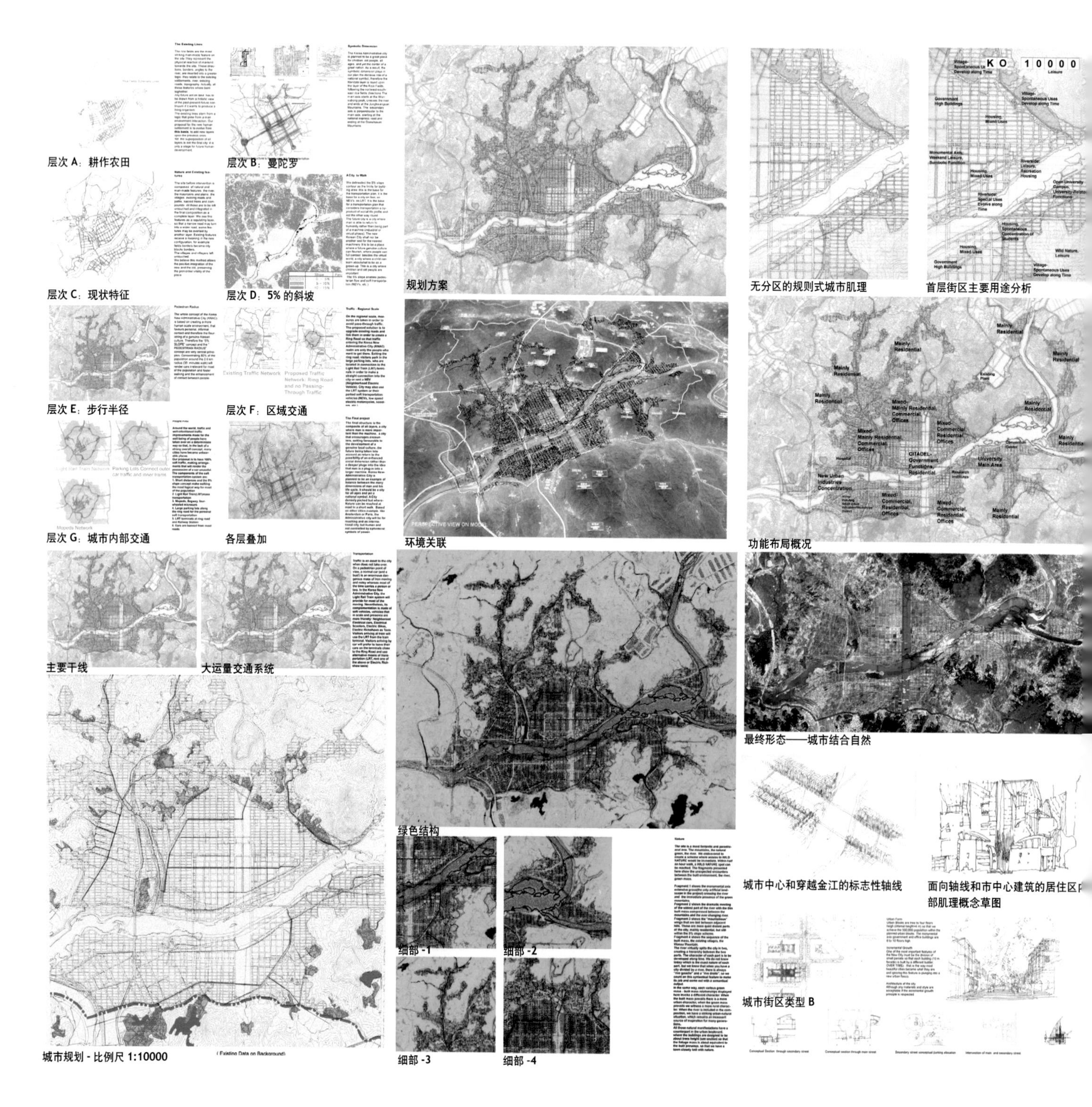

层次 A：耕作农田
层次 B：曼陀罗
层次 C：现状特征
层次 D：5% 的斜坡
层次 E：步行半径
层次 F：区域交通
层次 G：城市内部交通
各层叠加
规划方案
环境关联
无分区的规则式城市肌理
首层街区主要用途分析
功能布局概况
主要干线
大运量交通系统
城市规划 - 比例尺 1:10000
绿色结构
最终形态——城市结合自然
细部 -1
细部 -2
细部 -3
细部 -4
城市中心和穿越金江的标志性轴线
面向轴线和市中心建筑的居住区内部肌理概念草图
城市街区类型 B

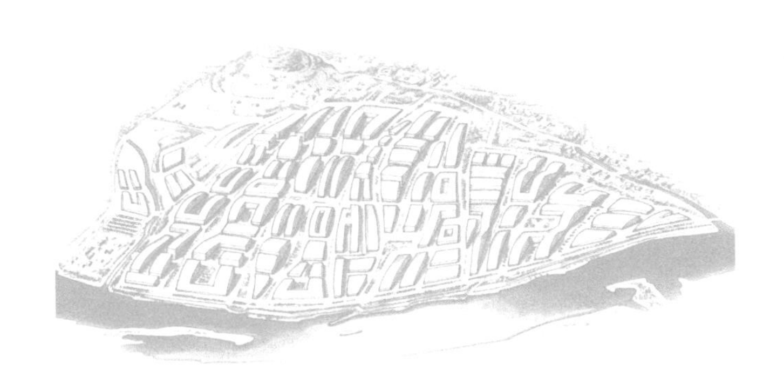

无题

Balsimelli, Andrea(意大利)
Schineano, Sallydagnaiz/Cappoqi, Pino/
Belloccisessa, Ilaria

DS 19125

韩国新复合型行政中心城市国际城市概念竞赛
概念规划

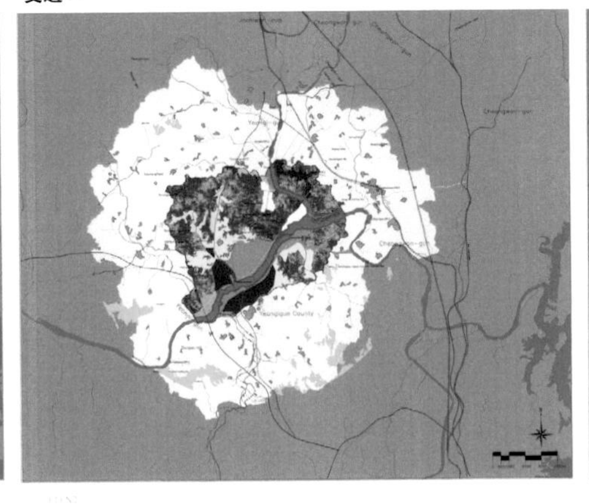

韩国新复合型行政中心城市国际城市概念竞赛
交通 A

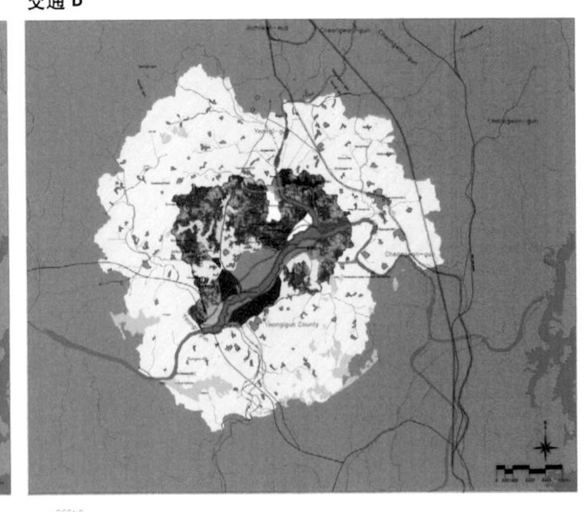

韩国新复合型行政中心城市国际城市概念竞赛
交通 B

韩国新复合型行政中心城市国际城市概念竞赛
山体保护及其景观

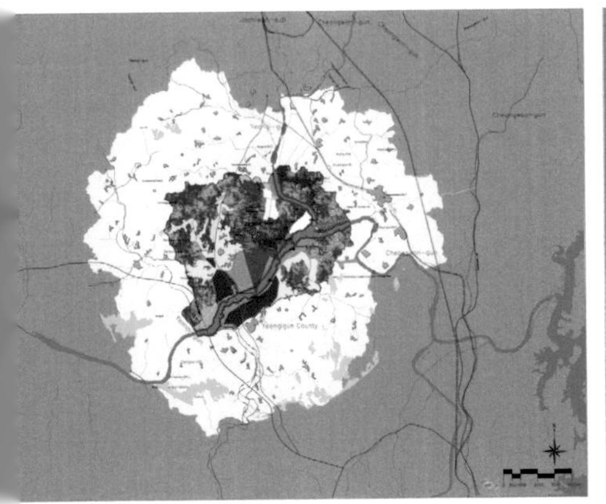

韩国新复合型行政中心城市国际城市概念竞赛
文化遗产、自然景观和旅游资源

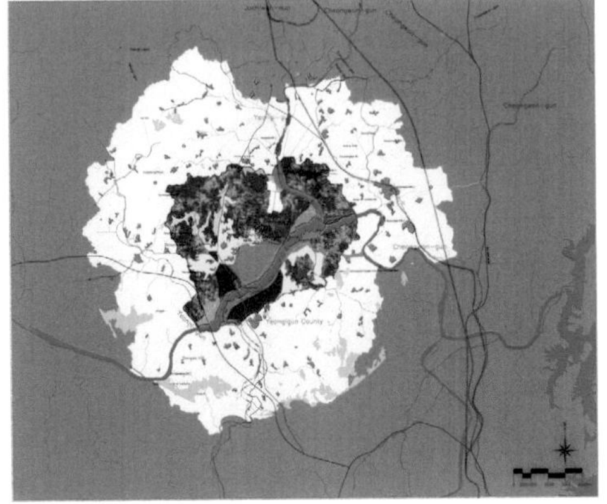

韩国新复合型行政中心城市国际城市概念竞赛

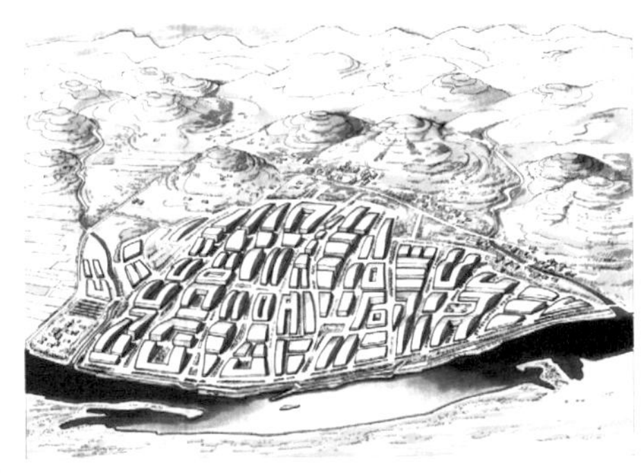

银杏之城

Kahng, Jang Wan（韩国）

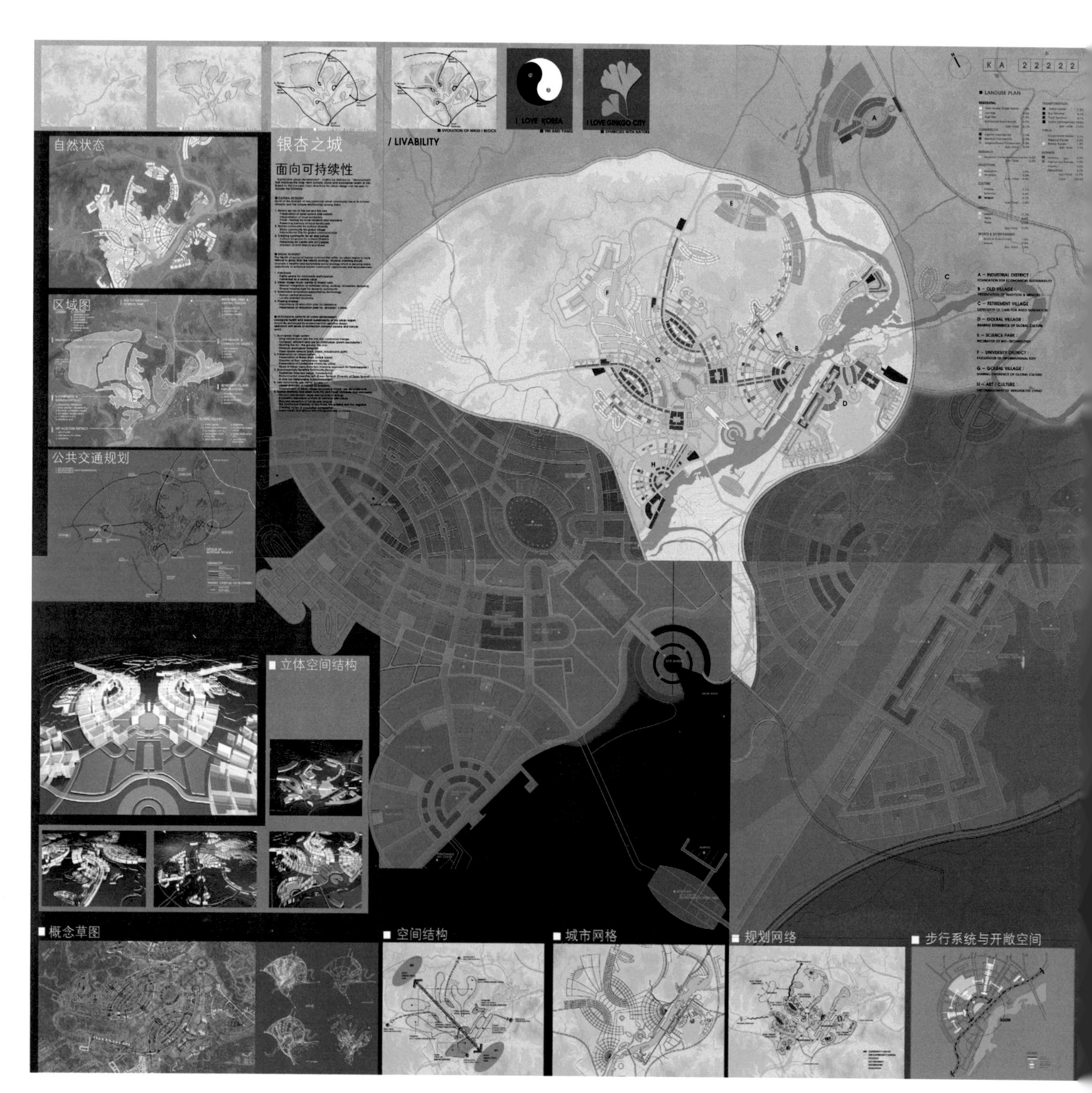

无题

Marx, Christian（德国）
Team 408

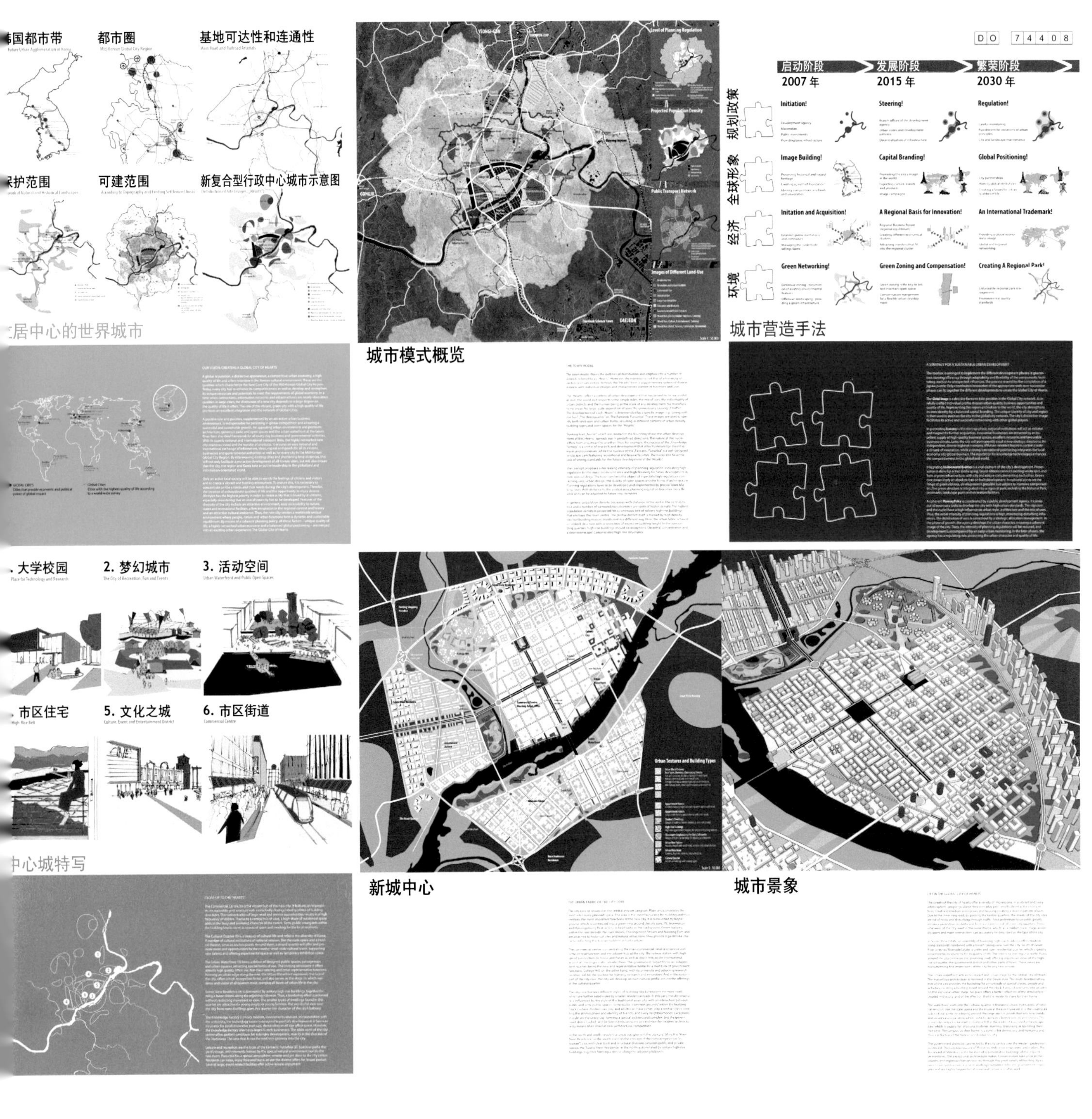

玲珑之城

Haimerl, Peter（德国）
Peter Haimerl Studio Fur Architecture

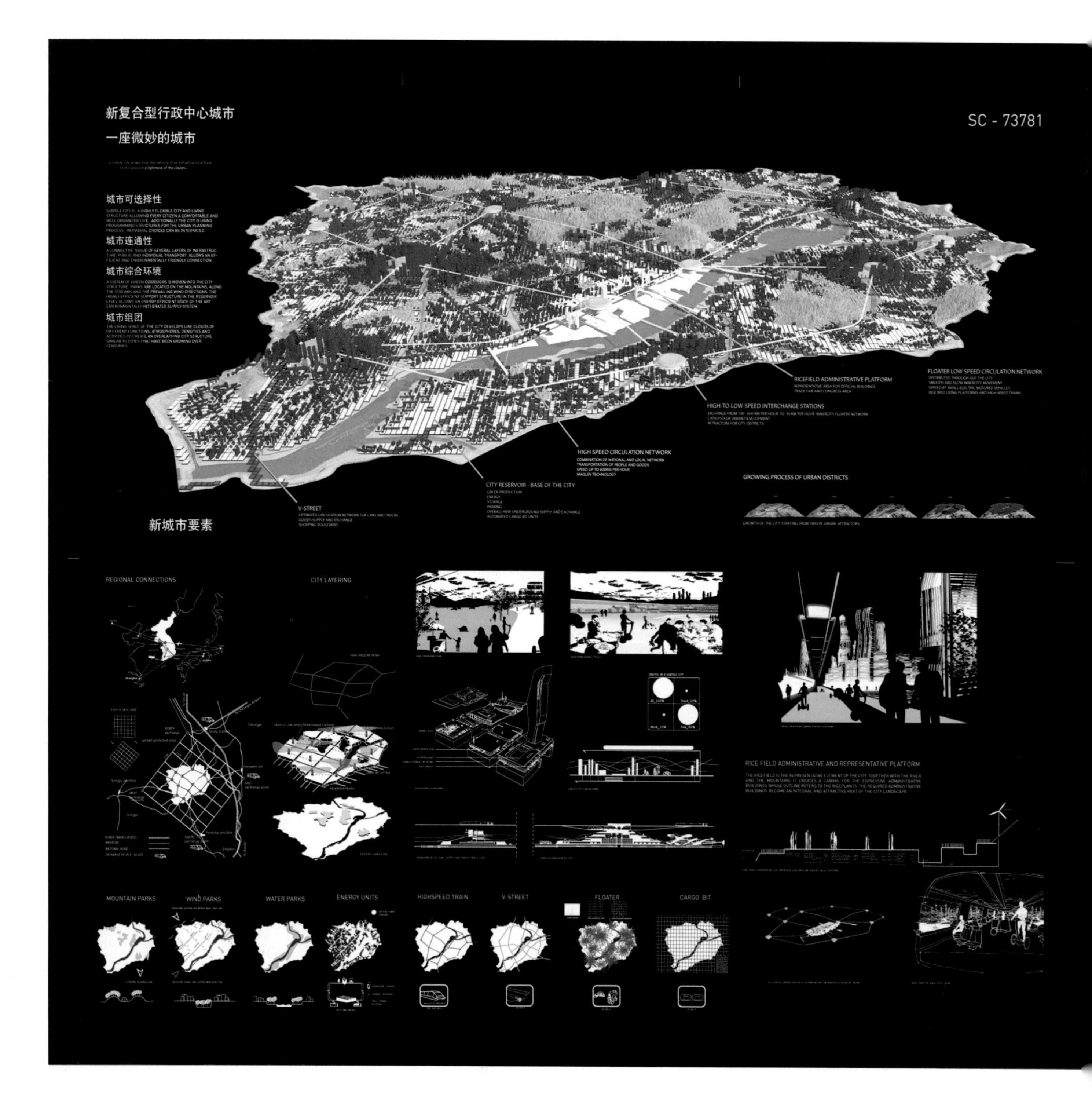

无题

Bomerski, Martin（德国）
Bomerski, Artur/Bomerski, Albert

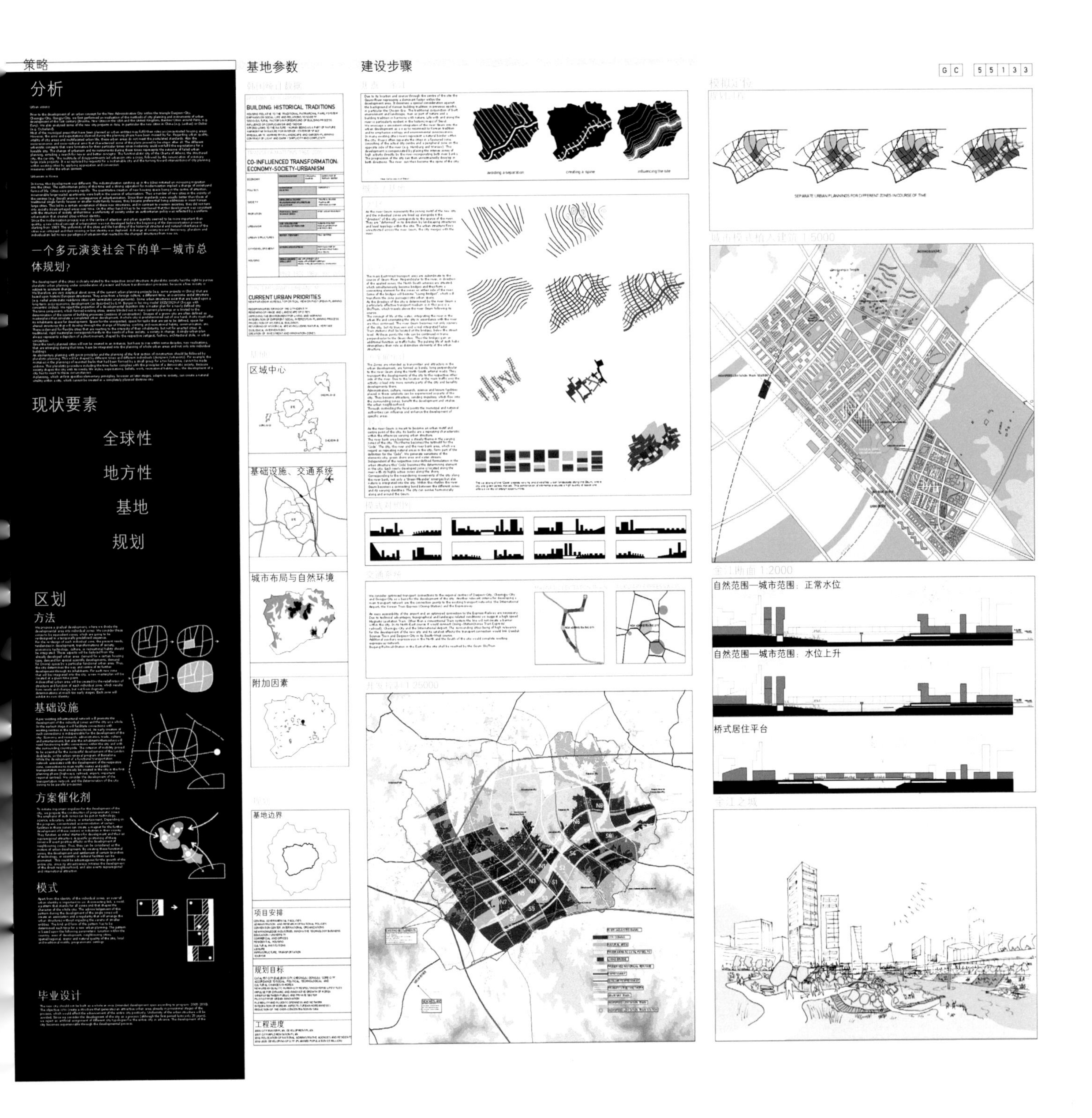

大型旅游城市

Yip,Ting Hin Michael（澳大利亚）
Fung, Si On/Njo, Victor/Horsting, Bas

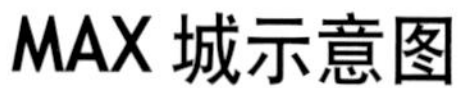

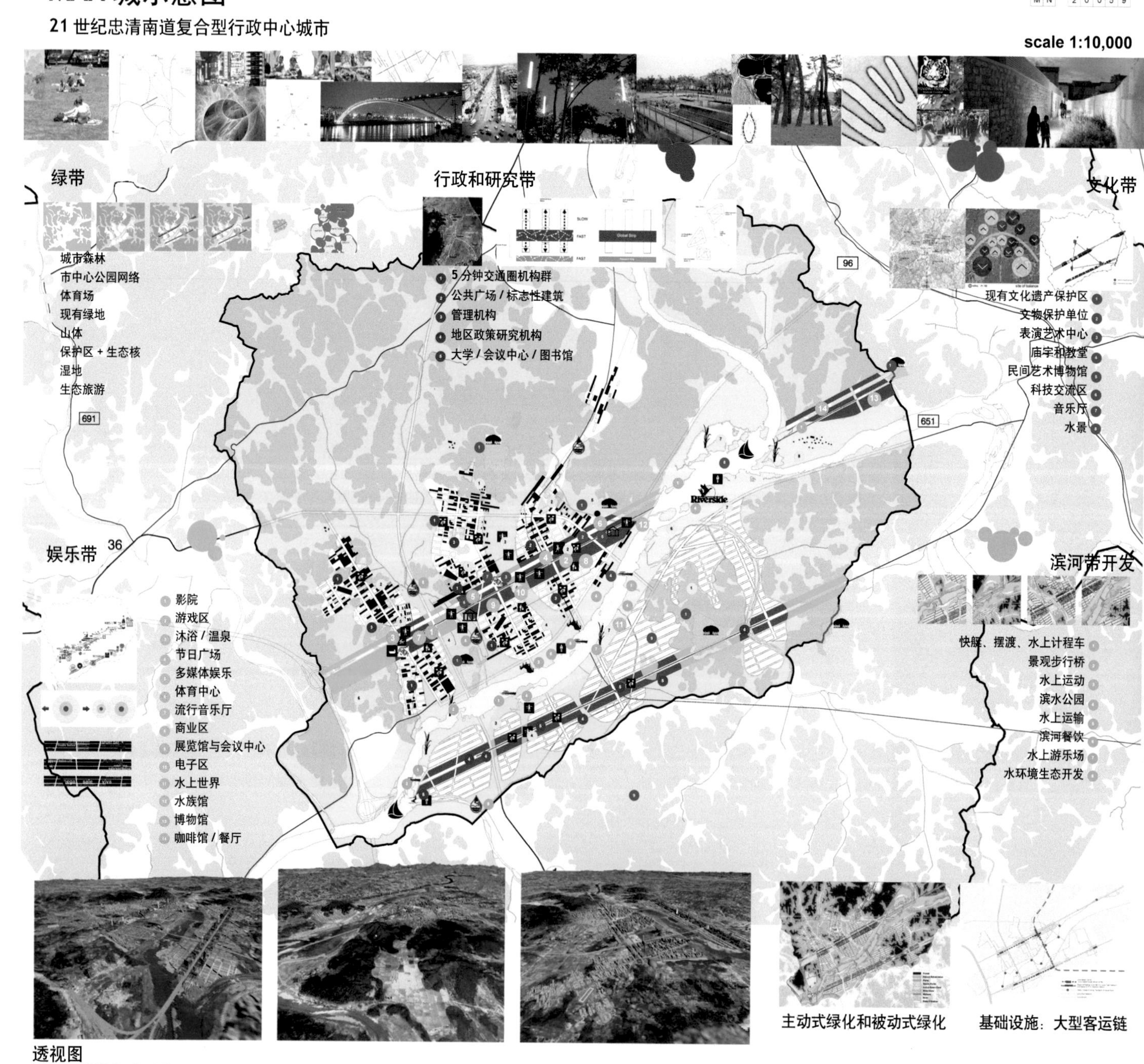

无题

Witthinvich, Jochch（德国）
Stelmach, Mathias/Braig,Monika

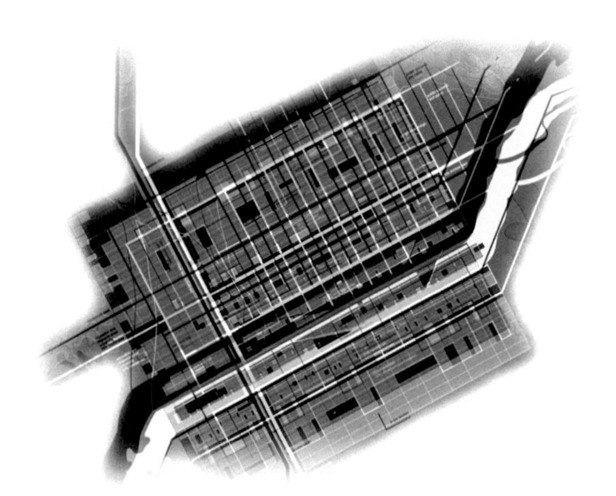

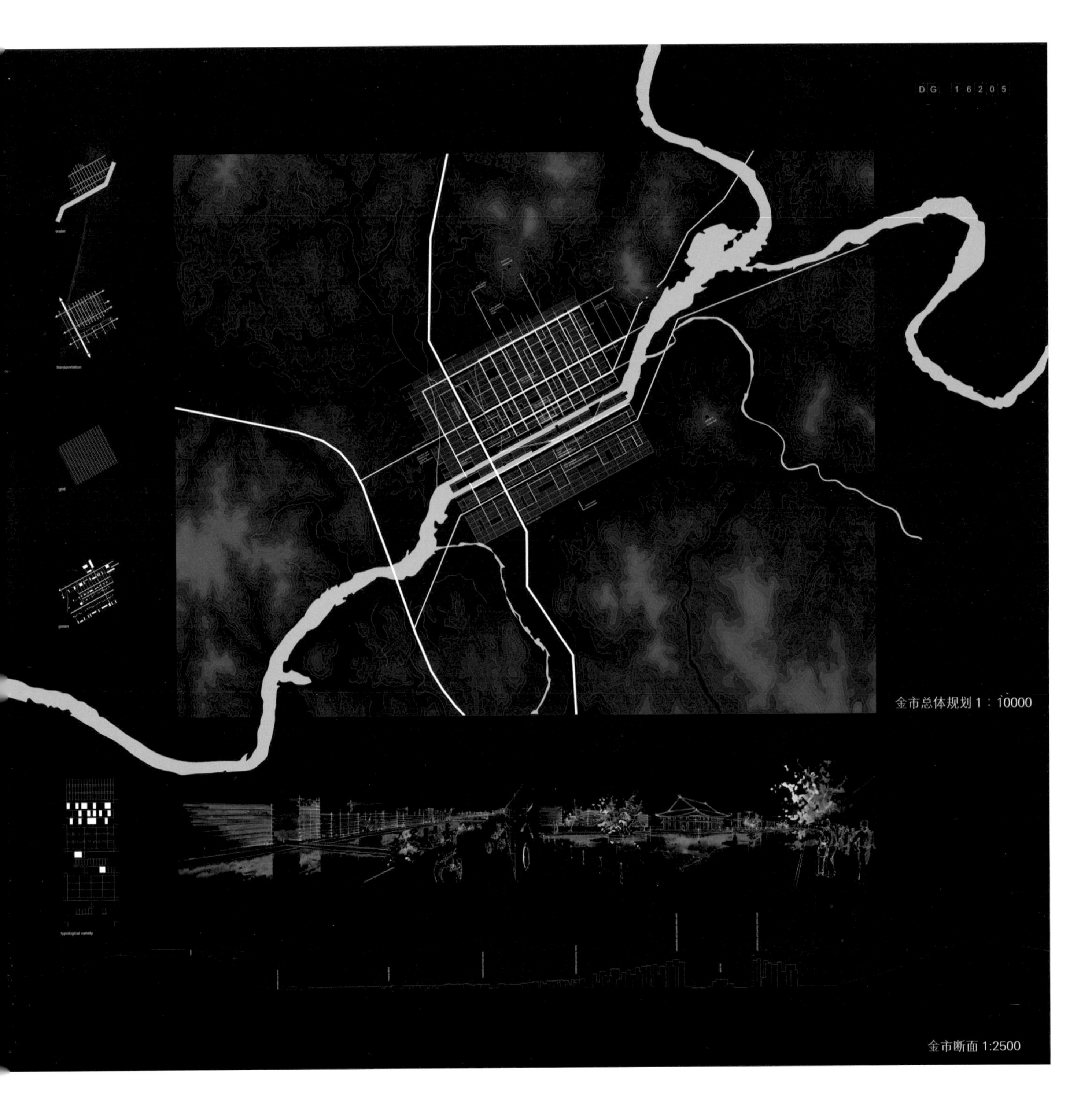

金市总体规划 1：10000

金市断面 1:2500

无题

Cho, Seong Ju（韩国）

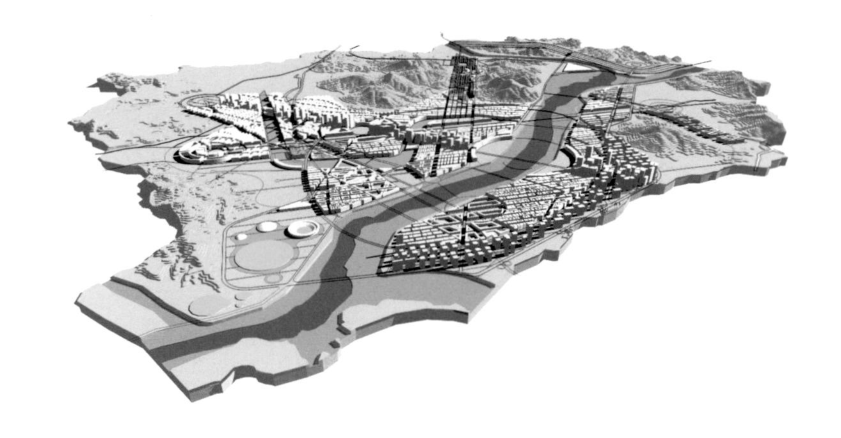

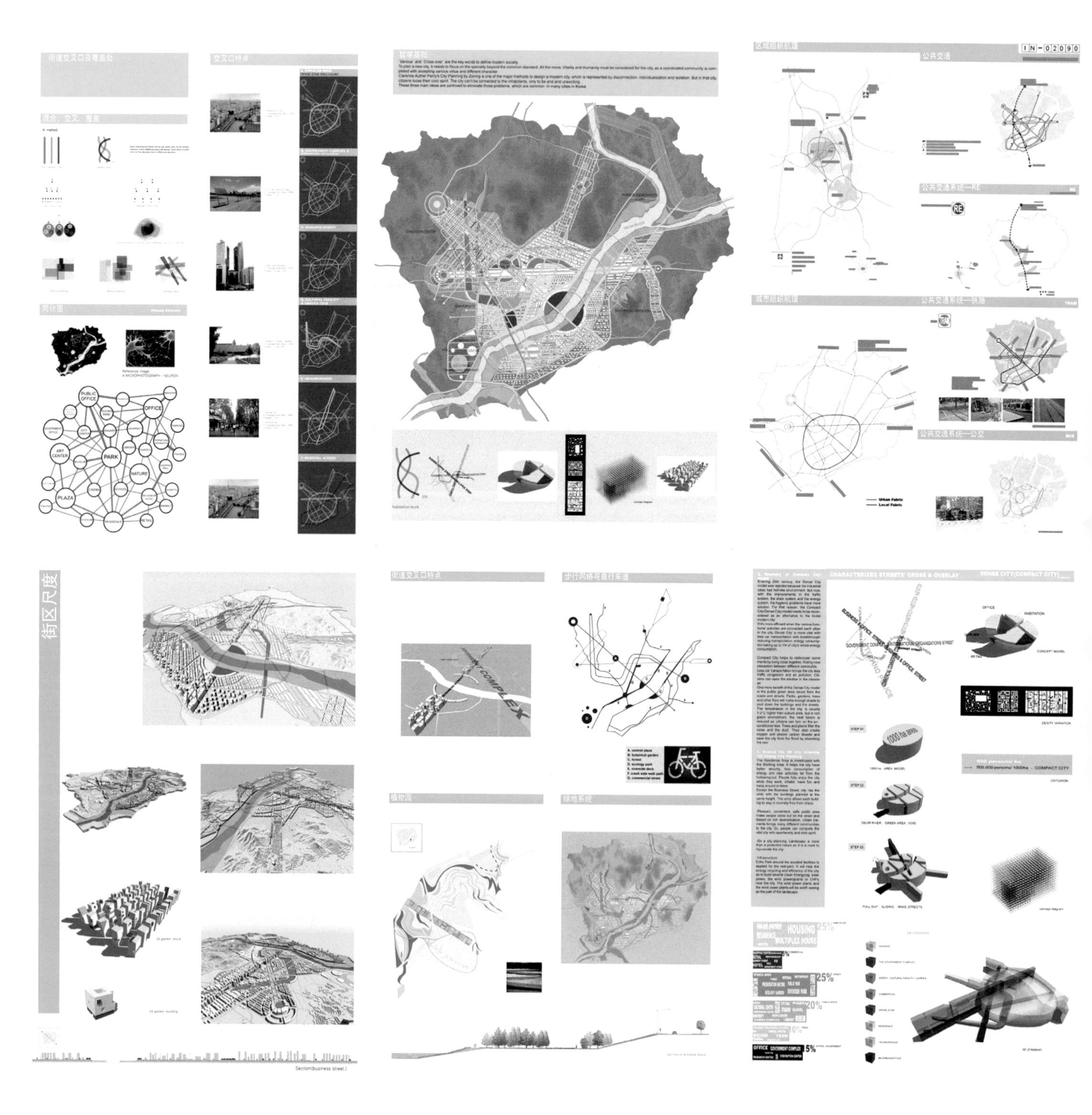

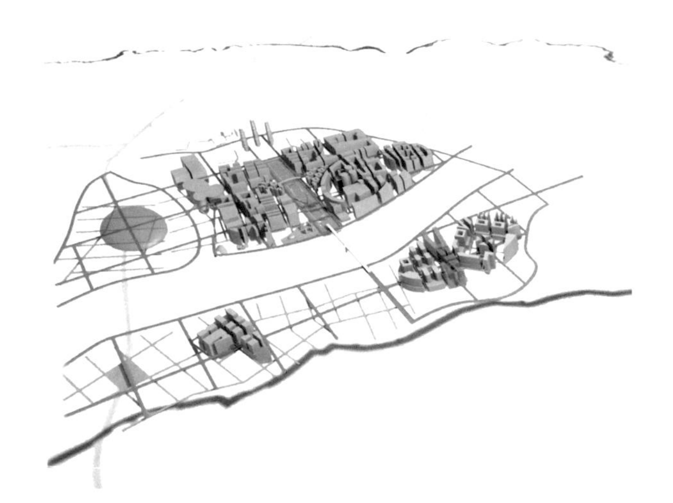

无题

Ahn, Jung Gil（韩国）
Inhousing Co., Ltd/Kang, Sung Joong/
Haein Architects&Engineers Co. Ltd

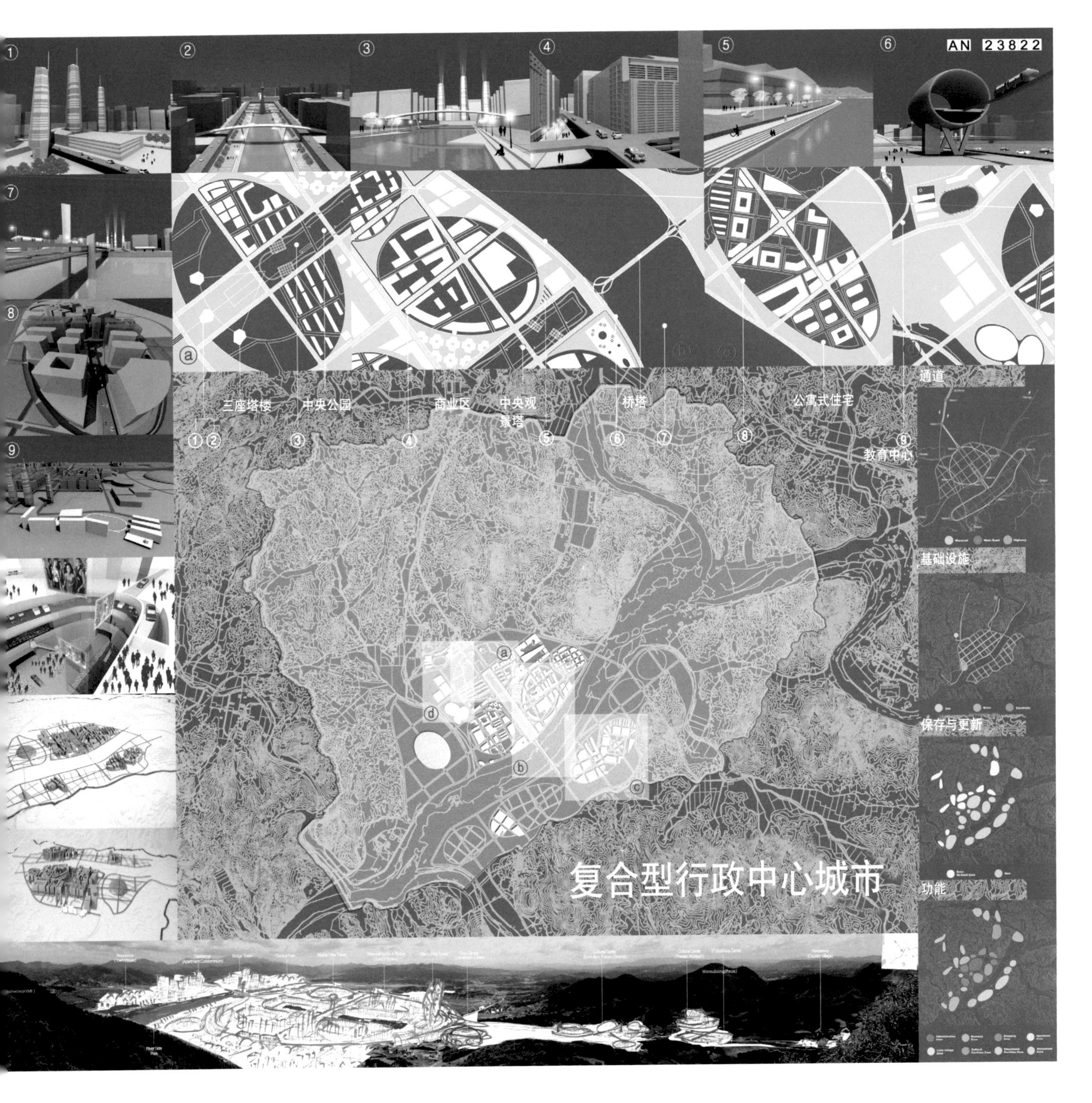

自我—连接—网络

Furuya, Nobuaki（日本）
Lee, Doo Yeol

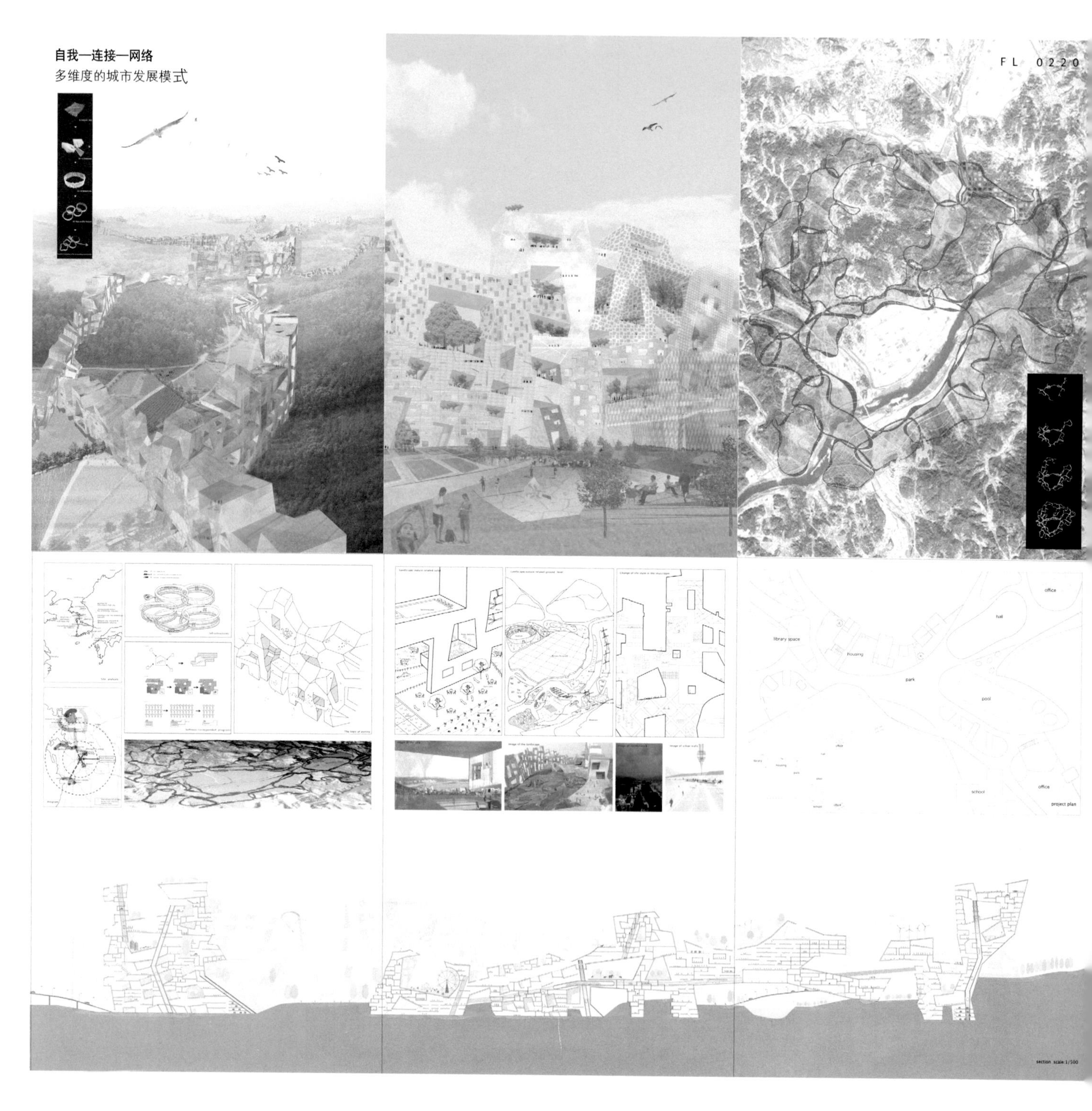

无题

Butler Sierra, Luis（新加坡）

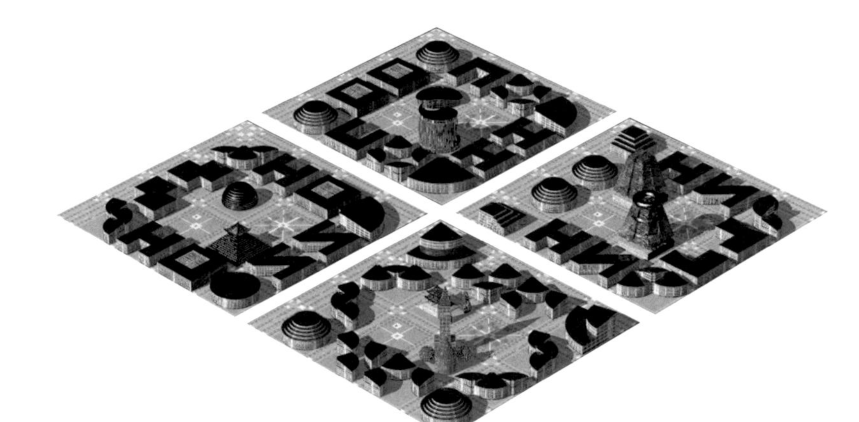

Daul

Lee, Su Man(韩国)

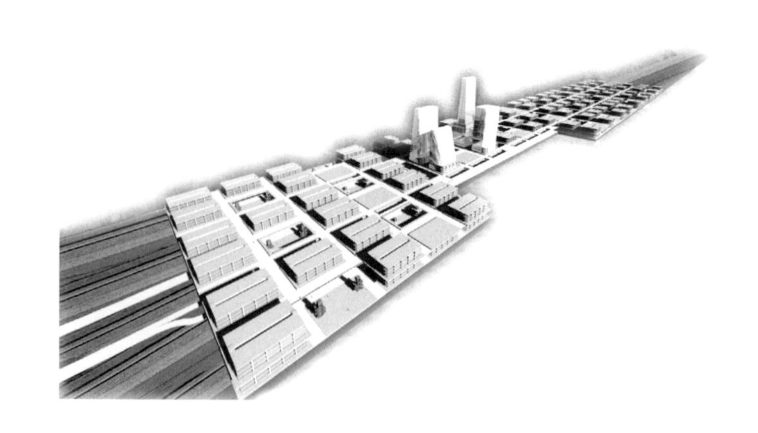

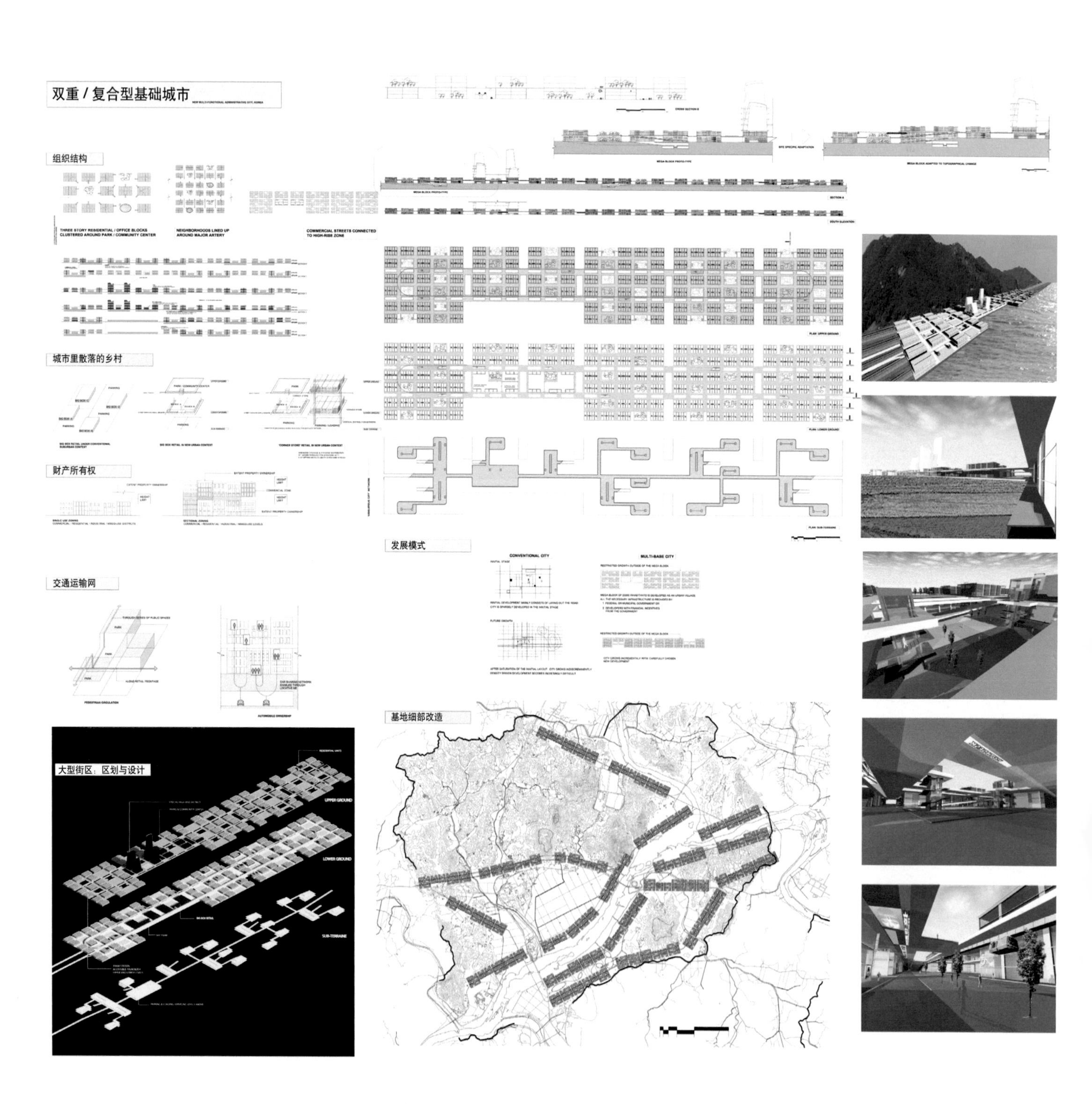

紧凑城市

Mundwiler, Stephan（美国）
Lee Mundwiler Architects/Lee, Cara/
Santos, Gustavo/Watanabe, Hiroyuki

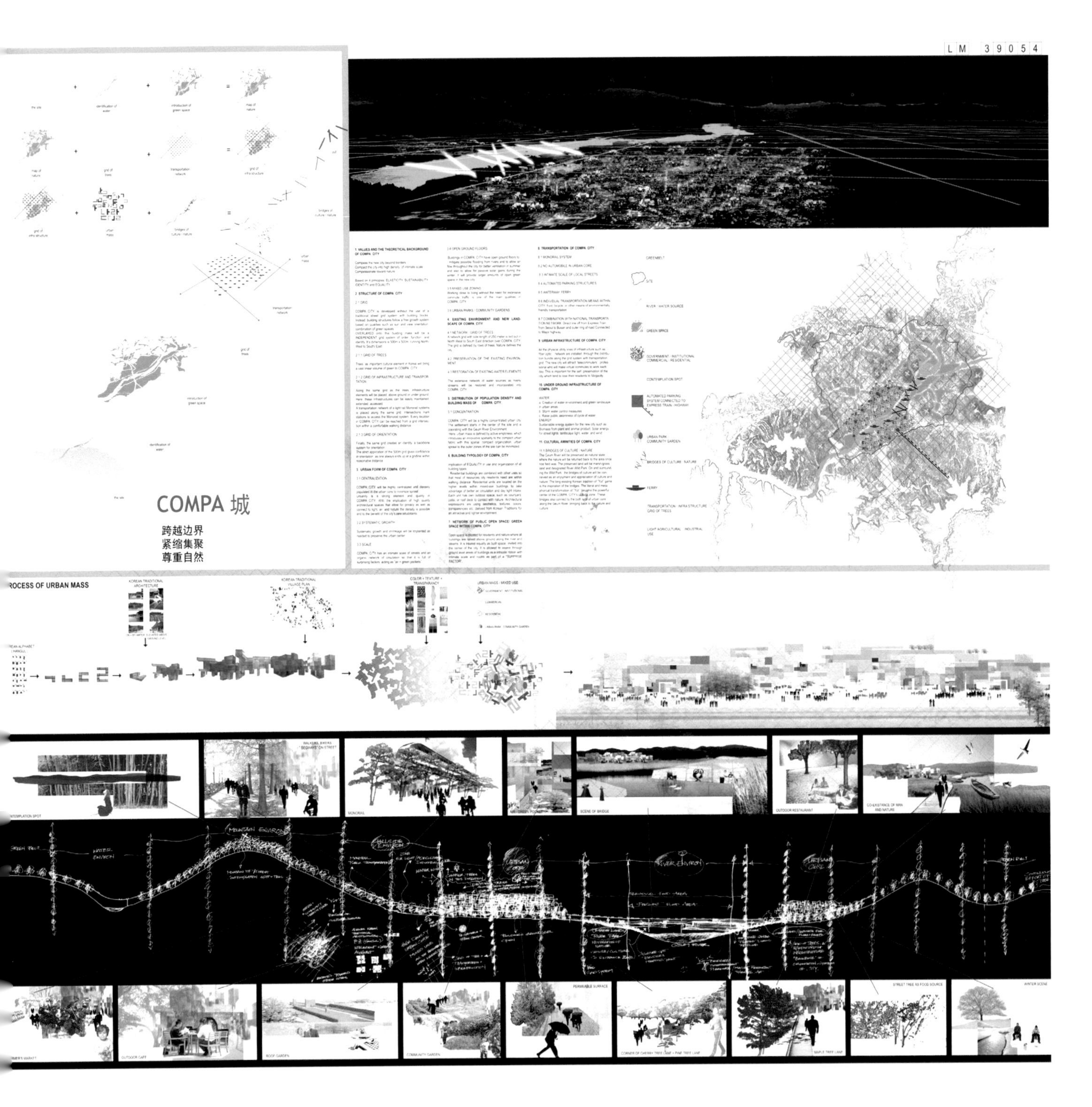

巨型城市

Petrosyan, Noure（亚美尼亚）
Zoroyan, Suartin

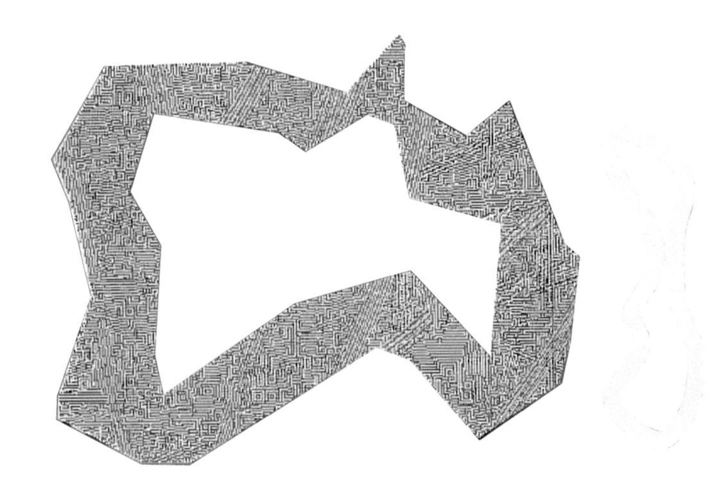

间隔地带

Haddad, Albert（法国）

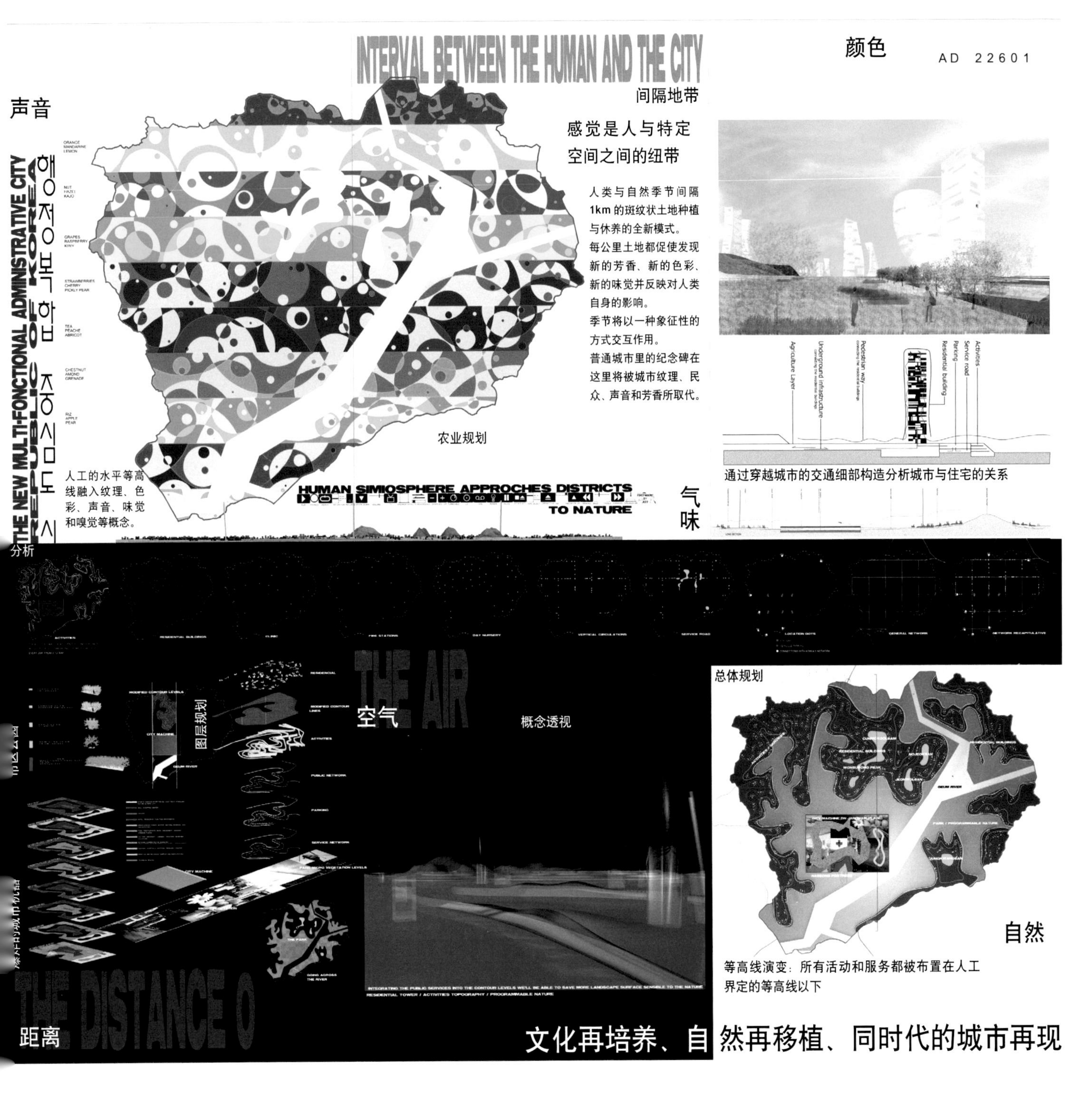

无题

Joyanovic, Bratislav（阿拉伯联合酋长国）

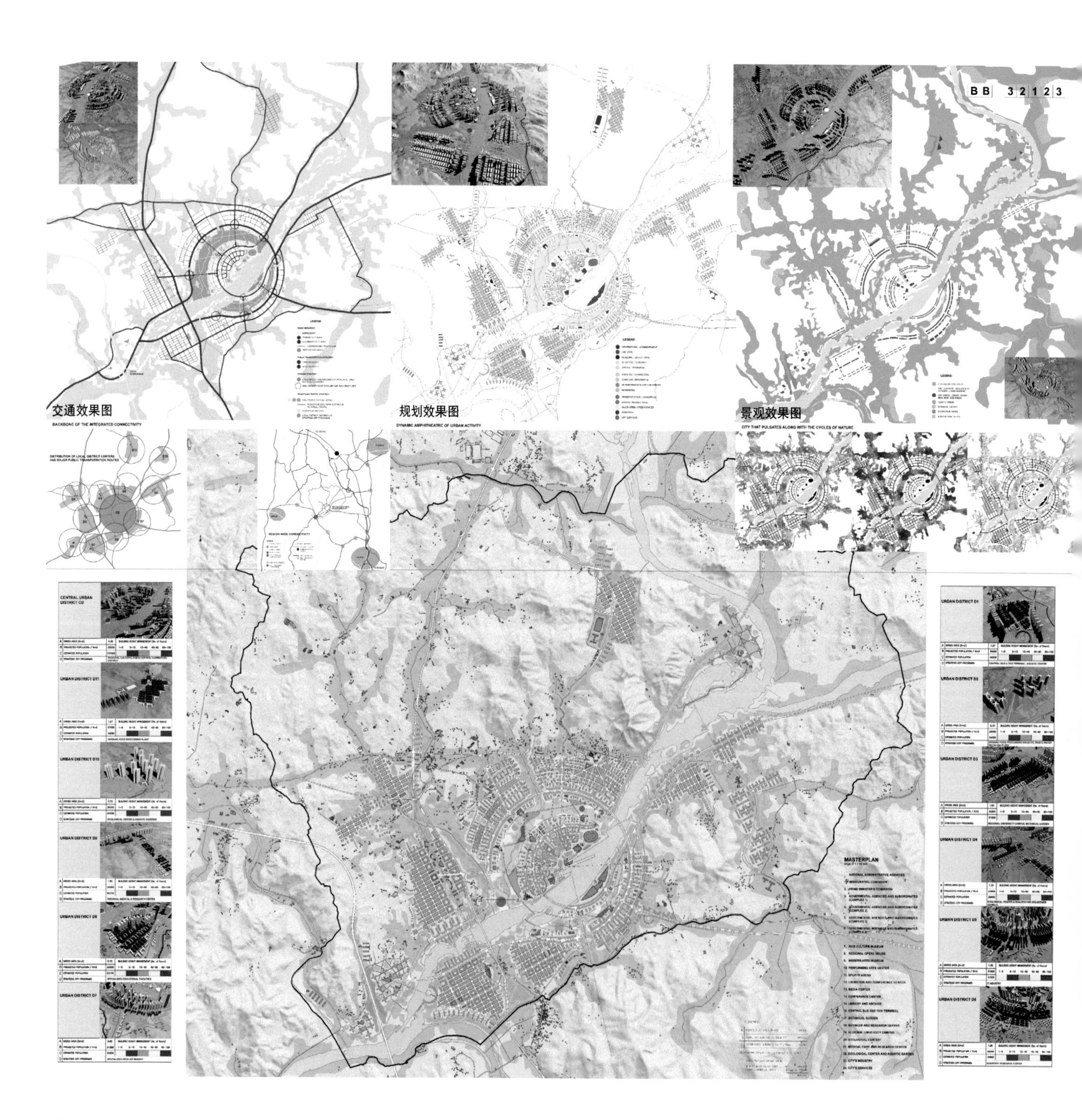

无题

Schaap, Ton（荷兰）
Venhoeven CS Architecten/Van Geissen, Cees/Kuiken, Rene

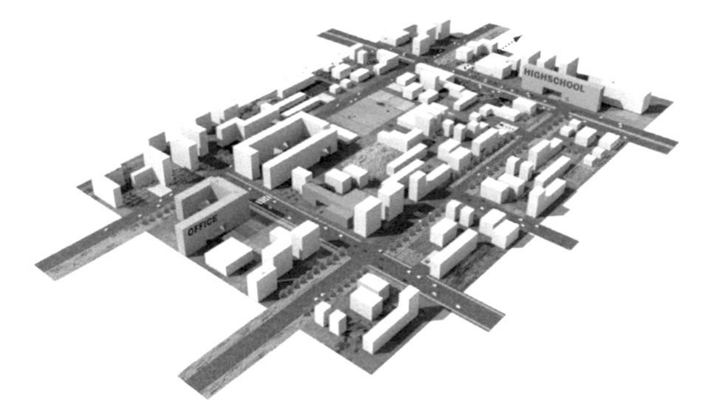

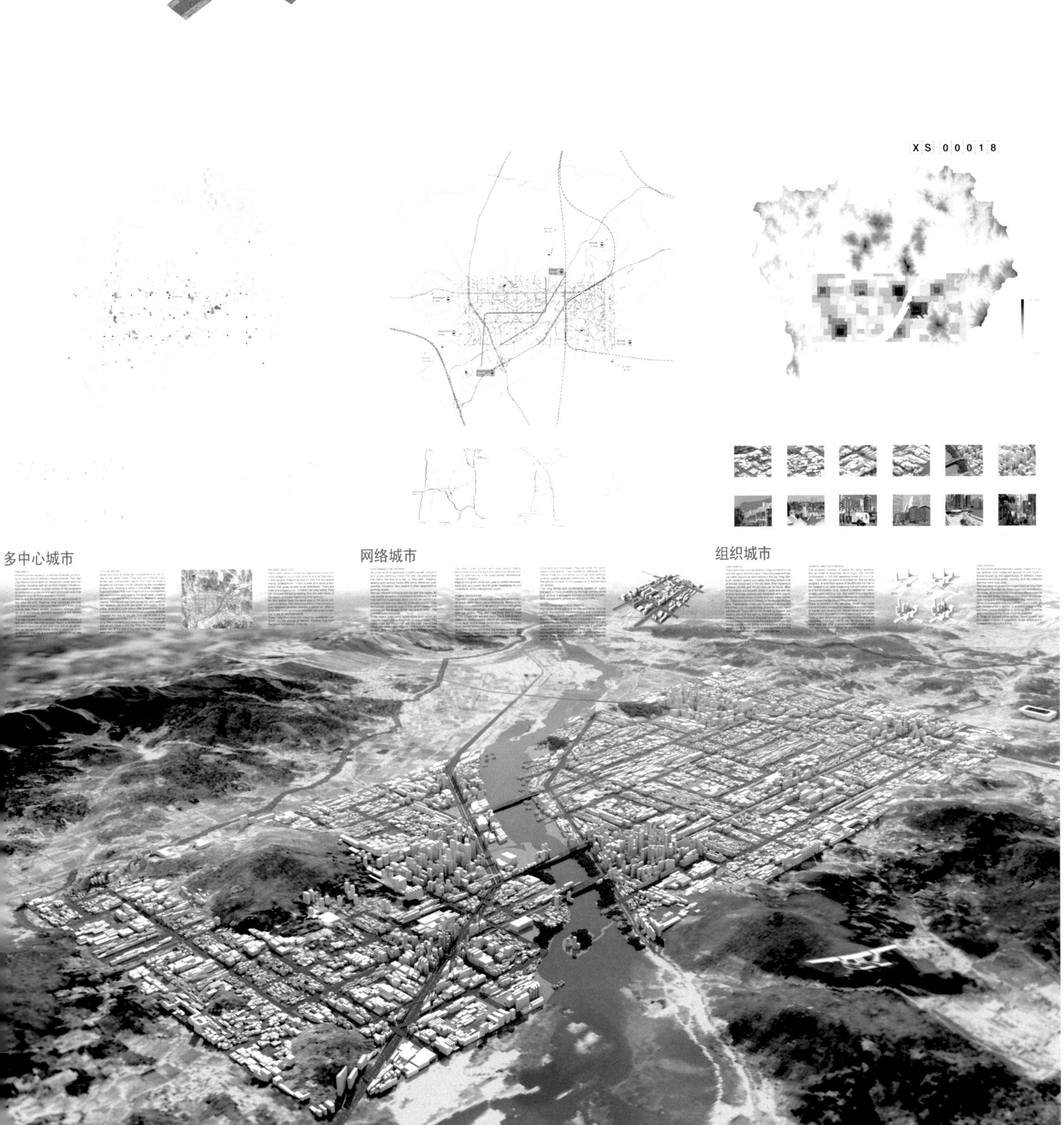

无题

Tooruerol, Paul（荷兰）

Disturbed Nature
(a metropolitan garden)

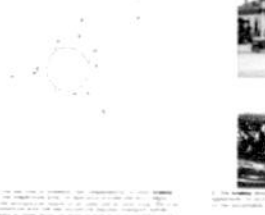

DN 05001

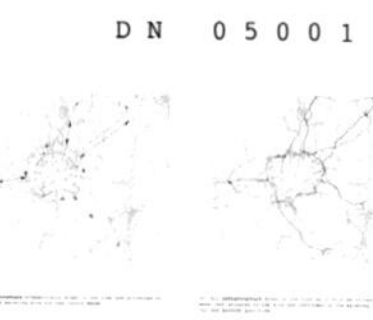

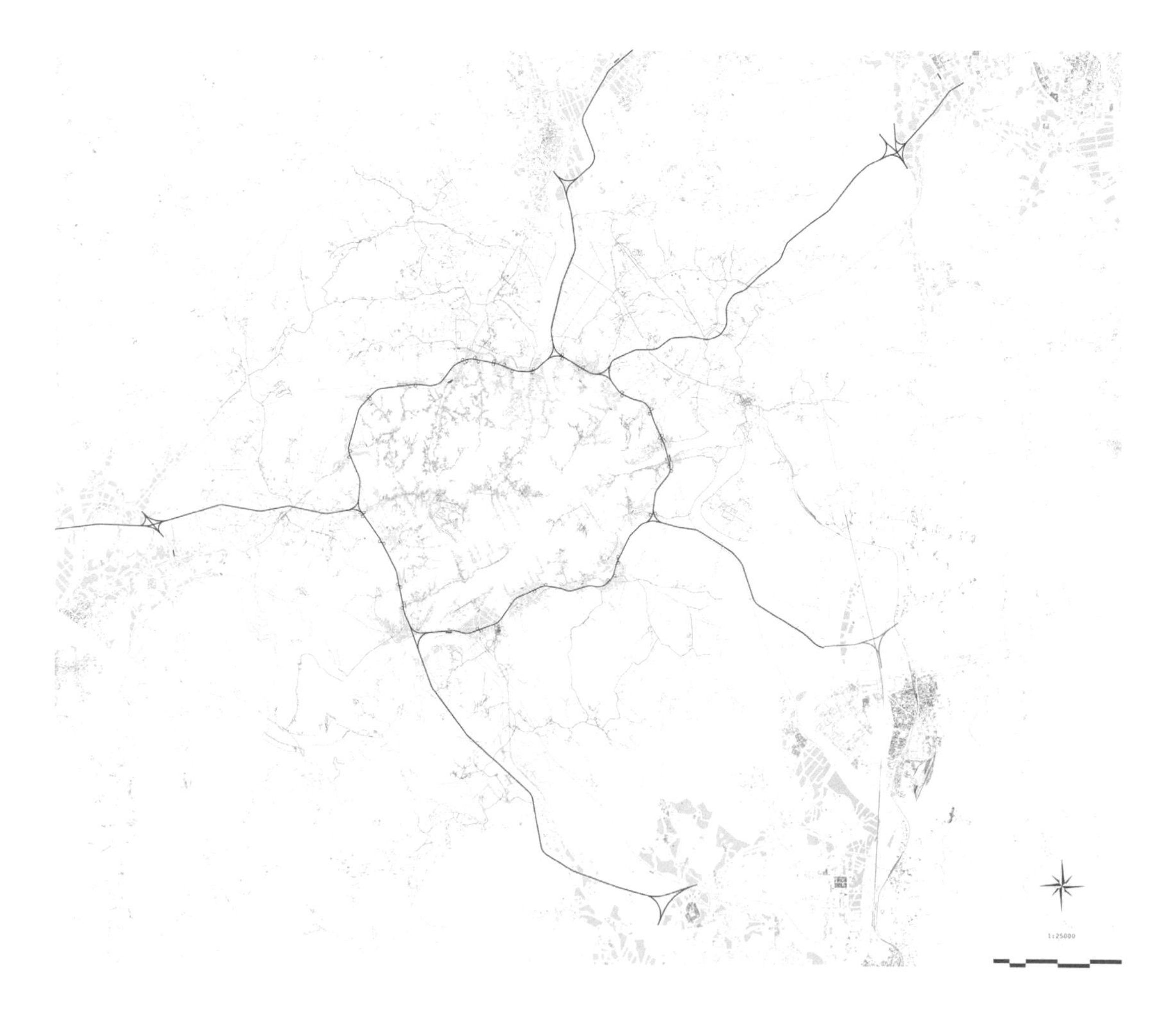

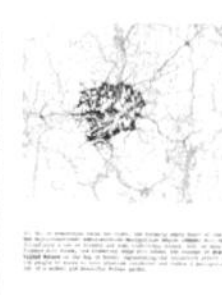

Disturbed Nature
(a metropolitan garden)

无题

Vladin, Petrov（英国）
Tsocaiev, Georgi

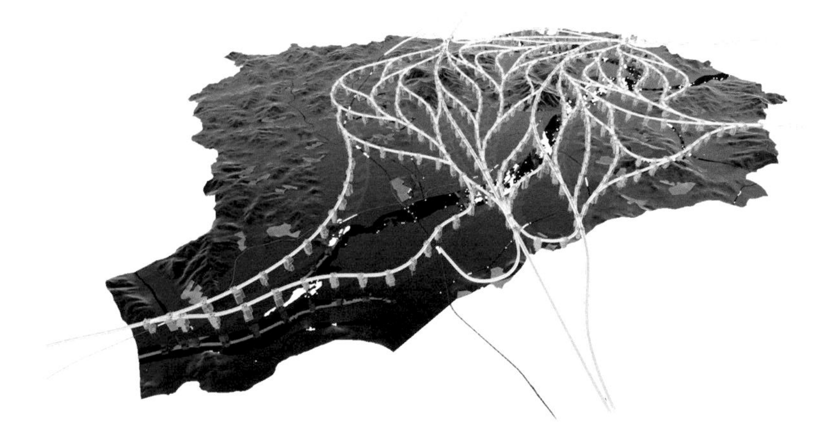

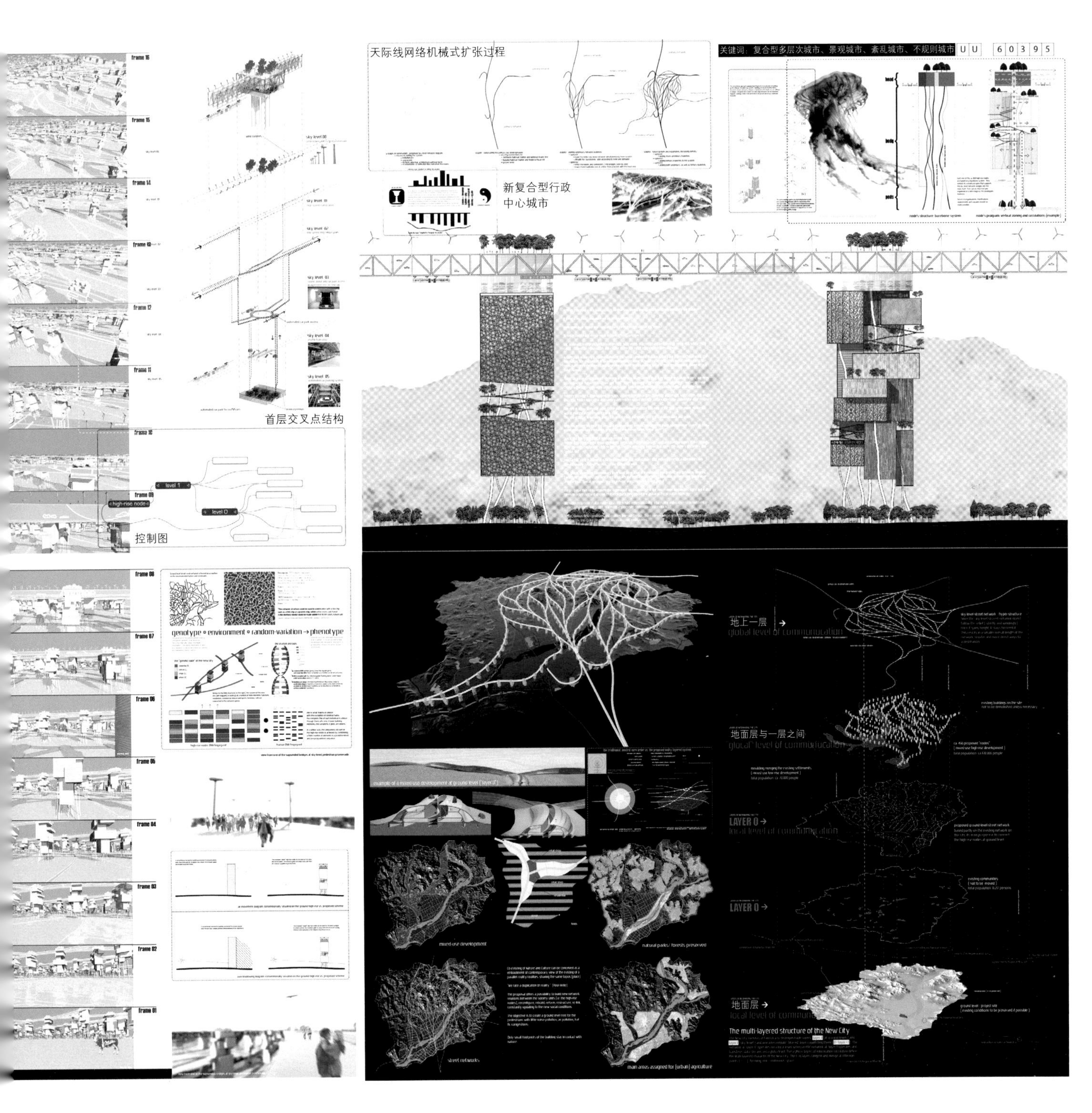

UBI离散城

Conti, Anna（意大利）

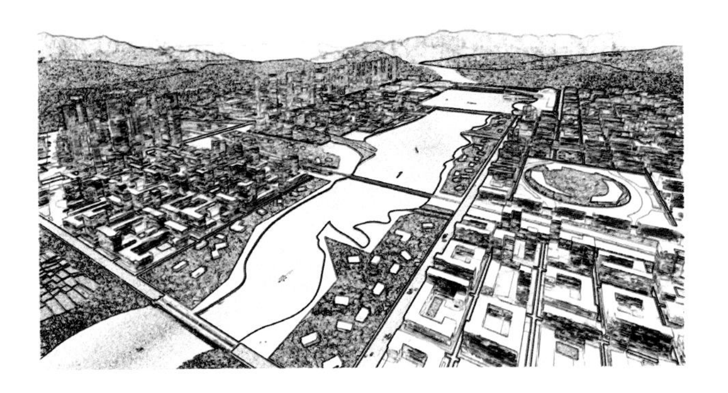

杨梅城

Dias, Carlos（巴西）
Sales, Pedro, M.R./Yamato Newton, Massafumi/Fehr,Lucas/Ursini, Marcelo L.

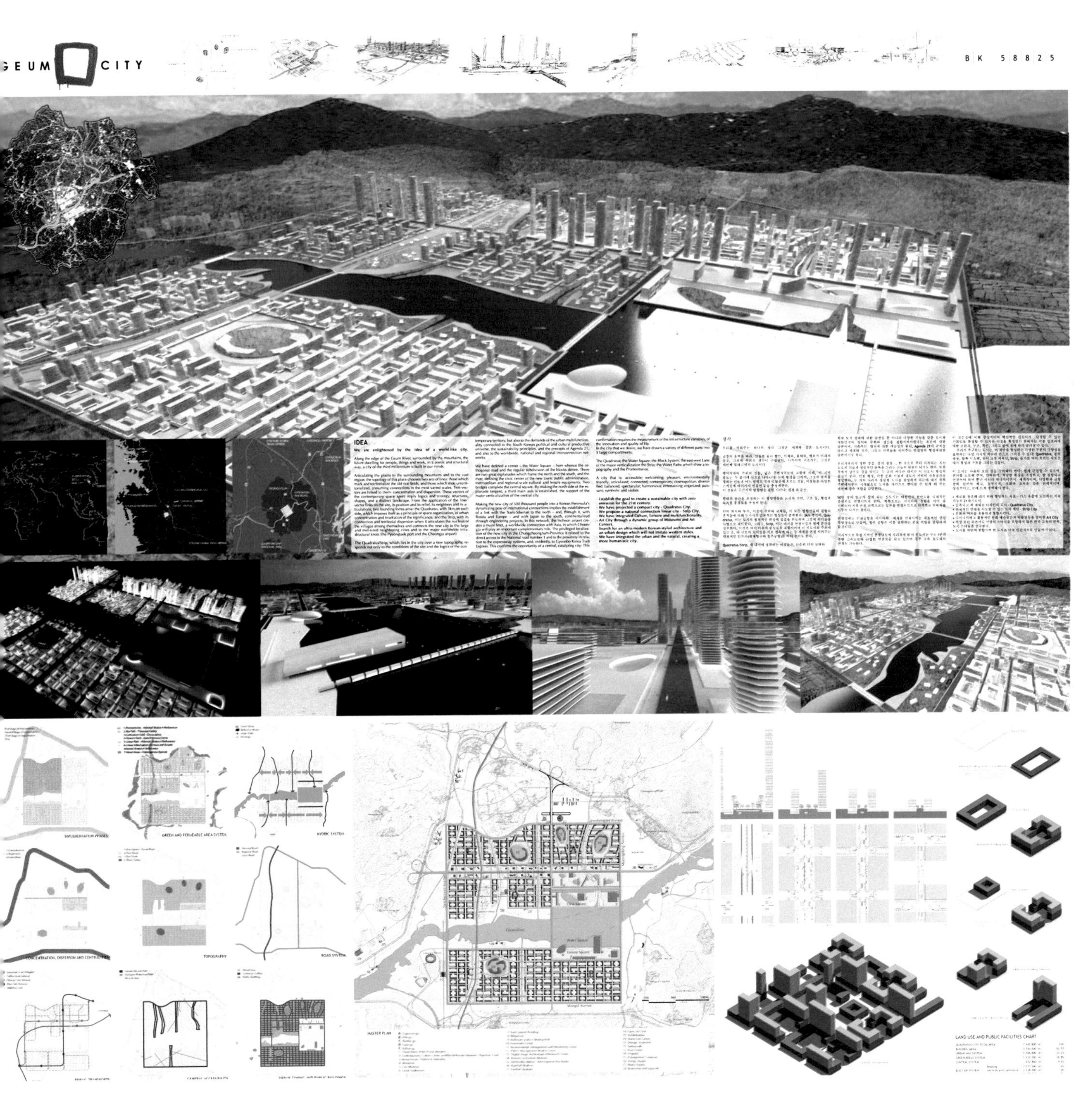

绿核城市

Spijker, wan't Jaakko（荷兰）

Bultstra, Henk/Deuten Bert, Karel/Steinarsson, Orri

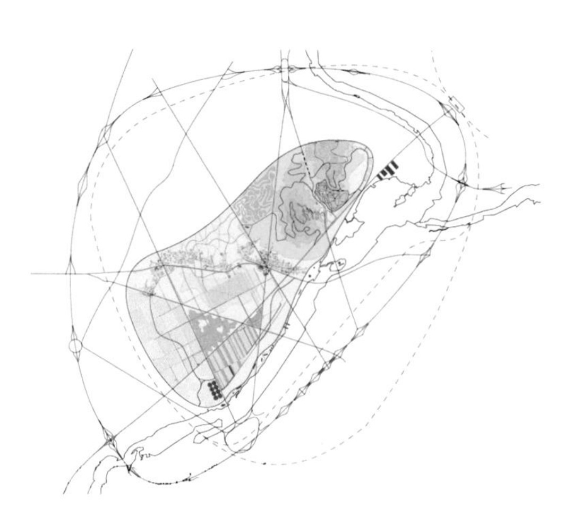

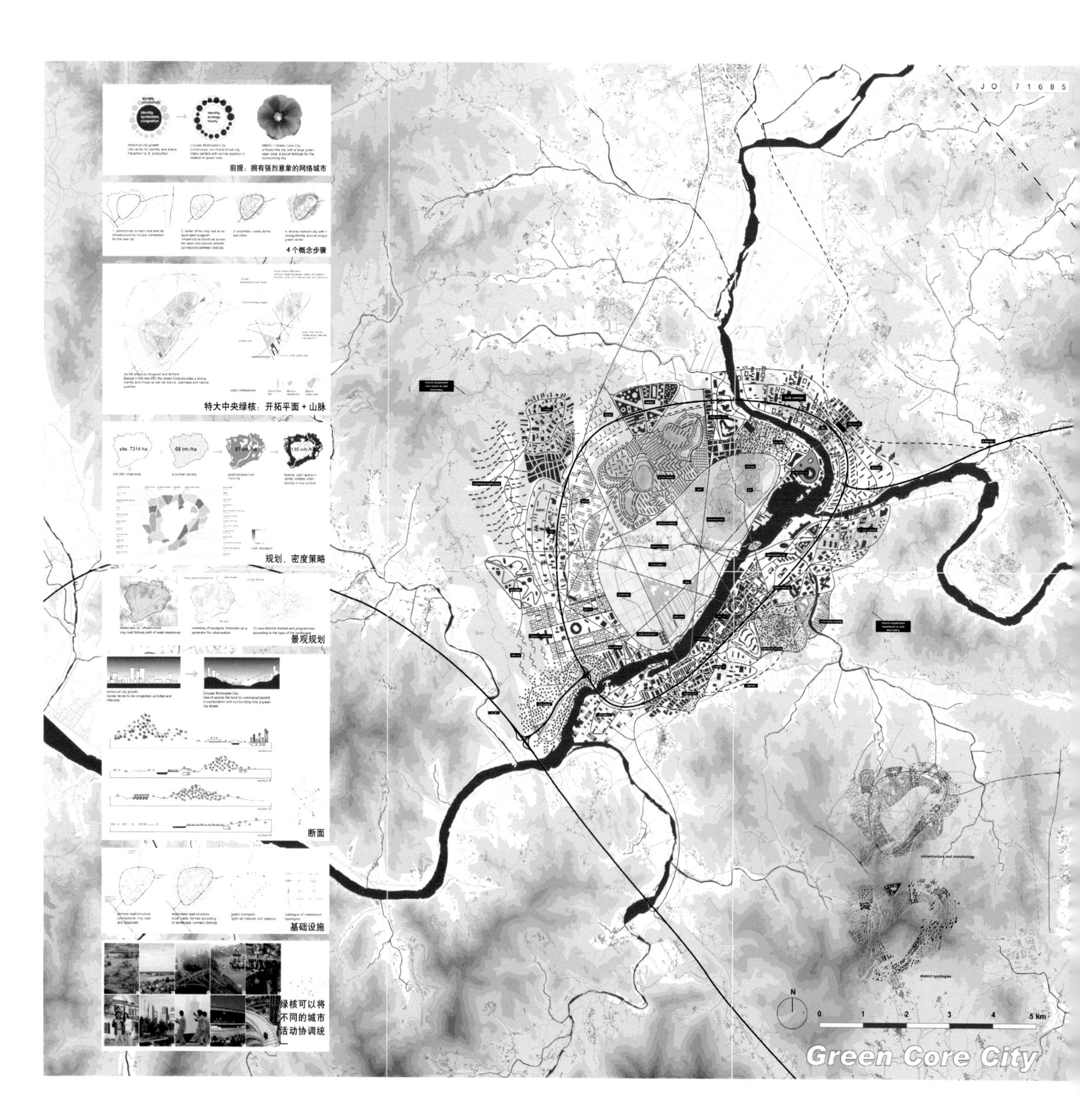

无题

Arellanes II, Michael（美国）

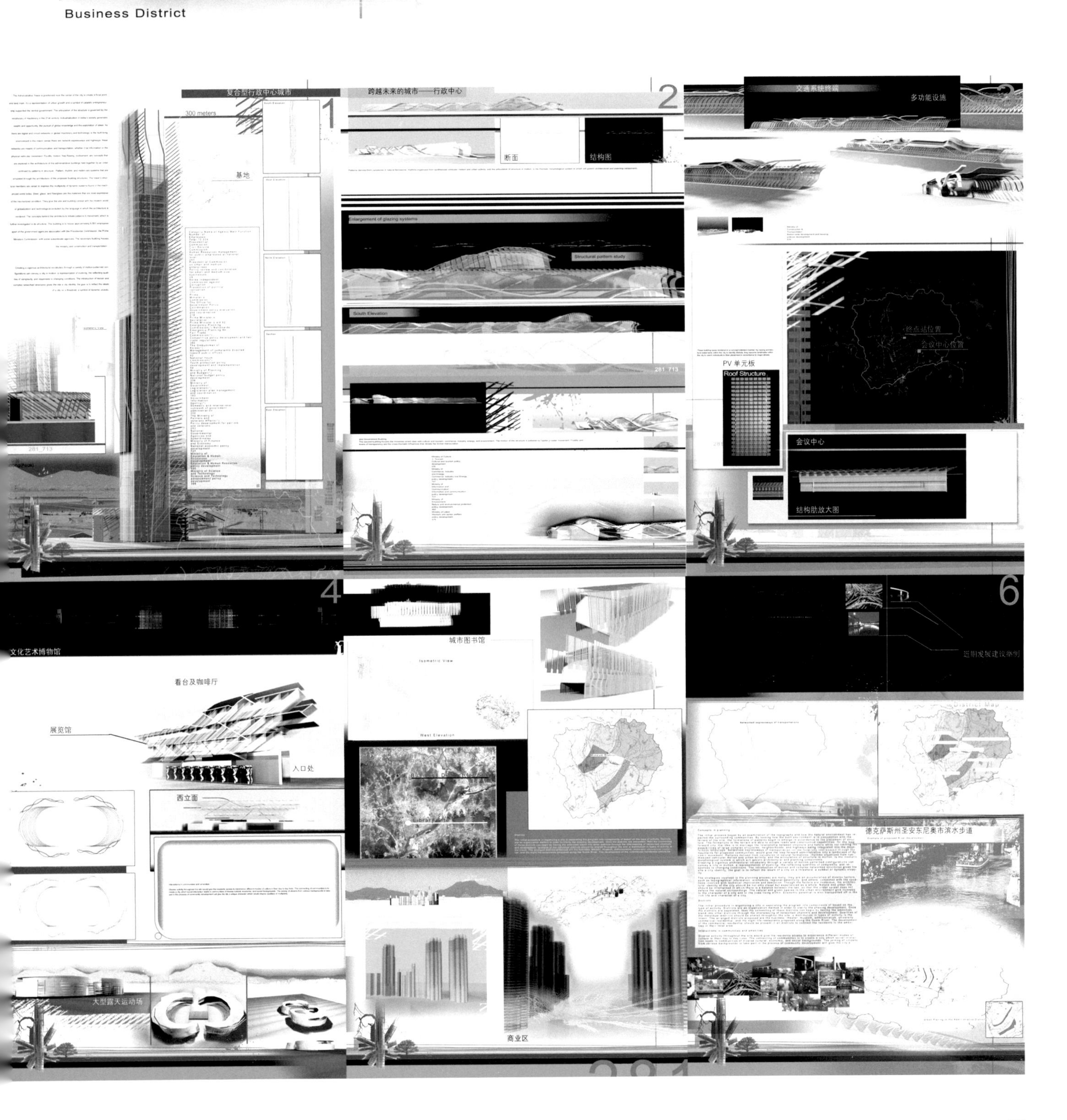

江南平原城

Hostench Ruiz, Oriol（西班牙）
Bernat Quinquer, Sergi/Caba Roset, Joan/
Crespo Solana, Rafel/Gonzalez Benedi, Sergi

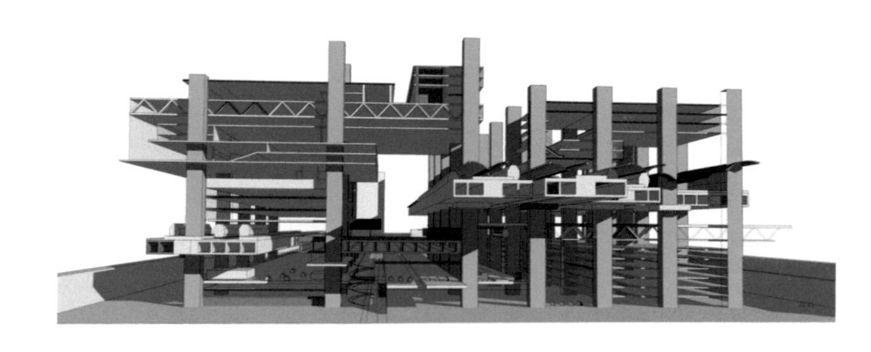

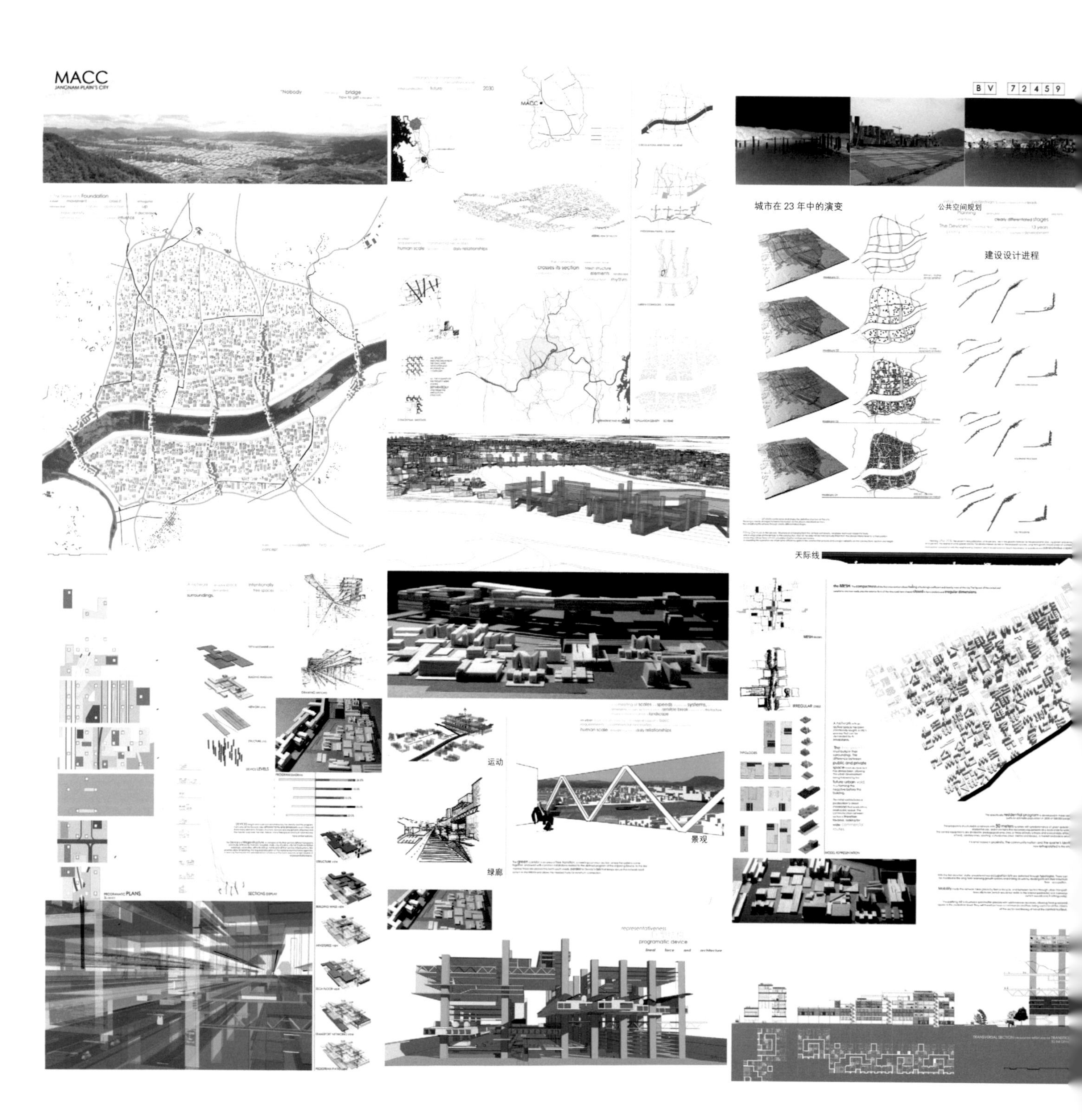

无题

Schelbert, Hannes（德国）
Hofer, Florian/Bucher, Sigfried

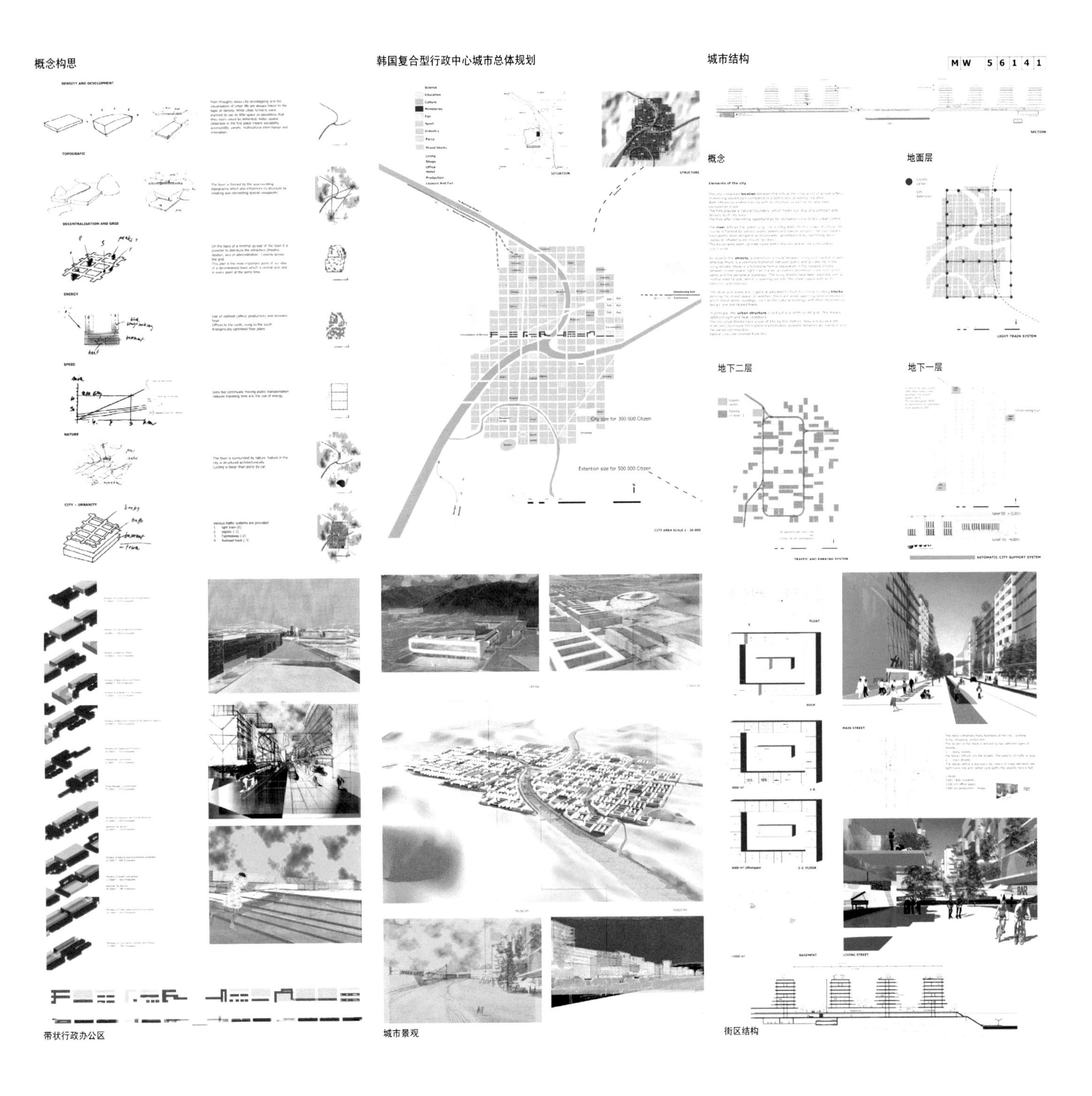

文城

Jones Evans, Dale（澳大利亚）
Dculys Landscape Architects /
Dale Jones Evans Pty Ltd. Architecture / Maki, Yamaji

无题

Varas, Alberto（阿根廷）
Estudio Alberto Varas Yasoc, Arqs/Montorfano, Hugo

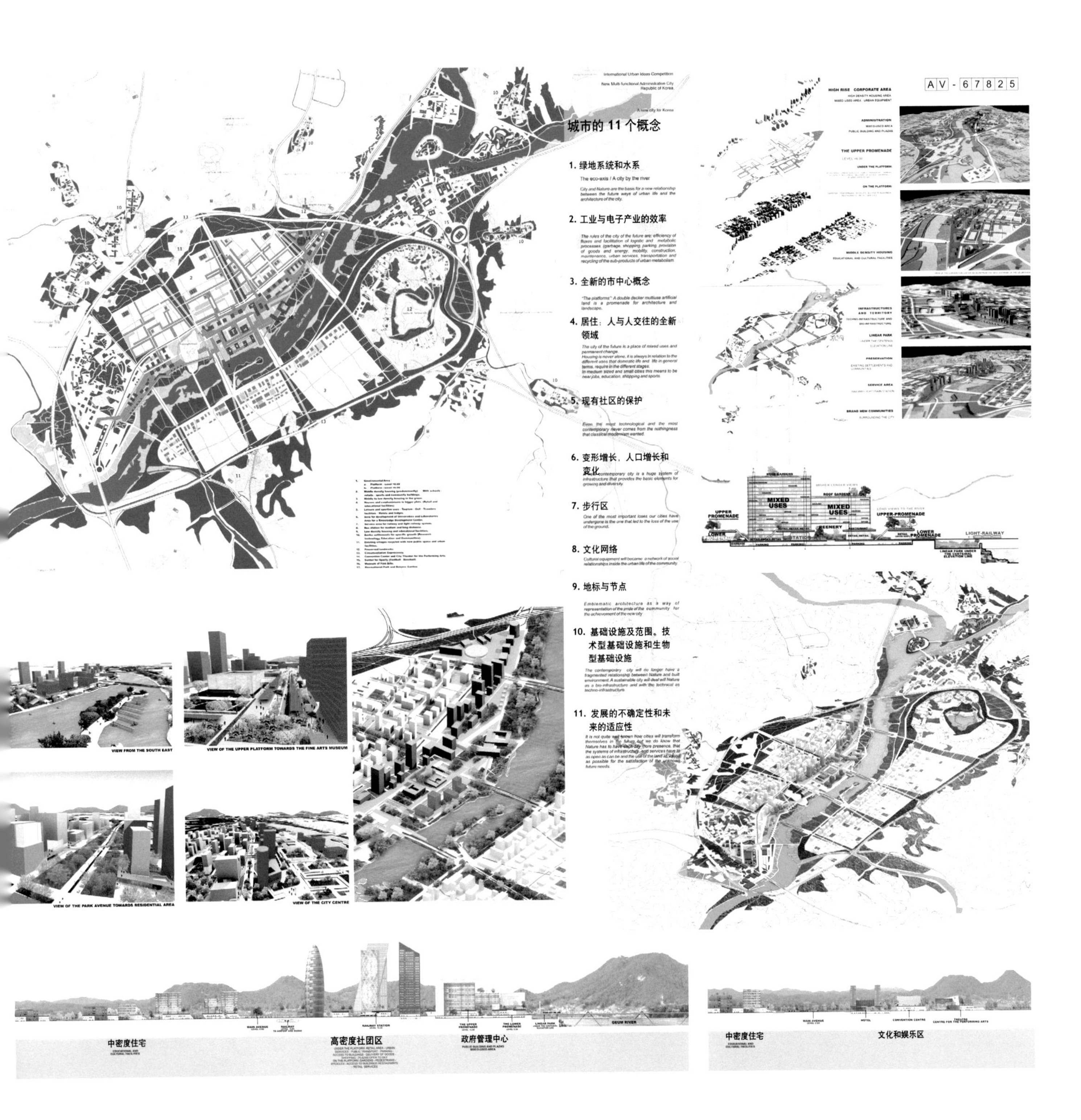

彩虹之城

Pasqualini, Isabella（瑞士）
Kuhnholz, Olof/Gran, Carianne

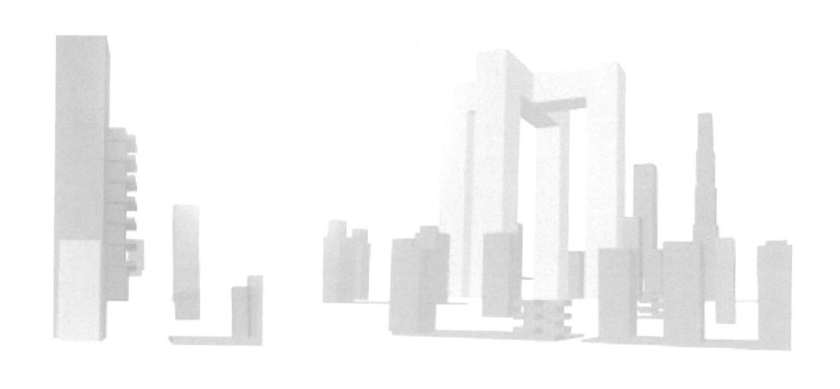

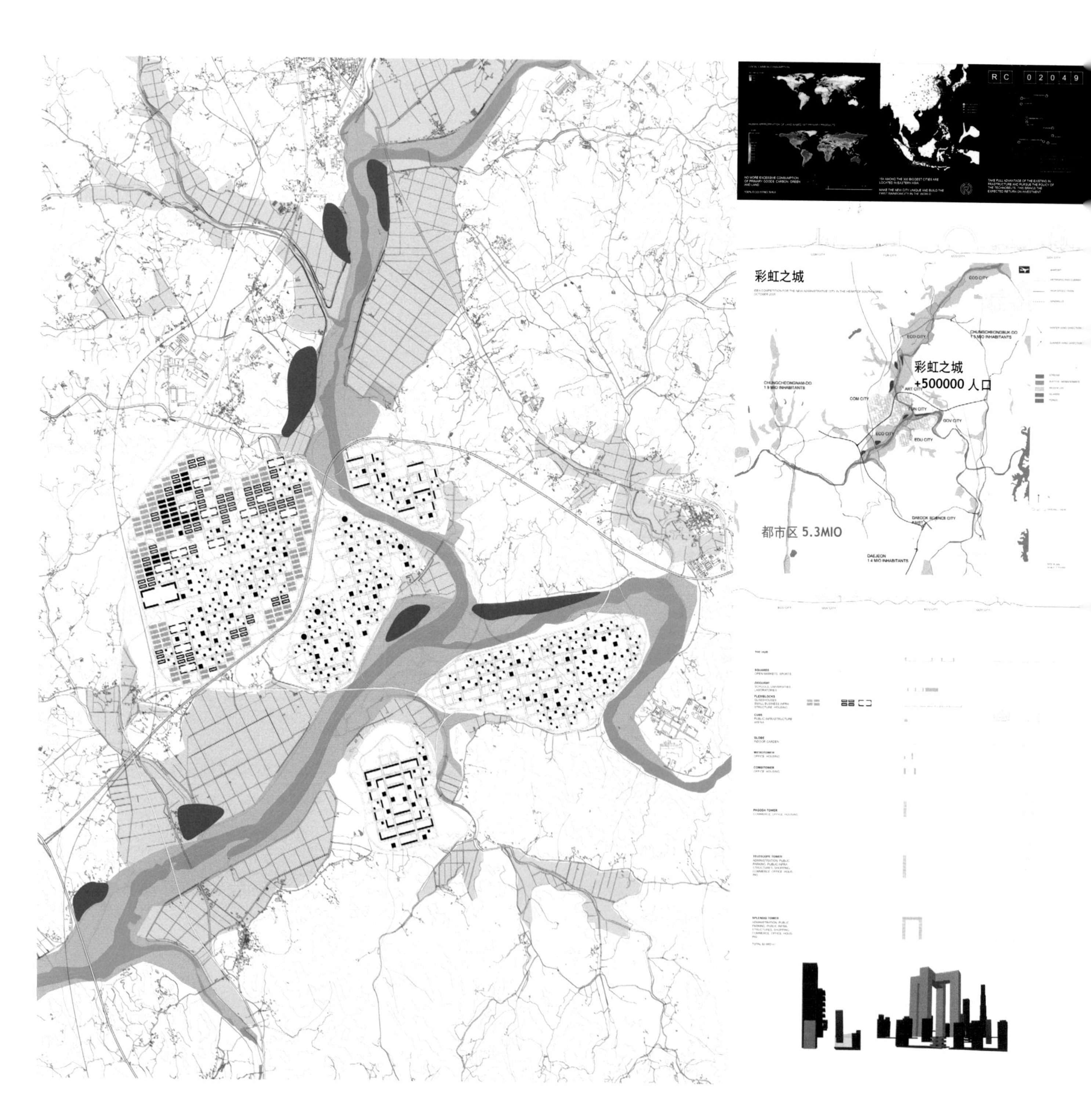

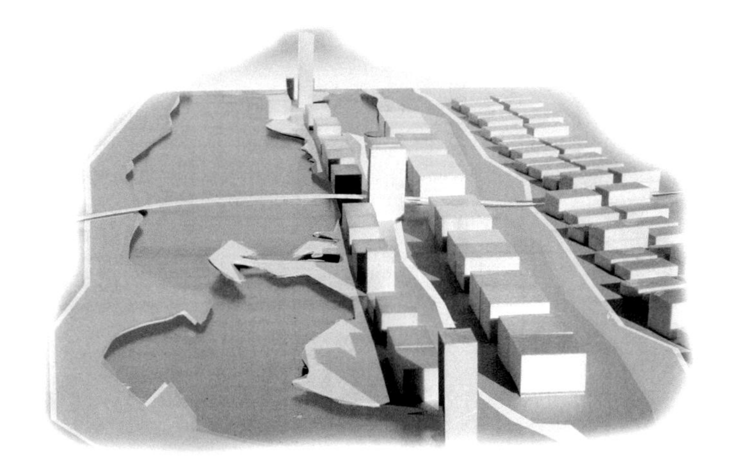

15个核心的城市

Fritzen, Andreas（德国）
Feiling, Oliver/Rung, Hannelore

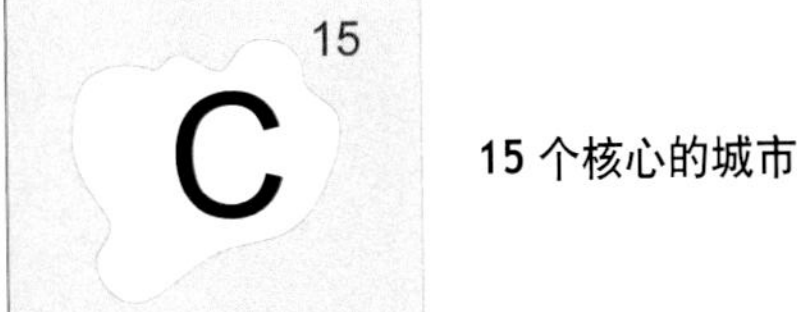

概念

“C-15”城市依托于 15 项核心工程，在其周边延伸发展。这种主动规划下的城市将有别于接近 30000 居民的地区。这些核心工程共同组成了一个多中心的、民主和各阶层融合的城市，确保城市的活力、创新、竞争与生机勃发，将为居民和世界创造一座具有普适性的复合型多功能行政中心城市。

当为公共权力机构规划核心工程并意识到城市应保持整体发展时，城市内部间隙空间便较少被管制和处理。这些内部间隙空间就像活动空间一样赋予城市活力，包容多元化、未知性和自发成长，通过最终的设计方案使其与临近的核心空间达到相互平衡与协调。这些残余空间可以采取公私合作的方式进行开发，并允许在城市和社会组织中存在必要的差异性。

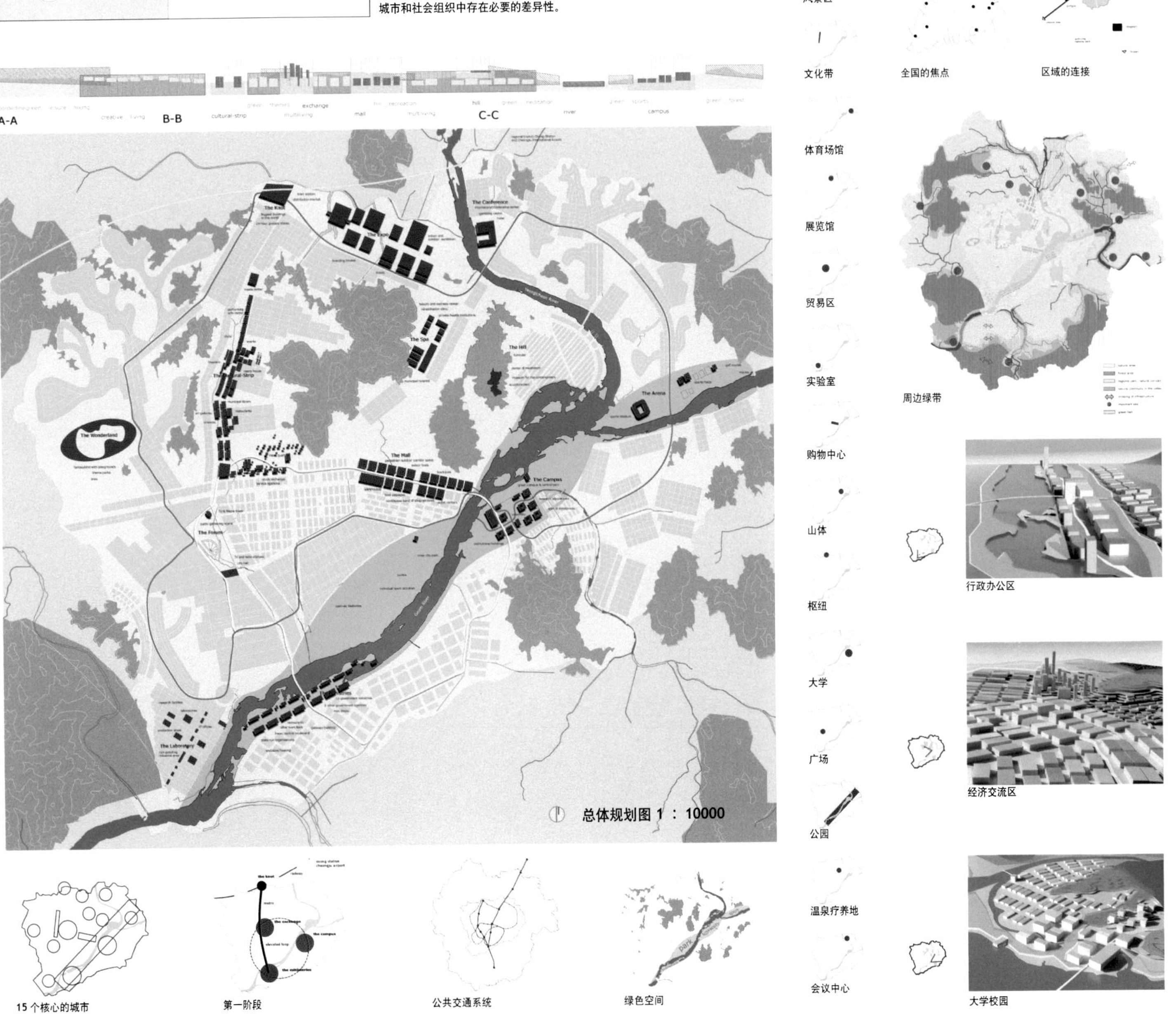

无题

Barbedo, Jose（葡萄牙）
Sottomayor, Nuno/Pinheiro Torres, Miguel/
Jigueiroa, Joao Luis/Alves Pinho, Albino

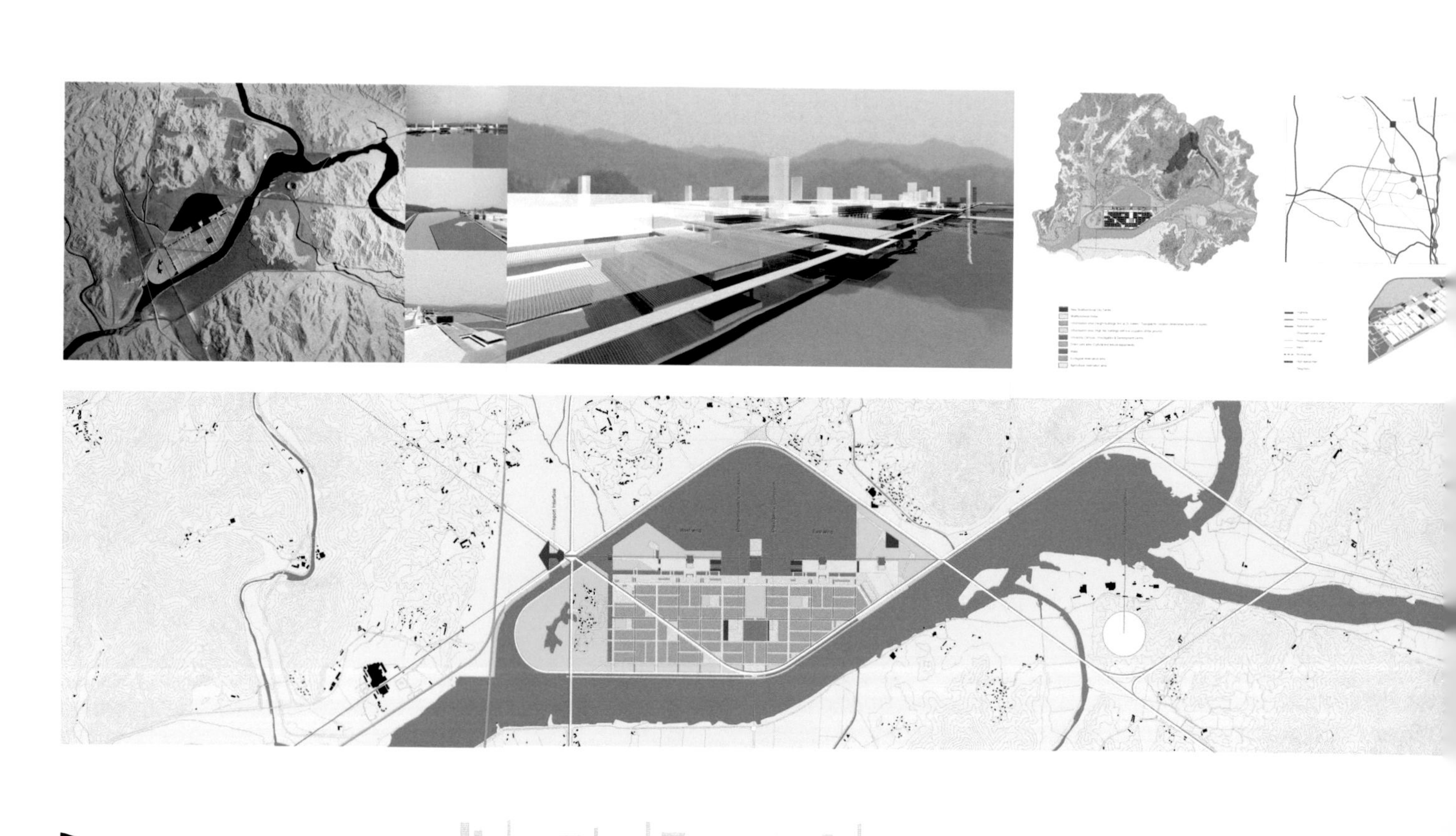

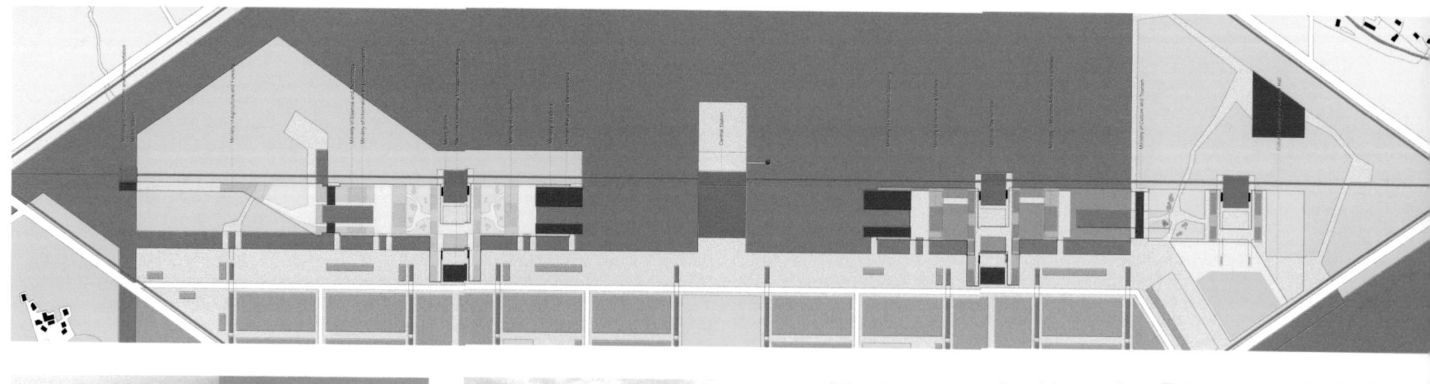

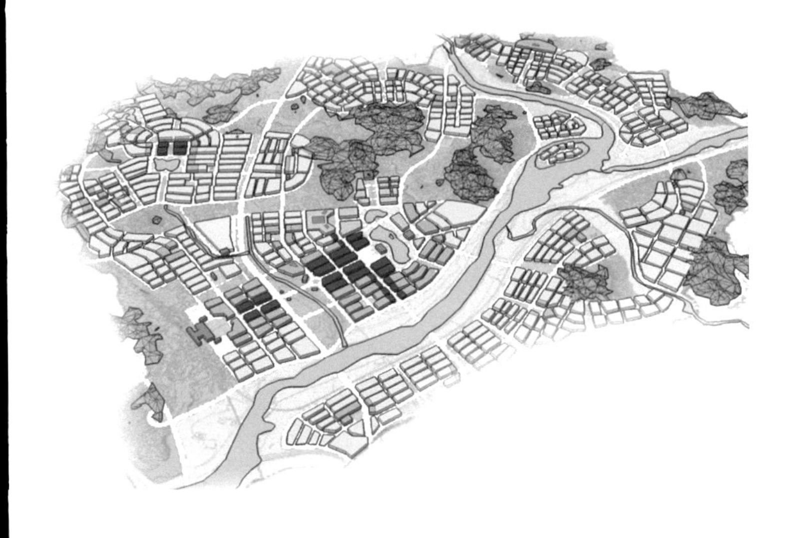

无题

Kamal, Zaharin（马来西亚）

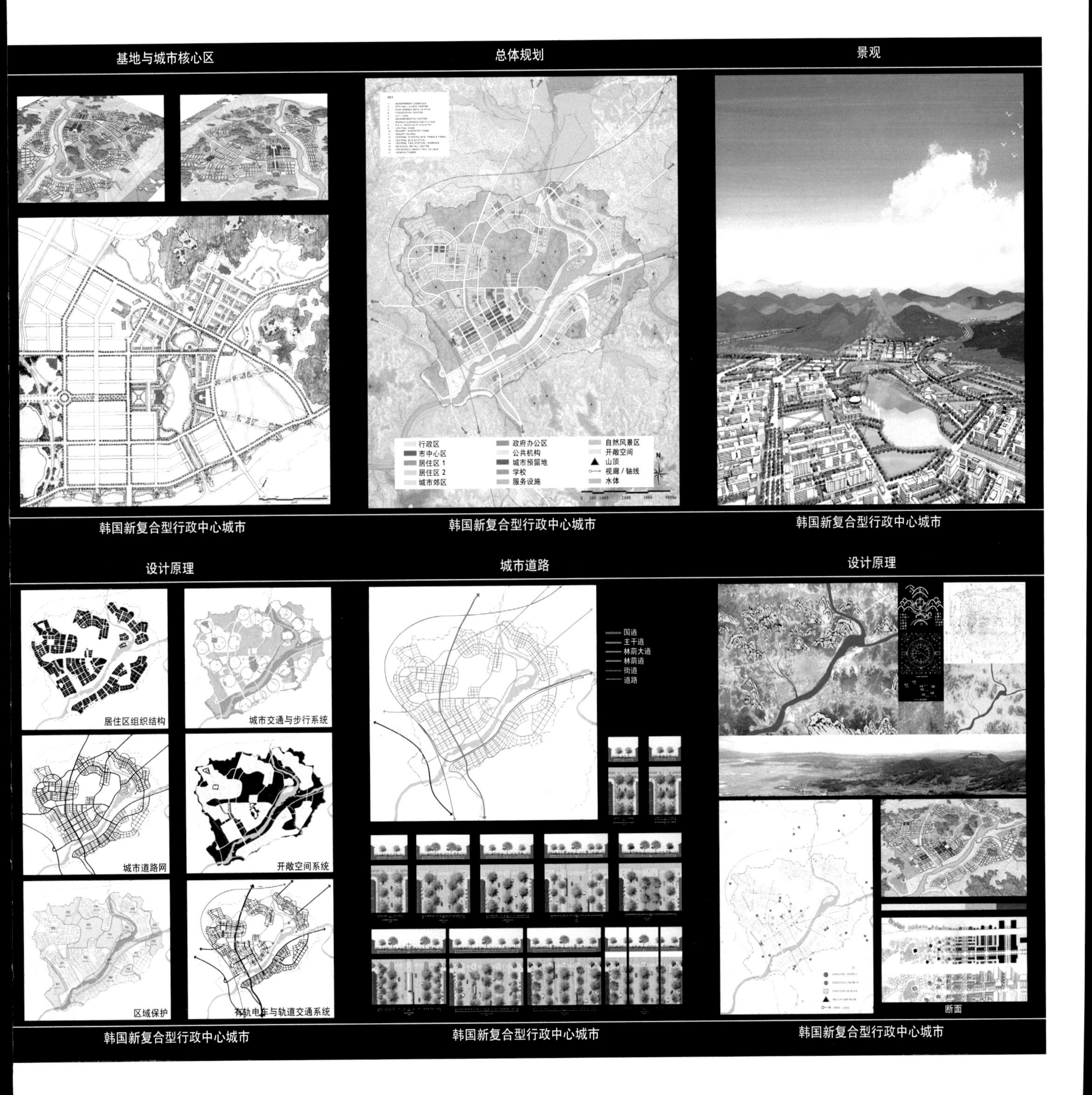

链状城市

Ngai, Ted（美国）
Schenk, Beat/Feng, Alice/
Kim, Chae Won/Feevajee, Ali

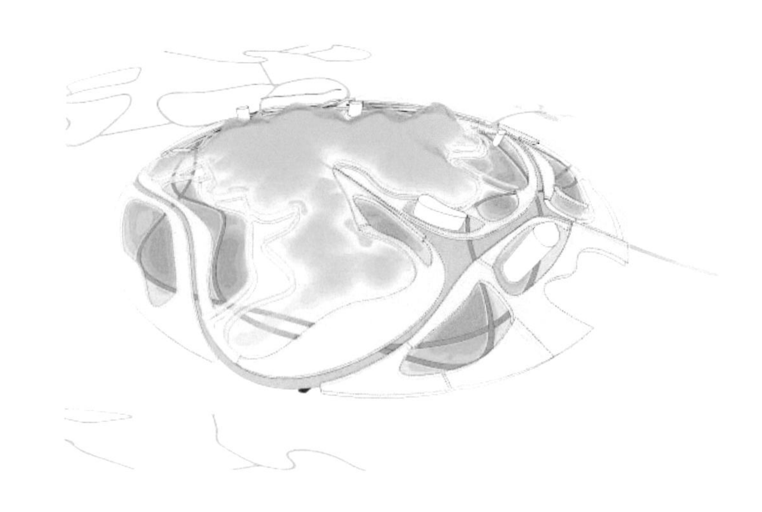

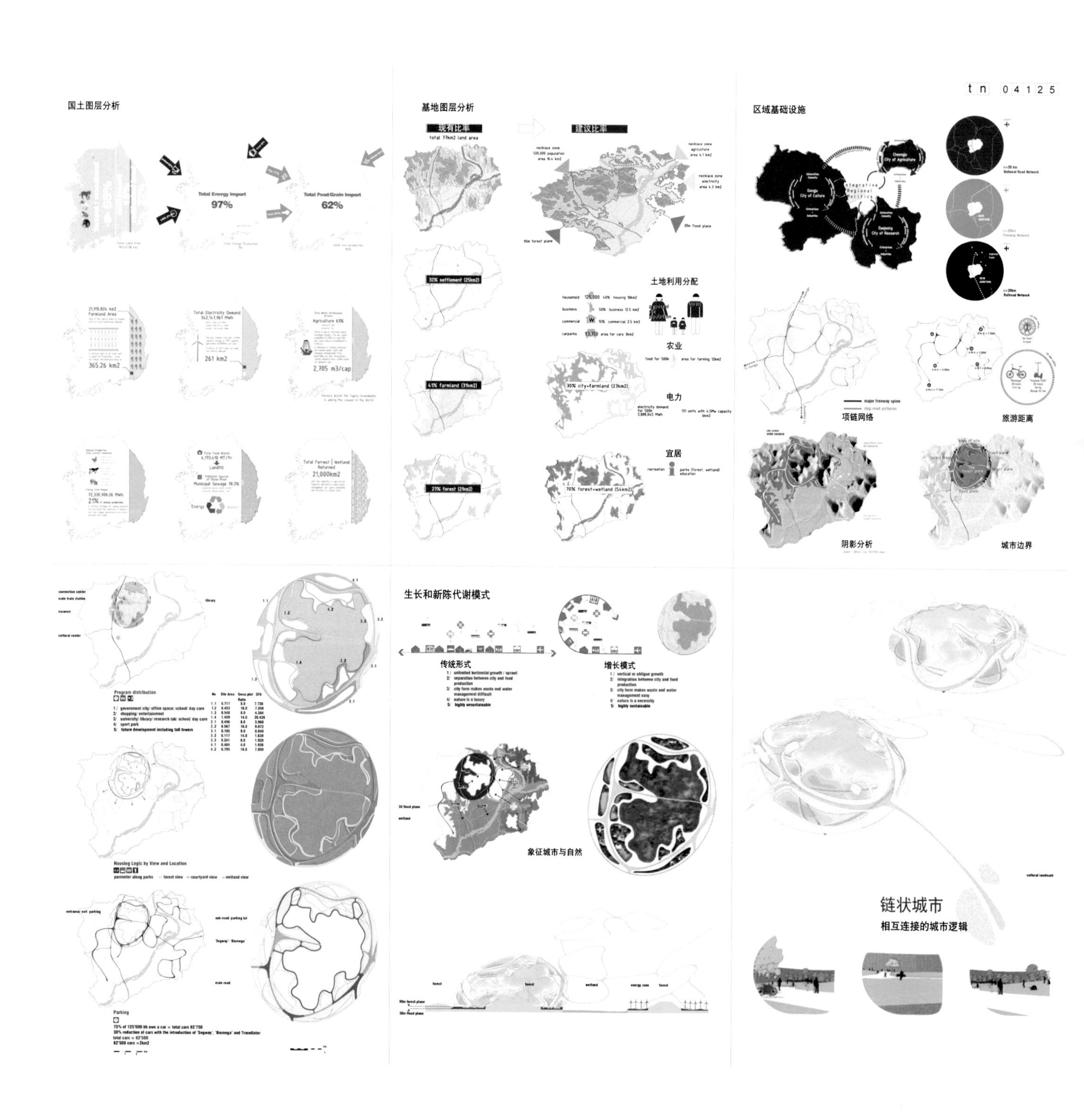

无题

Kim, Hee Seok（韩国）

1. 方案定位

背景

—通过建设新行政中心城市来疏散首尔地区过度集中的人口和财富

主要目标

—为行政职能营造一座概念城市

—为韩国的其他城市确立一个城市范例

主要原则

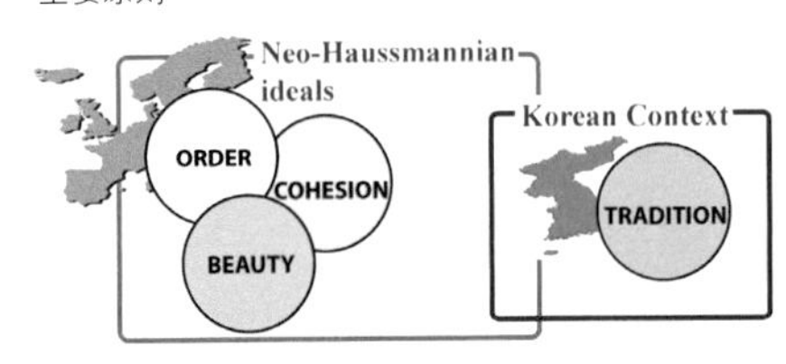

时序：尽量协调，避免混乱

集聚：形为整体，个性鲜明

景色：追求有序和凝聚的价值

传统：韩国传统建筑的再探索

韩国新复合型行政中心城市 2005 年 10 月

2. 区划与建筑

无分区制

—利用垂直分区替代水平分区

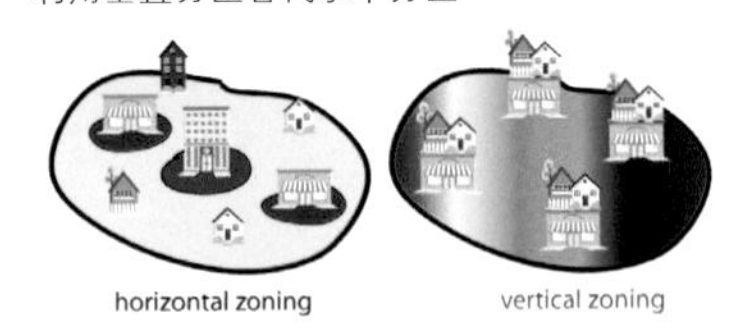

建筑

—两种建筑类型：常规建筑和特殊建筑

—常规建筑：城市的基本构成要素；风格一致但细节有别

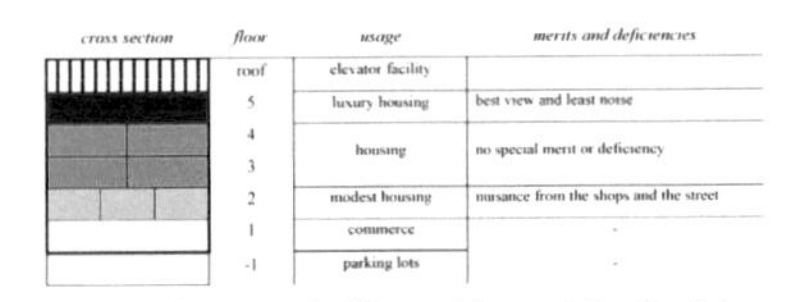

< structure of an unitary building and characteristics of each floor>

(Deeper color means more revenue of the tenants; A rectangle represents one household)

—特殊建筑：专门用途的建筑

—建筑学：传统韩国建筑的现代化

韩国新复合型行政中心城市 2005 年 10 月

3. 规划一：水

KN-45810

基地现状地理特征：

—被河流穿越

—随雨季形成的广大冲积平原

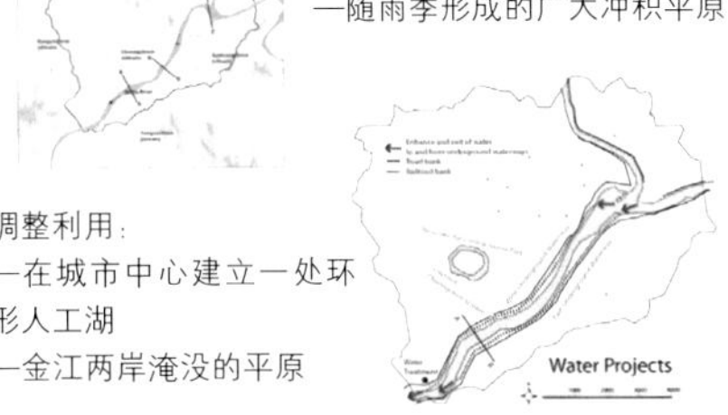

调整利用：

—在城市中心建立一处环形人工湖

—金江两岸淹没的平原

河岸冲积平原的开发利用

< cross section of the Geum River >

—在冲击平原上建设堤坝和场地

—空间综合利用：

1. 滨河公园
2. 堤坝中的交通主干道

—效果：

1. 便于河岸管理
2. 节约空间
3. 通过去除不明确的空间来美化河岸

韩国新复合型行政中心城市 2005 年 10 月

4. 规划二：总体规划和网络图

总览

—除了少量集中的特殊建筑外没有分区制

—道路自湖心岛公园呈网状发散

—城市中心位于湖心岛公园和金江之间

—冲积平原改造为公园

—将河漫滩要塞改造为公园

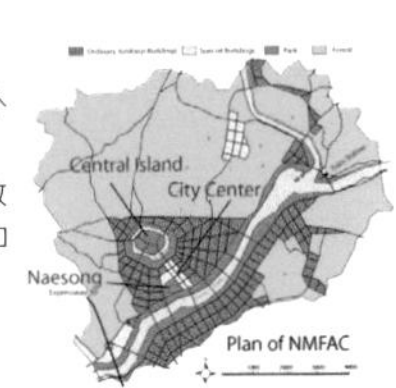

道路和铁路网络：公交优先

道路

—通过设立绕行主路使城市避免拥堵

—争取实用和美观而避免象征乏味的道路系统

—利用有轨电车、公交车和自行车组成主要的交通系统

铁路

—与大田市接通：延长大田市地铁 1 号线到 NMFAC

—与青州市接通：利用现有铁路网改造成城市高铁

韩国新复合型行政中心城市 2005 年 10 月

5. 规划三：公共空间

公园和森林

—每个公园和林地具有专门用途

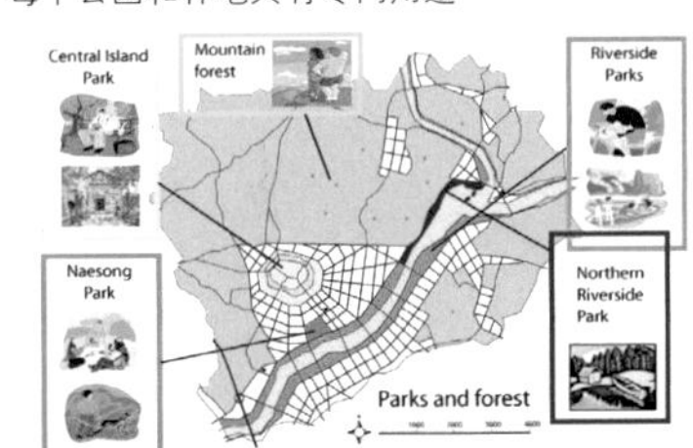

公共设施

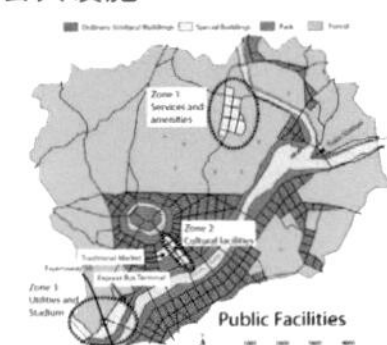

公共设施的两种类型

—大型和专业的设施集中于三处区域

—小型和普通设施在城市中均匀分布

标志性构筑物

—所有标志性构筑物将成为城市主要道路的对景

—市中心的标志性构筑物轴线：尽可能满足建筑美学的需要

韩国新复合型行政中心城市 2005 年 10 月

6. 环境氛围

街道及建筑鸟瞰图

河岸沿线正面

沿河步行道视野中具有韩国风格桥体

与沿街拱廊结合的有轨电车

韩国新复合型行政中心城市 2005 年 10 月

无题

Katainen, Juhani（芬兰）
Erra, Jyrki/Korpela, Pekka/Koivisto, Hannu/Haikio, Juho

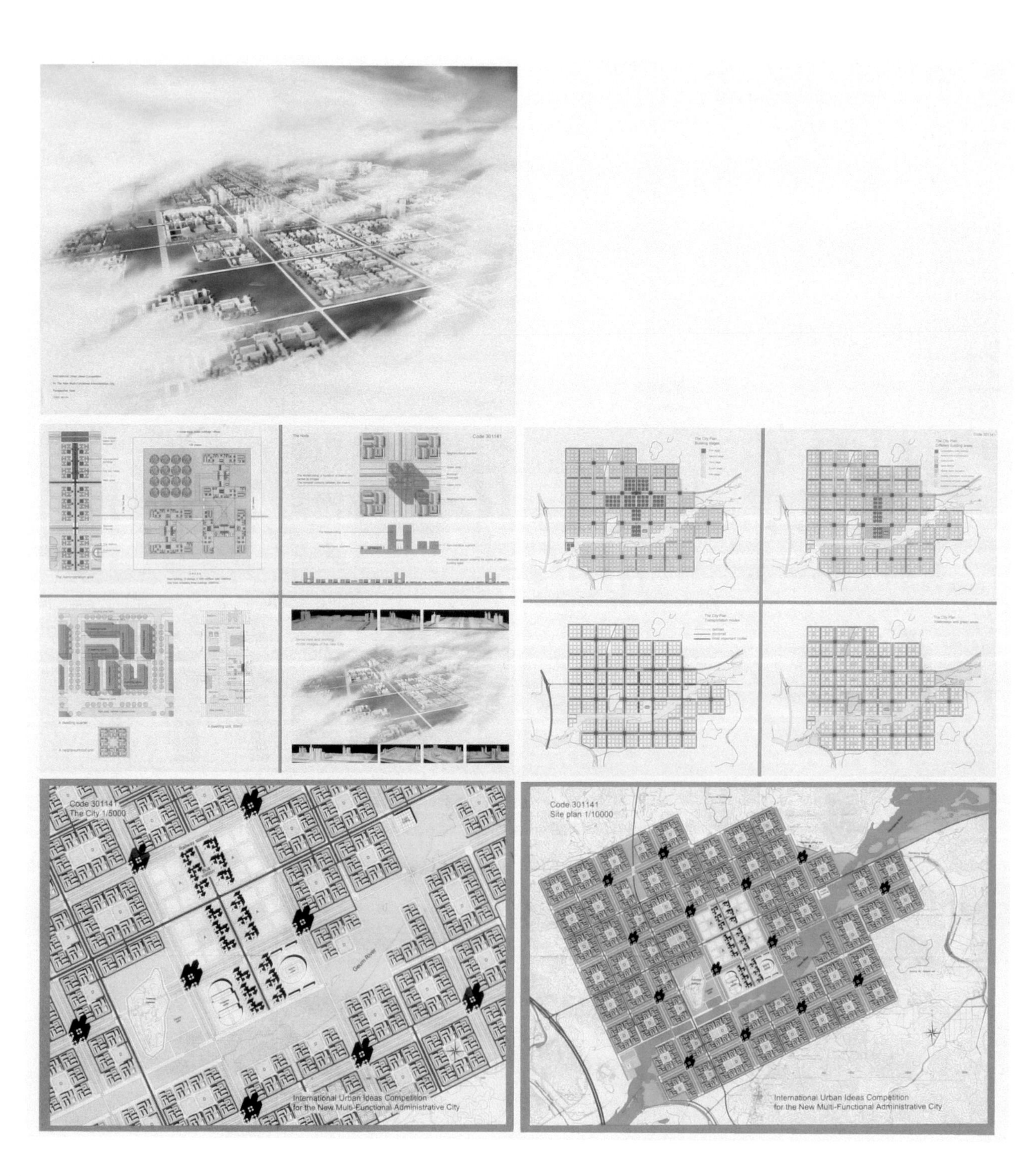

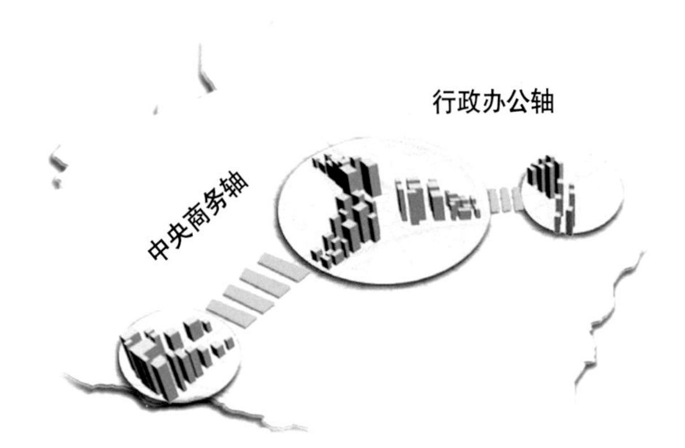

无题

Park, Sam Ho（韩国）
Chang, Hyun Jung/Lee, Jung Min/Park, Chul Woon

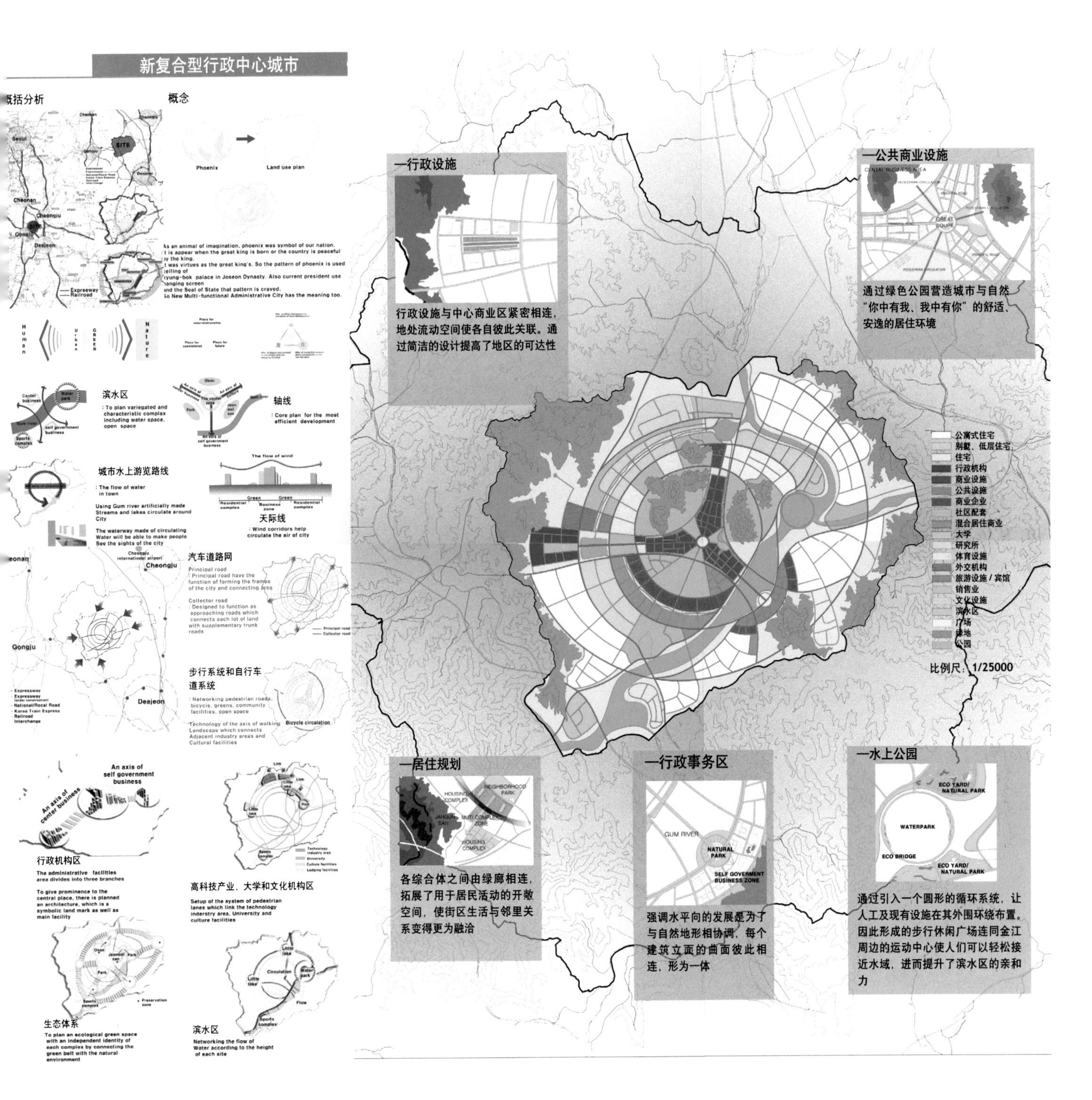

无题

Lee, Jae Yeol（韩国）
Noh, Gyeo Sun/Koo, Min Cheol

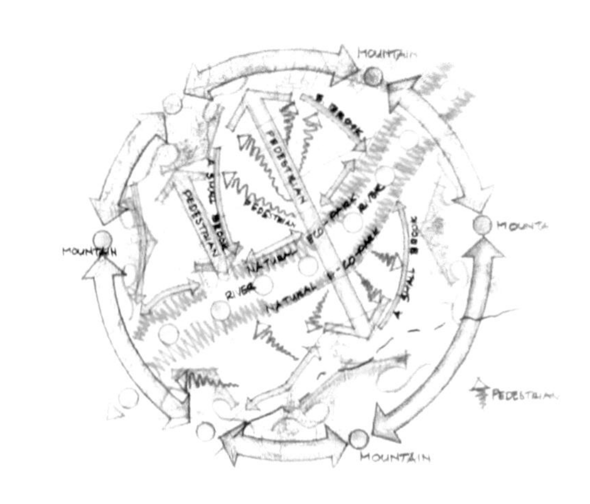

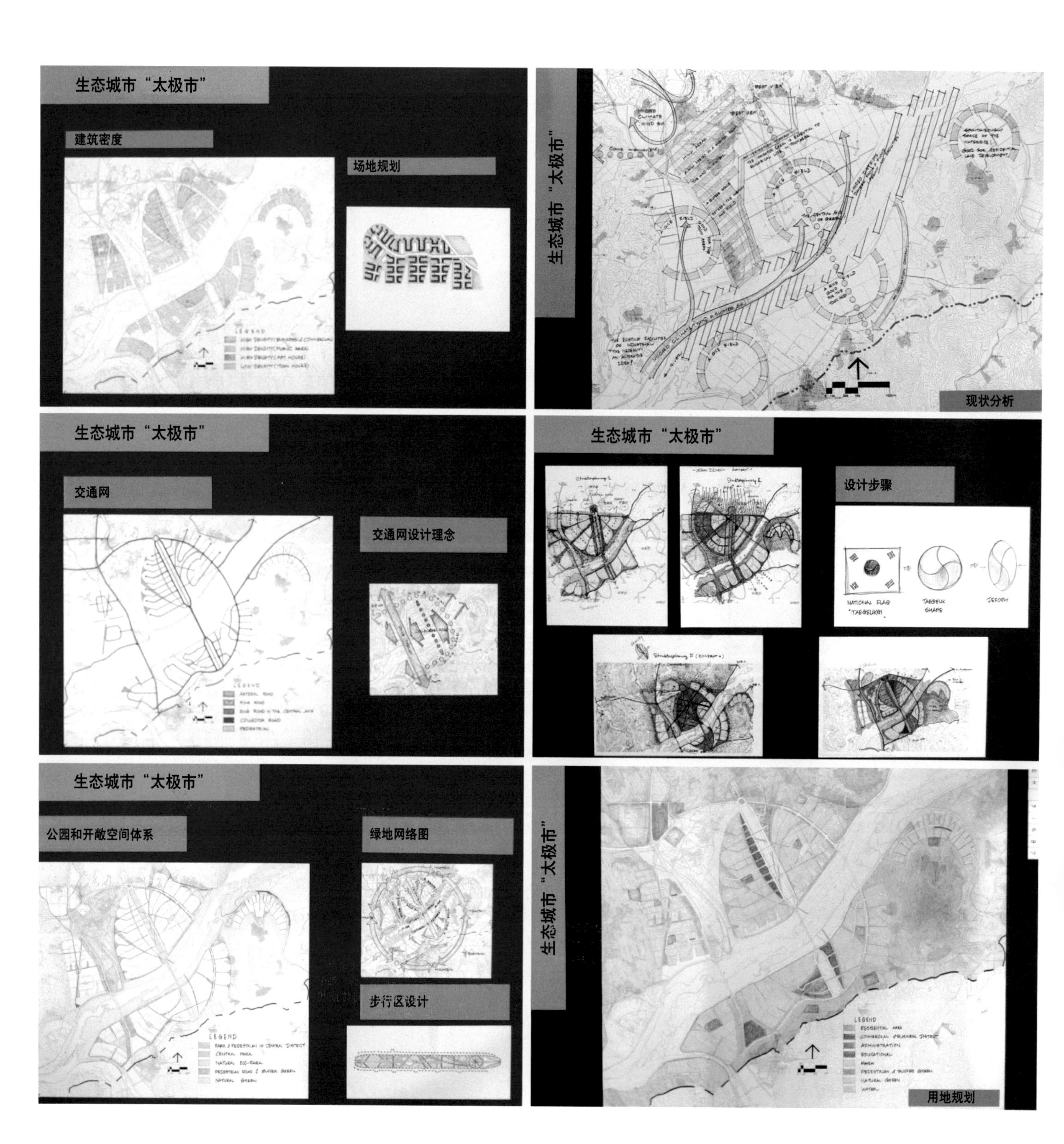

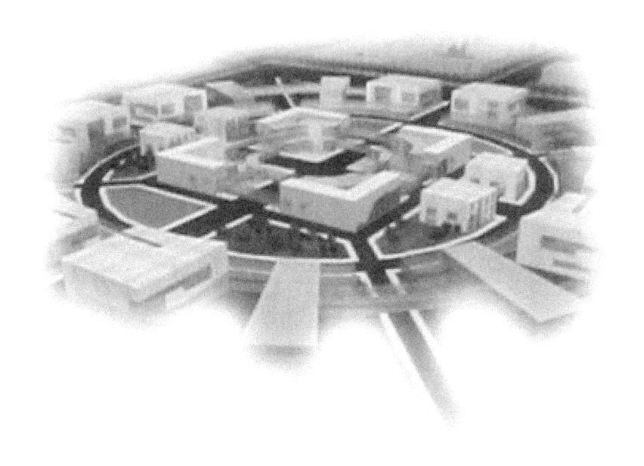

活力韩国　活力城市

Eun, Min Kyun（韩国）
Yoon, Ki Byung

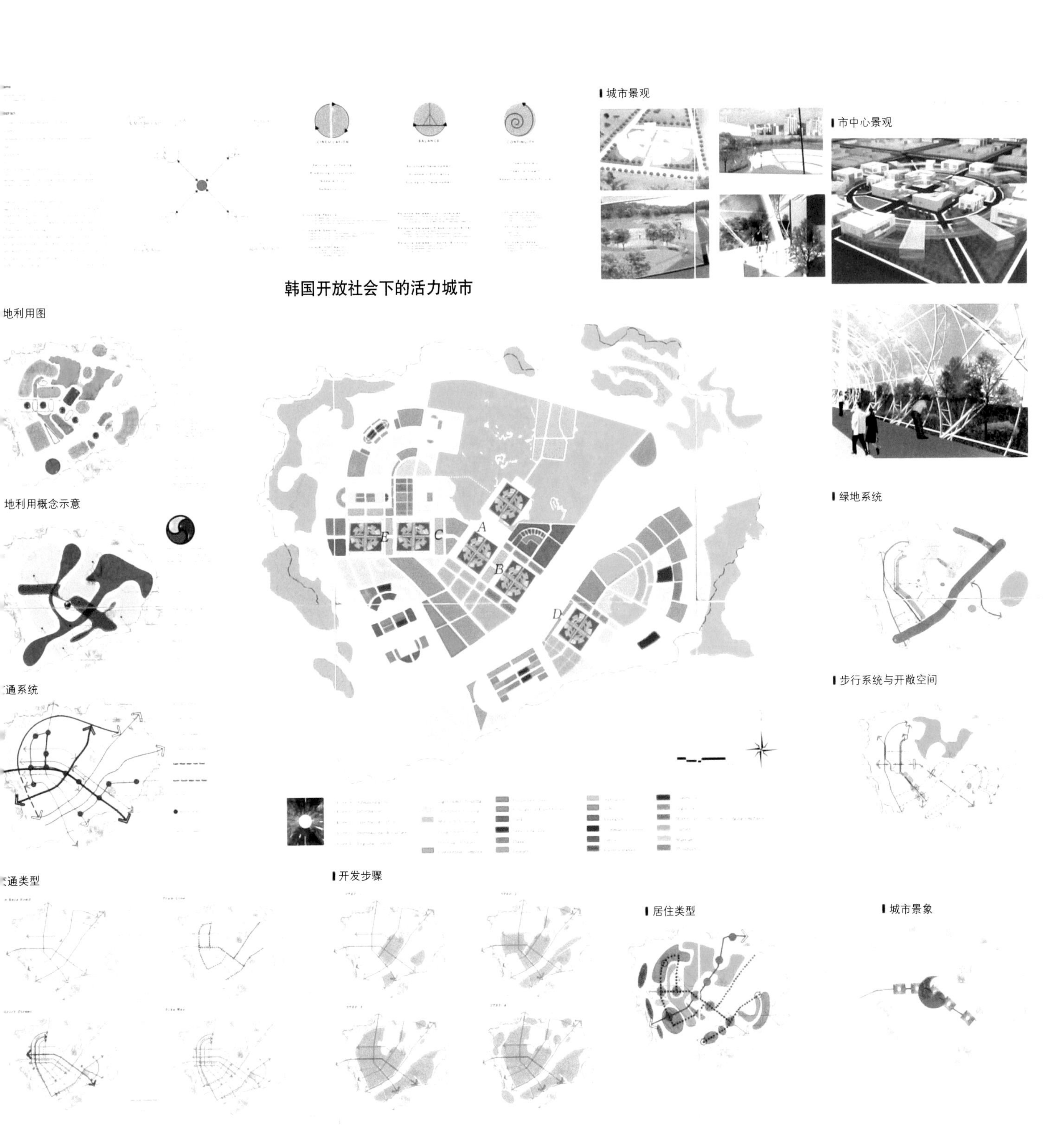

南部明星城市

Garratt, Dale A.（美国）
Garratt, Joy I./Garratt, Daejo S.

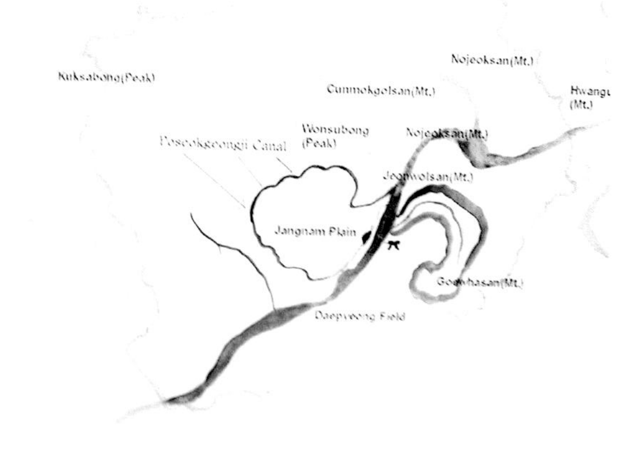

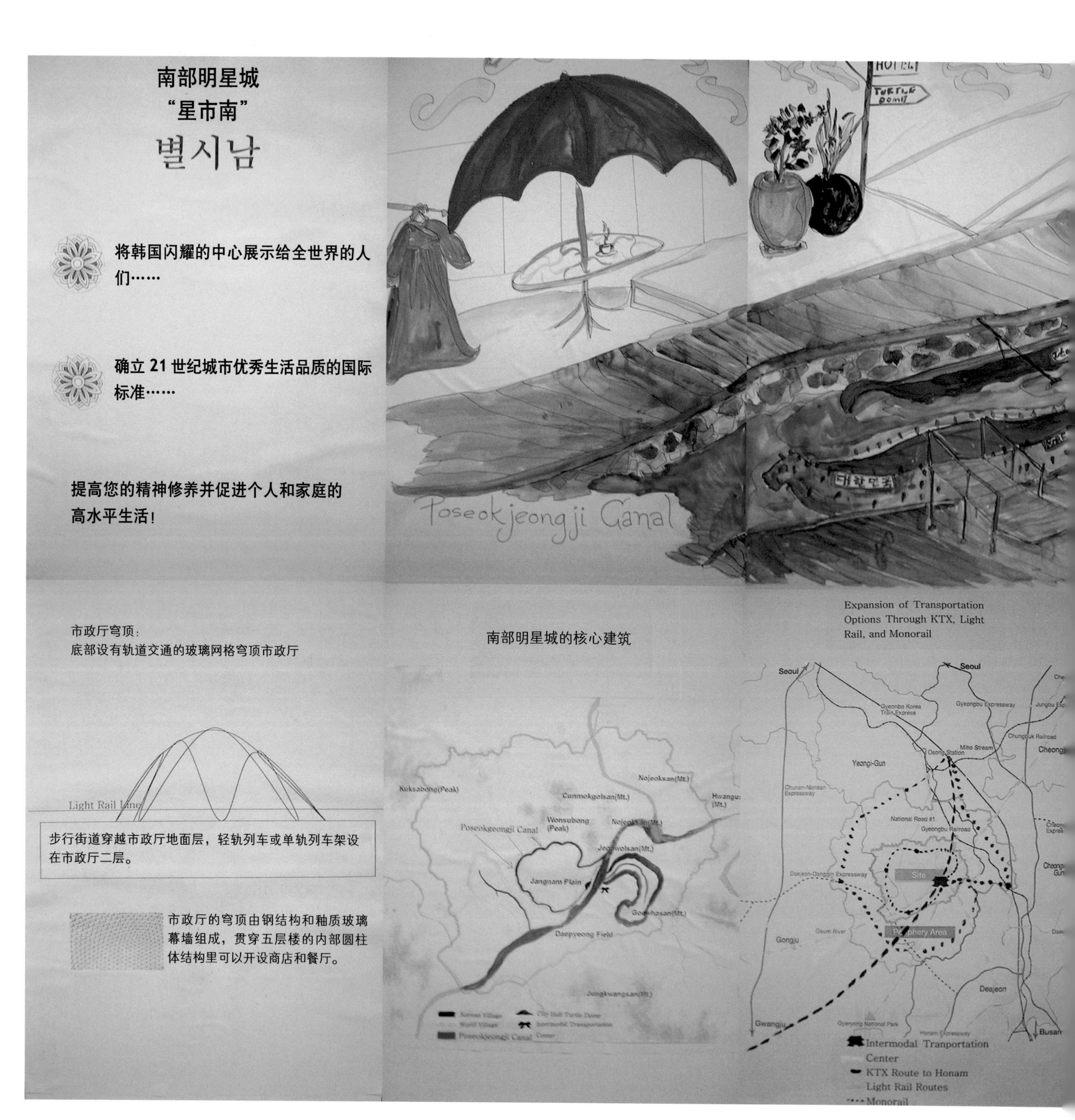

无题

Lorenzi Gianni（意大利）
Untersvlzner, Johannes

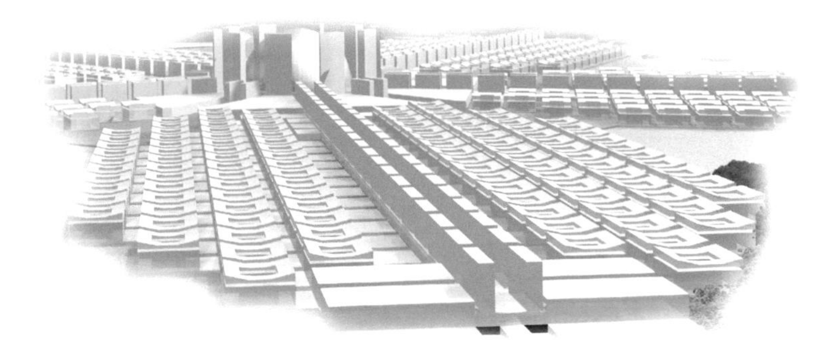

整个方案试图在检验这个花朵状的外观和结构能否在韩国复合型行政中心城市的建设中得以实施，并满足竞赛主办方所一直期待的最佳解决途径。更重要的是，可以通过将城市形态设计成韩国国花的形式来满足他们的这一要求。城市最终成型后，不论远近都能清晰地辨别出花朵状的造型。通过建筑布置、功能布局和行政机构安排等措施来实现城市设计的可识别性。

入围作品 / Entry Works

大都市测量控制

Lynch, Catherine（美国）
Metropolitan Planning Collaborative/
Todd, Lieberman/Benjamin de la Pena

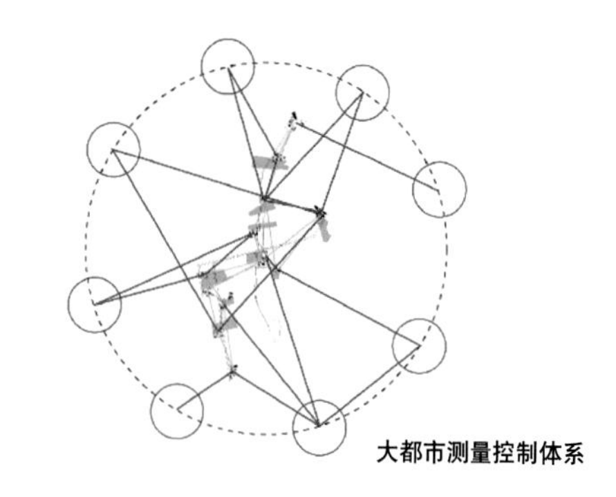

大都市测量控制体系

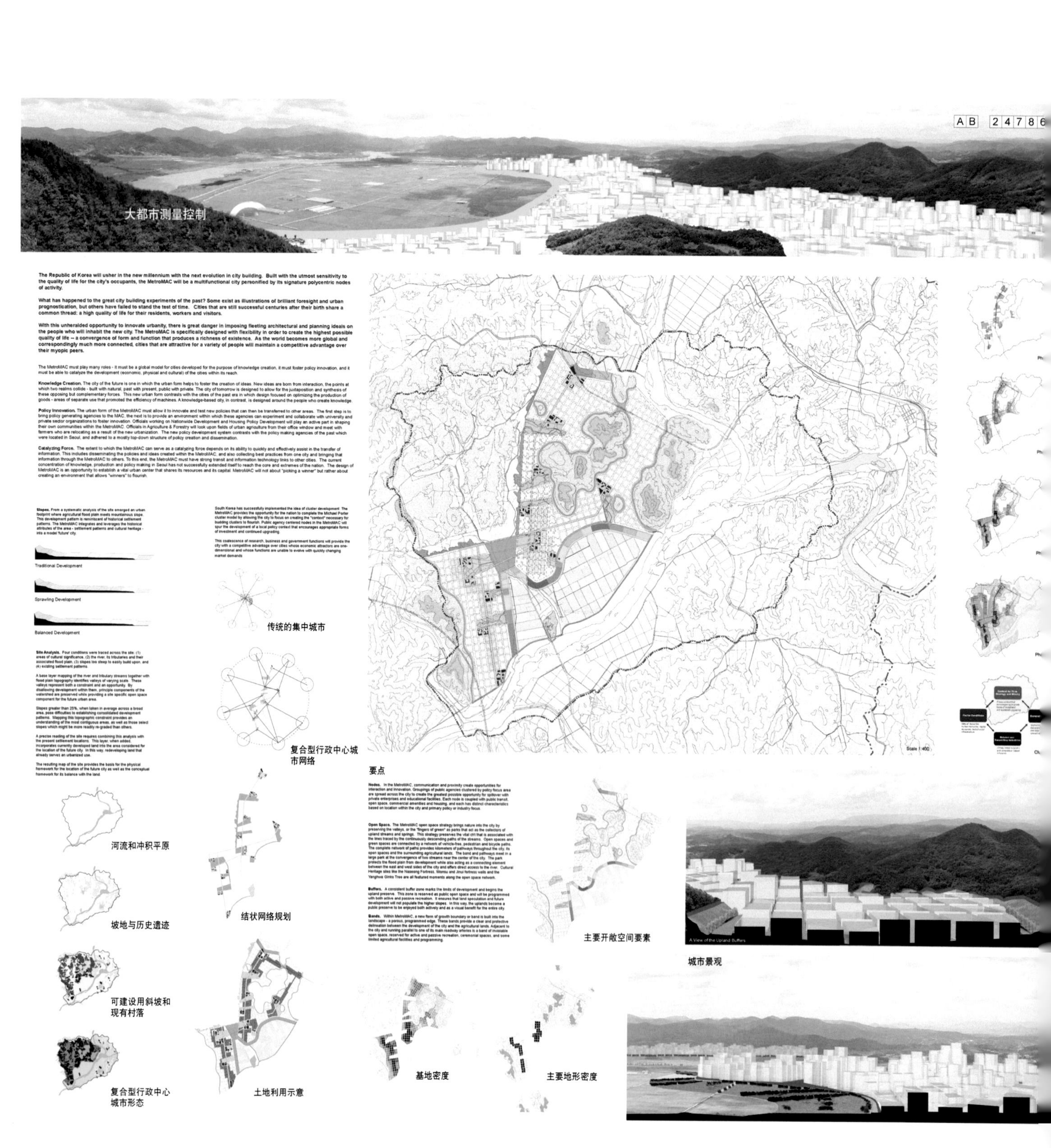

无题

Luck, Robert（英国）
Cvrtis, James/Pattni, Krishan

无题

Boender, Arnest（荷兰）

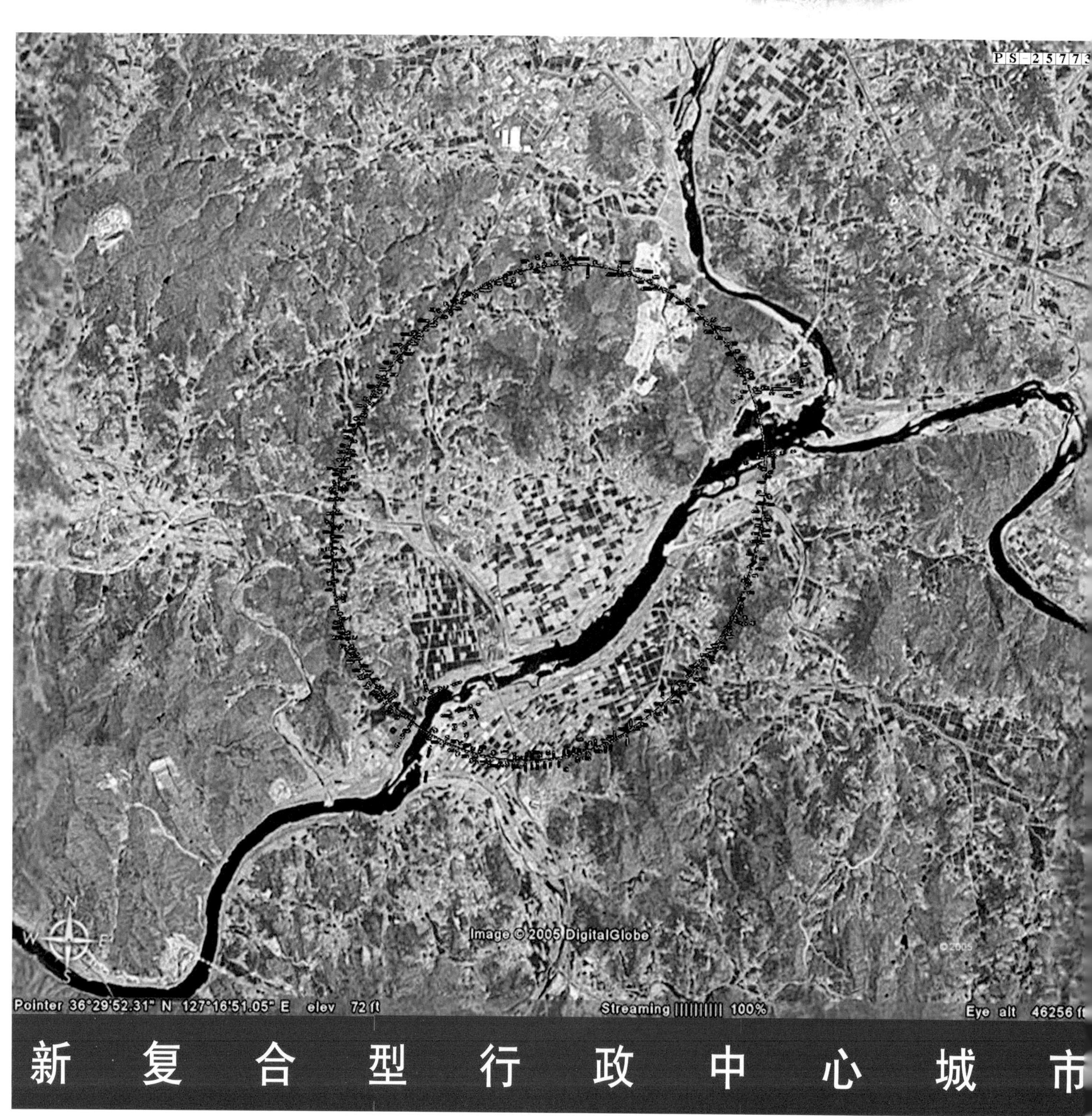

无题

Yang,Sung Goo（韩国）
Bae, Hyoung Du/Oh, Hyun Il

城市设计的构思并不是简单地开发自然或结合自然，而是力图将自然融入建筑中，进而形成一种复合结构。最终，城市将在自然需求和人类需求之间扮演一个“弥补者”的角色。在处理忠清南道的自然环境时，我们通过将自然与建筑相结合的方式来体现自然的价值。居住包围的景观是纯工环境，与城市周边的自然景色形成了鲜明对比。这是一个展示包括自然和建筑学在内的所有城市元素的理想方式，以及如何将建筑融入自然这自然景观时形成的另一种合理的方法论。

这种基本街区的初始构思源自于“庭院”这一韩国传统城市形态的构成要素。在韩国传统的村庄里，院落（一个家庭居住的多座住宅共同形成组团）中包含一种名为“庭院”的开敞空间，可以用来举办诸如婚礼、家庭礼仪和其他集会等的私人庆典活动。这就是为什么韩国传统村少存在类似于西方广场的集中公共开敞空间。在过去的时代里，这些“庭院”和院落共同组成了一个个村庄和城市。这种基本矩形单元的尺寸20m×120m”，基本与普通中学面积相等。

的建筑将保持 4～5 层高，临街建筑与街道的高宽比为 1 ：1，“口”字形的街区内部包含有自然生态的庭院，并可通过增加内部庭院的层数来观密度。有基于整体景观的背景和理念，街区的一部分会被模糊化处理，或者将几个街区相连后的院落连辟为城市中的带状步行公园。

近的巨型环是作为线性开敞空间、综合体建筑和步行天桥的复合功能构筑物。通过环绕滨河公园的方式增加城市内部的滨水面积，使其作为连接岸生态和交流的平台，并成为城市整体意象的首要构成元素。

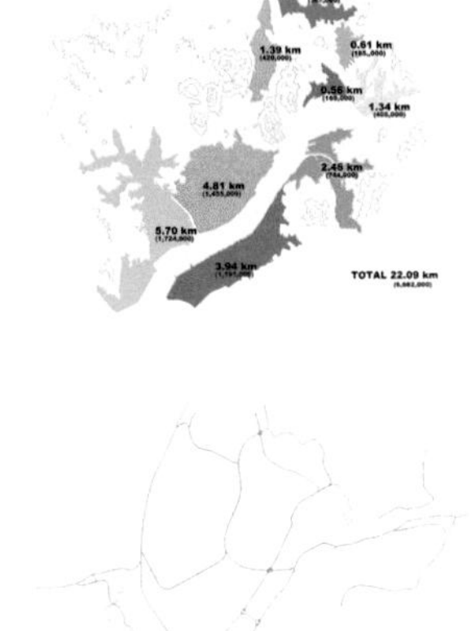

(m²)	Government	Housing	Education	Commercial Section	Cultural Facilities	Advanced technology Factory & Company	Fundamental Facilities: Road	Public Parks	water supply facilities	Etc.	The Sum	Total (each area)
A	751,060	1,711,000	150,054	480,899	150,461		823,561	157,905	16,952	17,891	995,308	4,838,782
B		2,255,500	1,158,406	569,922	166,277	420,413	926,696	216,057	20,631	17,059	1,180,443	5,780,961
C		1,609,000	971,648	405,864	155,220		625,222	115,667	15,945	18,202	774,528	3,959,060
D		1,113,575	488,000	245,867	71,300		328,628	115,451	14,851	11,585	474,515	2,395,257
E		598,475	140,608	138,946	40,294		196,022	57,279	4,423	5,333	263,957	1,382,280
F		150,000	80,000	127,401	27,091	684,087	157,699	27,281	16,695	60,101	261,716	1,350,795
G		284,900	135,968	60,968	17,680		81,517	28,395	2,887	3,993	116,792	614,308
H				130,421	19,028		127,352	408,305	593,259	31,836	1,160,752	1,310,201
I				44,560	8,000		50,415	265,559	168,055	8,697	492,666	545,050
Total (each function)	751,060	7,756,450	3,908,684	2,203,146	635,361	1,104,500	3,517,412	1,572,373	852,514	180,179	5,772,477	22,087,668

：是城市的中心区和主要的行政机构所在地，其主要职能是行政、文化和居住。所有建筑沿河而建，水成为了山体和河流间农田形成的供给者。沿河的带状公园将成为首都的一处象征空间和生态之桥的开端。

：这里紧邻 A 地区并有通往大邱和唐津的公路，将布置清洁的高科技工业城、水上乐园和办公楼等设施。

：覆盖了南部的大多数区域，通过生态之桥和滨水公园将南部相连。这里将建设两所大学、居住区和商业区等。

：包括了怪花山的水体部分，依山将建设小规模和生态性的住宅。水系将被改造成环形的滨水公园。

：包括春木谷山和露积山地区。此地环境优美，与市中心距离适宜，可以布置一些教育机构。规划中将在此建设一所大学和部分住宅。

：该地区与 E 地区隔美湖川相望，能够方便地到达高速铁路站，可以布置一些清洁的高科技工业园。

：位于 F 和 I 地区之间，将作为居住相关用地。

地区：I 地区的北侧部分将建有体验自然水景的生态公园和教育机构。通过生态廊和步行道与市中心的环形建筑相连，并与南部的水系共同组成城市的外围生态循环系统。

无题

Min, Yong Il（韩国）

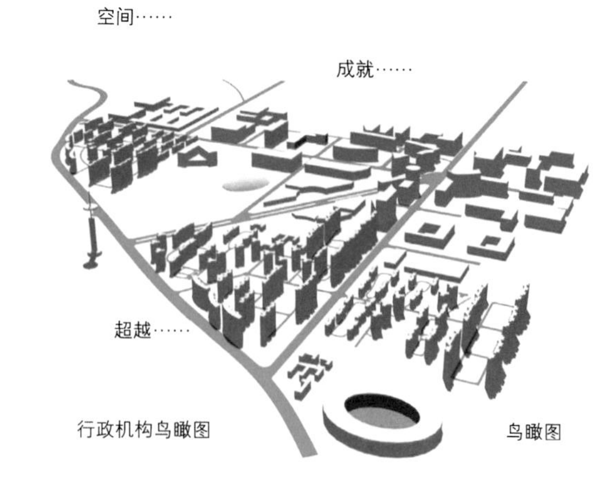

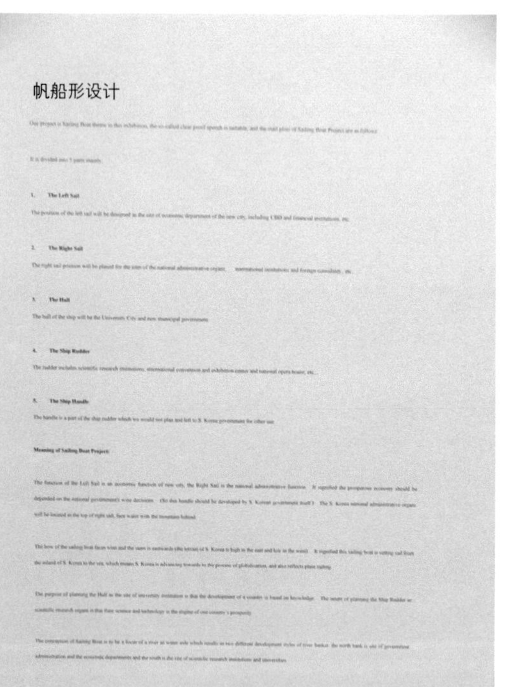
帆船形设计

1. The Left Sail
2. The Right Sail
3. The Hull
4. The Ship Rudder
5. The Ship Handle

Meaning of Sailing Boat Project

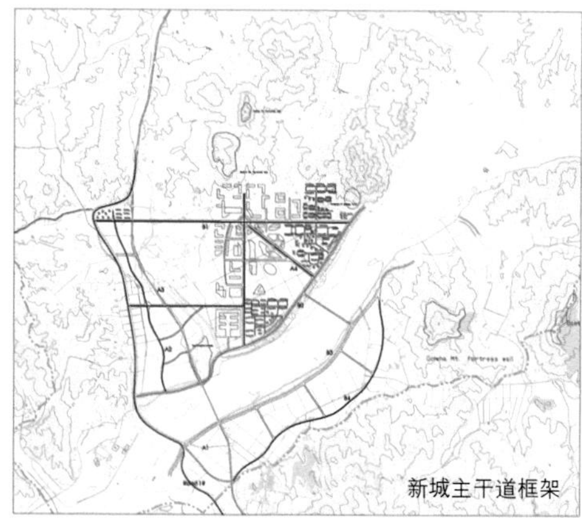

新城主干道框架

新城功能分区

中央商务区规划 鸟瞰图

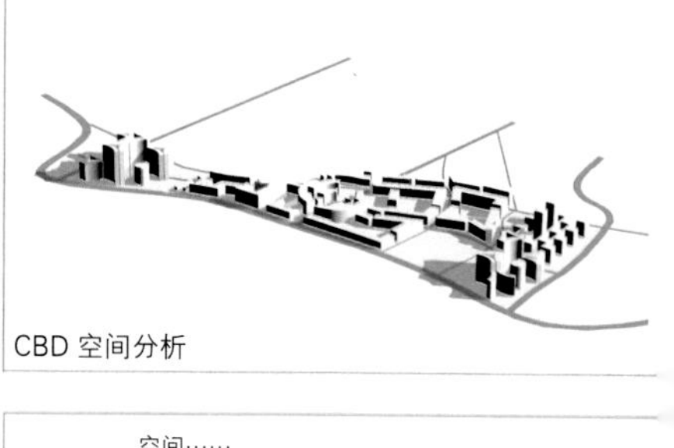
CBD 空间分析

行政区规划设计

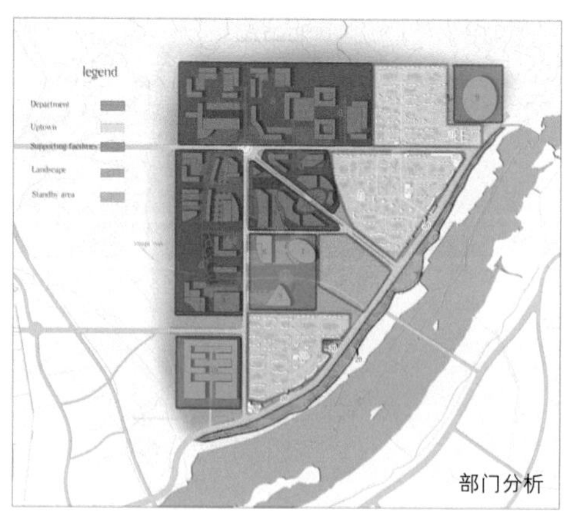

部门分析

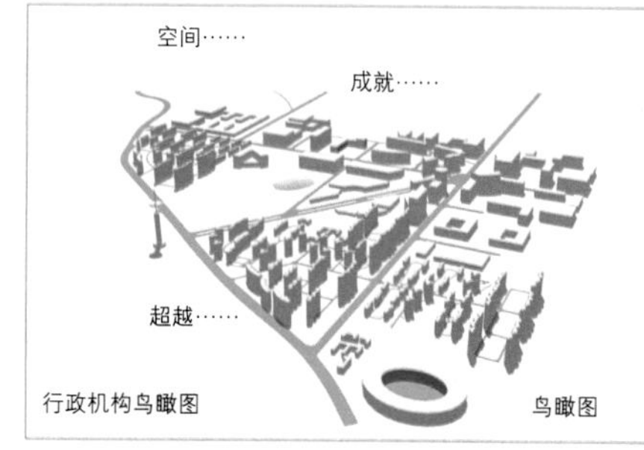

无题

Mattuiss, Bouw（荷兰）

NL Architects/Matton, Ton/Malkit, Shoshan

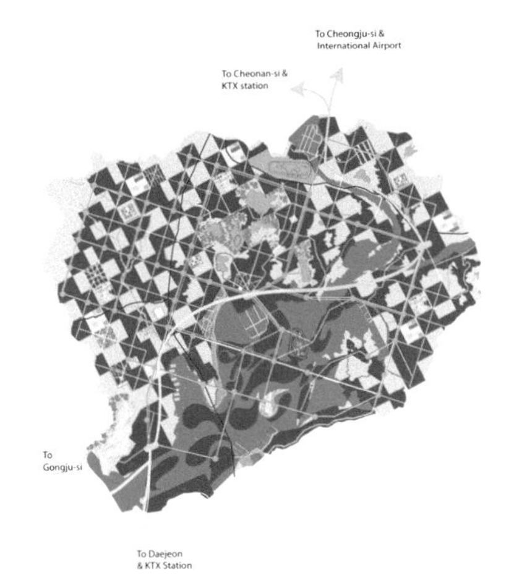

韩国行政中心城市

高效、开放、健康

MB 0605454

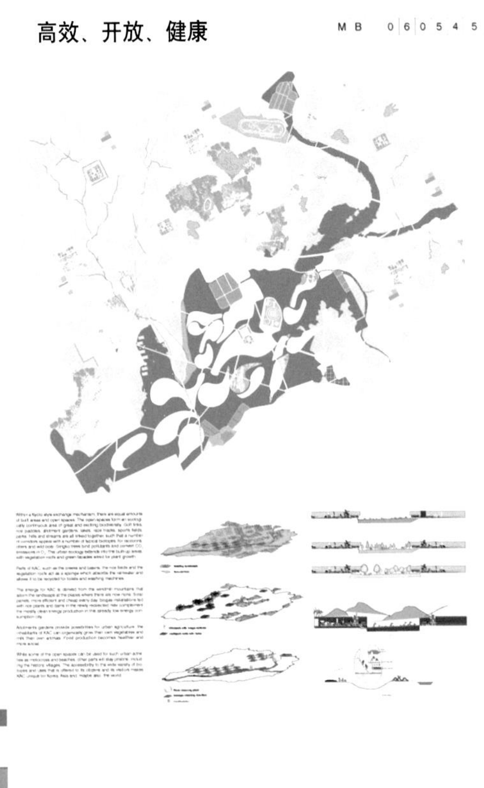

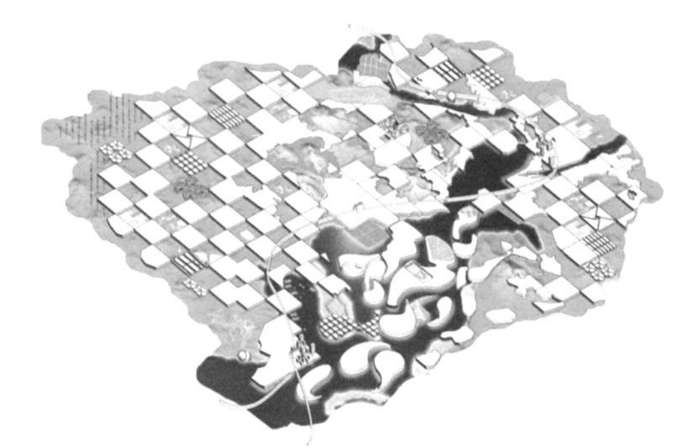

建成区的开敞空间构成了生态补给系统

无题

Canovas, Andres（西班牙）
Nicolas, Maruri/Amann, Atxu/
Lopez Fernandez, Ana/Bravo, Mauro

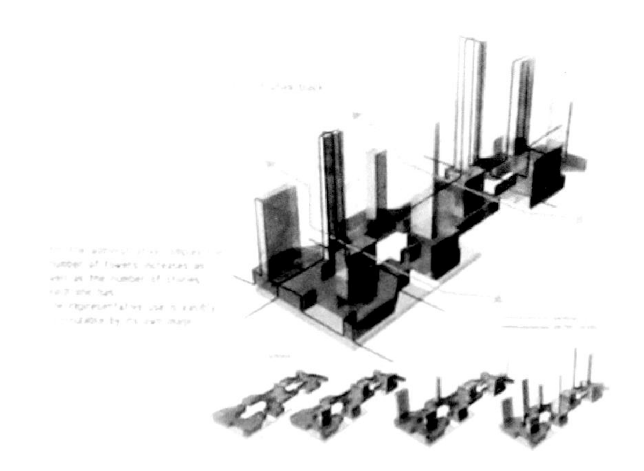

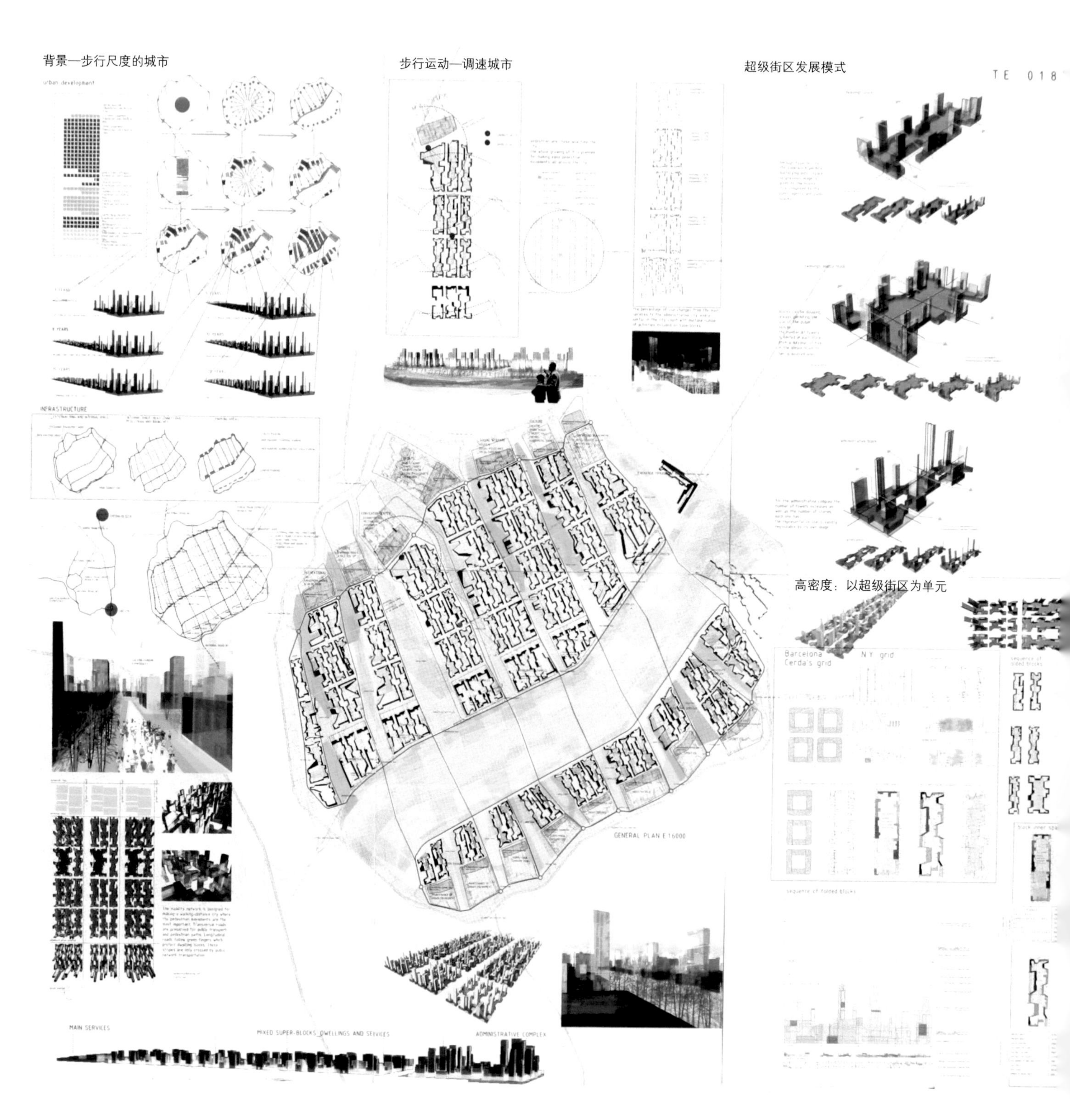

无题

Sommer, Richard（美国）
Miller, Laura/Kwon, Mee Hae/
Modesitt, Adam/Dan, Adams

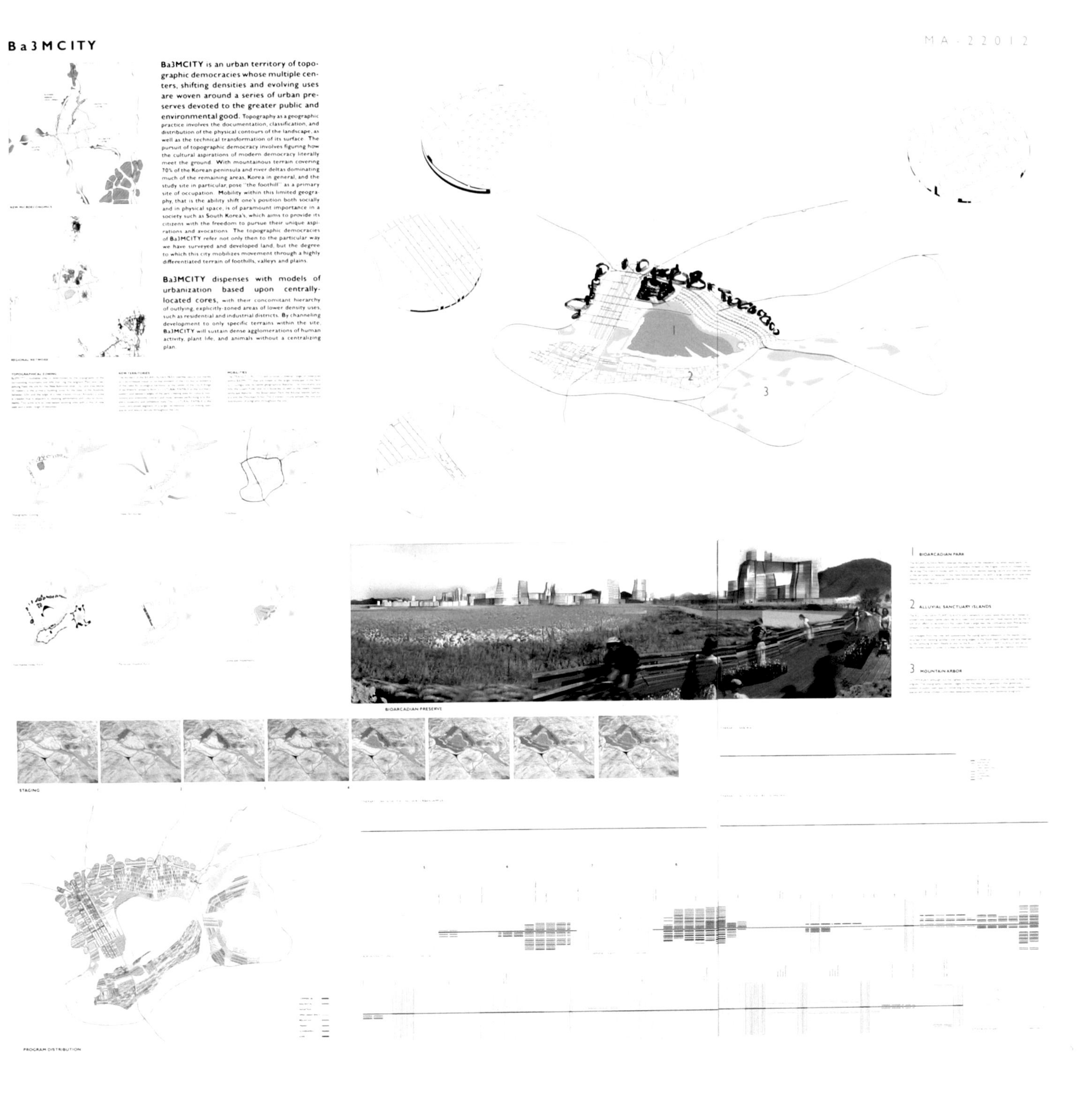

Ba3MCITY
MA-22012
Ba3MCITY is an urban territory of topographic democracies whose multiple centers, shifting densities and evolving uses are woven around a series of urban preserves devoted to the greater public and environmental good. Topography as a geographic practice involves the documentation, classification, and distribution of the physical contours of the landscape, as well as the technical transformation of its surface. The pursuit of topographic democracy involves figuring how the cultural aspirations of modern democracy literally meet the ground. With mountainous terrain covering 70% of the Korean peninsula and river deltas dominating much of the remaining areas, Korea in general, and the study site in particular, pose "the foothill" as a primary site of occupation. Mobility within this limited geography, that is the ability shift one's position both socially and in physical space, is of paramount importance in a society such as South Korea's, which aims to provide its citizens with the freedom to pursue their unique aspirations and avocations. The topographic democracies of Ba3MCITY refer not only then to the particular way we have surveyed and developed land, but the degree to which this city mobilizes movement through a highly differentiated terrain of foothills, valleys and plains.
Ba3MCITY dispenses with models of urbanization based upon centrally-located cores, with their concomitant hierarchy of outlying, explicitly-zoned areas of lower density uses, such as residential and industrial districts. By channeling development to only specific terrains within the site, Ba3MCITY will sustain dense agglomerations of human activity, plant life, and animals without a centralizing plan.
1
2
3
1 BIOARCADIAN PARK
2 ALLUVIAL SANCTUARY ISLANDS
3 MOUNTAIN ARBOR
BIOARCADIAN PRESERVE
STAGING
PROGRAM DISTRIBUTION

无题

Solid Arquitecnra S.L.（西班牙）
Maroto Raros, Frsncisw Javies/
Soto Aguirre, Aluaro

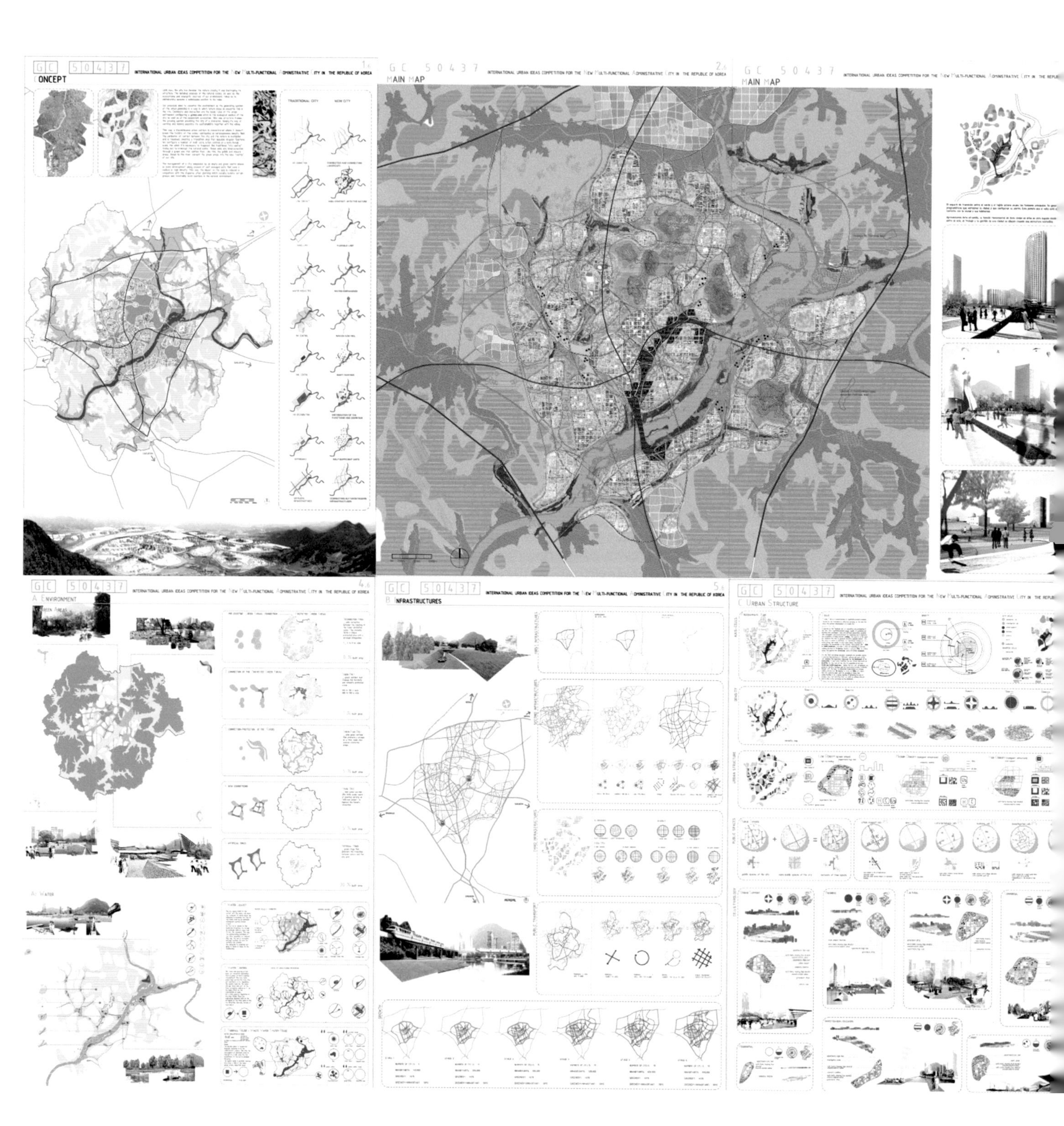

韩国首尔迁都规划竞赛作品集
International Urban Ideas Competition for the New Multi-functional Administrative City in the Republic of Korea

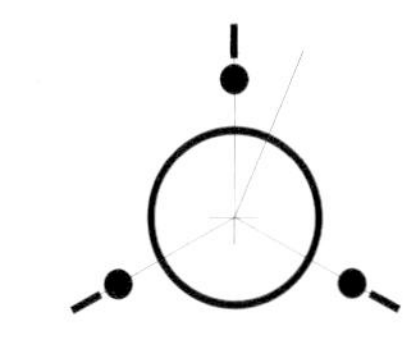

附录
APPENDIX

国际城市概念竞赛规章

介绍

大韩民国位于新兴的东北亚地区，目前正规划建设新复合型行政中心城市来促进国家在21世纪的持续发展和创新增长。这一构思既有利于实现国家“协同发展”和“跨越前进”的政策目标，又有助于建设韩国的美好未来。

在20世纪的下半叶里，韩国经历的快速城市化、工业化和经济增长导致了国内政治、经济、产业以及人口向首尔都市圈的过度集中。然而，与超负荷的首尔都市圈形成鲜明对比的是国内其他地区异常缓慢的发展。这种与日俱增的发展不均衡性加剧了地方民众的不满情绪，并导致了国家竞争力的日益下降。为了扭转这种局面，政府推出了三项主要政策来促进首尔都市圈与地方的协同发展：(1) 将核心的公共机构布置到其他地区；(2) 赋予地方政府更多的自治权；(3) 建设新复合型行政中心城市。

建设新复合型行政中心城市是以上三项政策中最为核心的部分，这必将作为全新的国土整治模式而被载入史册。这座占地73.14km^{2}能容纳50万人口的新城将由作为城市中枢的中央政府部门及其周边的产业、教育与文化等设施共同组成。

我们希望这座新城能够创造出良好的实际效果并对其他城市产生积极的影响，使其成为城市营造中的创新之举和典范之作。怀着这样的期望，特此举办国际城市概念设计竞赛来征集可以达到甚至超过我们期望的既具创造性和想象力又便于实施的城市概念设计方案。

这座新城为设计师们提供了一展才华的绝好机会，同时您的创意和灵感可以开拓国际城市规划与设计的全新视野。我们诚邀您前来支持并参与此次竞赛，以此展现您的创意和激情。

第1章
竞赛目标

■ 针对新复合型行政中心城市举办此次国际城市概念设计竞赛的目的是为了获得优秀和富有创新精神的城市规划与设计理念，并使它成为21世纪新城建设的的范例。

■ 我们希望通过此次竞赛可以激发在城市设计创新和城市文化等领域的积极探讨，同时热切期望竞赛中能够涌现出适应未来全球城市发展的各式方案。

第2章
竞赛规则

1. 竞赛类型

■ 这是一次公开的国际城市概念竞赛。

2. 语言表达

■ 此次竞赛的官方语言是韩语和英语；

■ 翻译中如若发生异议，将以韩语表述为主，英语表述为辅；

■ 所有提交的文件均应采用英文书写。

3. 组委会和主办方

■ 此次竞赛的主办方是复合型行政中心城市建设推进委员会，联系方式如下：

- 联系地址：韩国首尔钟路区世宗路77−6号　中央政府综合办公楼615号　复合型行政中心城市建设推进委员会　邮编：110−760
- 联系电话：+82−2−3703 −3569
- 传真号码：+82−2−3703−3492

4. 竞赛日程

■ 竞赛发布日期：2005年5月27日

■ 报名时间：2005年6月1日～7月11日

■ 参赛细则发布日期：2005年7月12日

■ 咨询与解答

• 提交问题日期：2005年7月12～25日

• 问题答复日期：2005年7月29日

■ 作品提交日期：

2005年10月18～25日

• 通过国际邮政或快递渠道提交作品时，邮戳日期不应迟于2005年10月25日，并于2005年10月31日17点（GMT+9）之前投递至参赛办公室方为有效。

■ 评审委员会评选日期：

2005年11月11～14日

■ 获奖方案公示日期：

2005年11月15日

5. 评审委员会

5.1 评审委员会成员

■ 评审委员会由7名委员（包括3名韩国人士和4名外籍人士）组成。

■ 当评审委员会的常务委员因故缺席时将由两名候补委员（包括1名韩国人士和1名外籍人士）接替其职位。

■ 主办方邀请的评审委员会成员名单如下：

• 7名常务评委：

Min, Hyun Sik—建筑师／教授（韩国）

Ohn, Yeong Te—城市规划师／教授（韩国）

Yoo, Kerl—建筑师／教授（韩国）

David Harvey—教授（美国）

Arata Isozaki—建筑师（日本）

Winy Maas—建筑师（荷兰）

Dominique Perrault—建筑师（法国）

• 两名候补评委：

Park, Sam Ock—教授（韩国）

Nader Tehrani—建筑师／教授（美国）

5.2 评审委员会会议

■ 评审委员会须在首次评审会中选举出负责人。

■ 如果有常务评委缺席首次评审会即失去评委资格，由候补评委予以替代。

■ 评审委员会的决议须通过投票的方式进行表决，并采纳占多数票的意见。当有常务评委短时间缺席时将由候补评委临时接替其职位。

■ 相关专业顾问将在评审期间全程配合评委们的工作，但没有投票表决权。

6. 专业顾问

■ 专业顾问简介

• Ahn,Kun Hyuck博士，首尔国立大学国内城市与地理系统工程学院教授

• 联系电话：+82－2－880－8200

• E－mail：ahnkh@snu.ac.kr

■ 专业顾问担负着确保竞赛如期完成、报名指导、咨询答疑以及将参赛作品匿名等职责，同时还监管技术委员会，协助评审委员会的工作。

■ 技术委员会负责对提交的作品进行验收，确保作品符合参赛规则的要求。

7. 参赛资格与奖项设置

7.1 参赛资格

■ 任何符合条件的个人或公司不论是以个人名义还是团体名义均可报名参加此次竞赛。

■ 为确保赛事公平，属于下列情况的个人将不具有参赛资格。

• 竞赛主办方的成员或职员；

• 评审委员会成员（包括常务评委和候补评委）、专业顾问以及技术委员会成员；

• 为赛事提供技术支持的韩国国土研究院职员；

• 韩国土地会社（赛事推广及管理企业）职员。

7.2 奖项设置

■ 为参赛方设置的奖项名额最多为5个（包括个人和团体），团队成员组成形式如下：

奖项设置的最大名额

组别	个人	团体
个人	5	—
个人＋团体	3	2

■ 如果参赛者以个人+团体的形式参赛，那么他/她可以以个人名义获得奖项。

■ 所有参赛团体必须选出一名负责人，团体获奖时其负责人可以以个人名义获得特别奖。

8. 报名

■ 报名日期：2005年6月1日上午10点~7月11日下午5点(GMT+09)

■ 报名地点：竞赛官方网站

■ 报名方式与费用：

参赛方须参照报名须知在互联网上进行信息录入，并通过信用卡缴纳报名费100美元(100000韩元)。参赛者以本人邮箱地址作为用户名，获取密码后可于2005年7月12日通过竞赛官方网站下载竞赛细则。

9. 竞赛详细信息发布

■ 所有已报名的参赛方可于2005年7月12日上午10点(GMT+09)之后凭用户名和密码通过竞赛官方网站下载以下信息：

①（必要的）竞赛附加条款；

② CAD格式的规划底图（比例尺1:25000）；

③图片格式的其他有关图纸（韩国地图、忠清道区域图、周边环境图、项目基地图）；

④项目基地照片：航拍图、模型照片、不同位置的人眼视角照片等；

⑤项目基地视频影像；

⑥项目基地信息；

- 区域概况：通达和环绕该地区的交通（高速公路、国道、高速铁路、铁路、国际机场），基础设施干线和主要地形地貌等；
- 自然环境气候（季节特征、风、温度、湿度、降水量）、地形、植物种类、地质状况、水系等；
- 人文环境：人口、经济区和历史遗迹等。

⑦登记表（随作品一起提交）。

10. 咨询与解答

■ 咨询时间：2005年7月12日上午10点~7月25日下午5点(GMT+09)。

■ 答复时间：2005年7月29日上午10点（GMT+0 9)。

■ 咨询方式：竞赛官方网站(http://competition.macc.go.kr)。

■ 任何有关竞赛事宜的咨询须提交至官方网站的“咨询与解答”页面。

■ 所有咨询须以英文或韩文的形式提交。

■ 所有咨询的解答均通过官方网站予以回复。

■ 所有问题的答复可被视为对已发布竞赛规则的补充。

11. 作品提交

11.1 成果要求

■ 以下为提交作品的必含内容：

①设计说明书［不超过50页的A4(210mm×297mm)纸打印文件20份］；

②作品展板［六张A1(594mm×841 mm)尺寸的展板］；

③内含设计说明书和不低于300dpi分辨率作品方案的CD光盘一张；

④登记表（含复印件1份）

■ 所有作品文件须用英文表述，测量单位采用公制。

11.2 成果内容

11.2.1 设计说明书

■ 设计说明书用A4纸打印，包括封面和分隔页在内不得超过50页，一共提交20份。如果因图示需要而含有A3的纸张，那么每页A3纸将按两页A4纸来计算。

■ 设计说明书须沿左侧装订。

■ 虽然参赛方可以在设计中自由表述创意构思，但我们建议参赛方能够在设计说明的阐述中包含以下内容：

- 新行政中心的价值取向及其理论依据是什么？
- 新行政中心应该拥有怎样的城市形态和结构？
- 为实现作品提出的城市形态和结构形式，现有环境将作何种改变？
- 描述新行政中心的总体景观意象。

人口密度与建筑群如何布置与控制?

- 城市采用何种居住形式?
- 居民如何沟通交往?
- 如何组织公共开敞空间系统?
- 主要交通运输方式及其运营管理模式是什么?
- 为实现城市的可持续发展应建设怎样的基础设施系统?
- 采取怎样的城市规划策略来展现文化传统?
- 对城市地下基础设施有何建议?
- 方案执行阶段应首先考虑什么问题?
- 针对新行政中心的其他意见。

■ 设计说明书的封面上只允许标注个人身份代码（图 1），不得有任何其他字母、数字或插图出现。

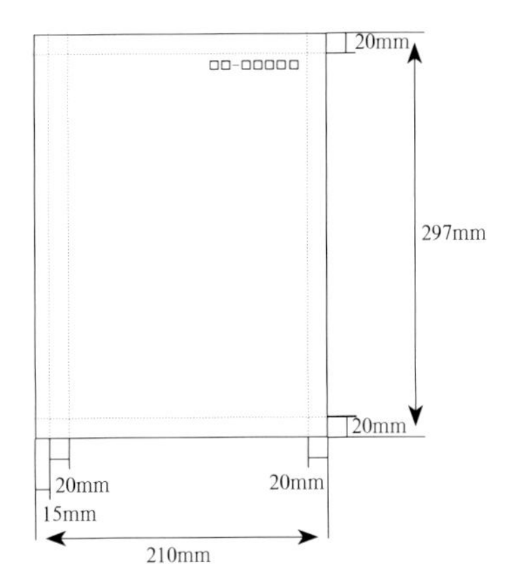

图 1　设计说明书封面形式

■ 除封面以外的正文和图表格式可由参赛方自行确定，但要注明页码。

11.2.2　方案展板（6 张 A1 纸）

■ 需要提交 6 张 A1 纸张（594mm × 841mm）大小的展板。

■ 展板内容没有特殊的版式要求，参赛方可以自由选择能够准确表达构思的版面形式，例如可自由布置规划总平面图、概念图和示意图等。

■ 展板本身要求轻质、抗压、坚固和平坦，边缘处不得绘制边框。

■ 展板需用高强度纸张进行外包装以保证作品安全，参赛方的个人身份代码应在包装纸上标明。

■ 评委们在评审方案时，展板将按照（图 2）的排列形式进行组合展示，参赛方的个人身份代码应标注在 3 号板上方的恰当位置，每张展板的序号要用阿拉伯数字写在板的背面。

11.2.3　CD 光盘

■ 为方便出版和使用，以下电子文件应附于提交的 CD 光盘中：

①显示 6 幅展板全貌的电子图片 1 张；

②设计说明书：A4 纸张大小的 Ms-Word 和 Adobe Acrobat 格式电子文档；

③展板上的所有图像文件。

■ CD 光盘应装于一个塑料光盘盒中，并按（图 3）所示标明参赛方的个人身份代码。

■ 图片格式应能被目前常用的图形软件识别（如 JPEG、TIFF、PSD）。图片分辨率不应低于 300 dpi（用于打印 A4 纸张大小的出版物）。

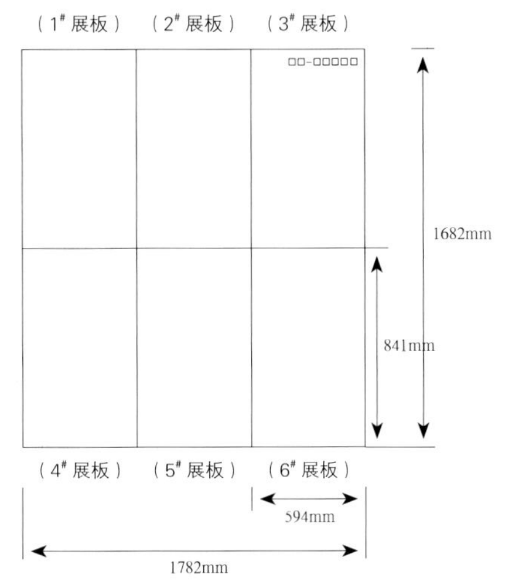

图 2　展板排列形式

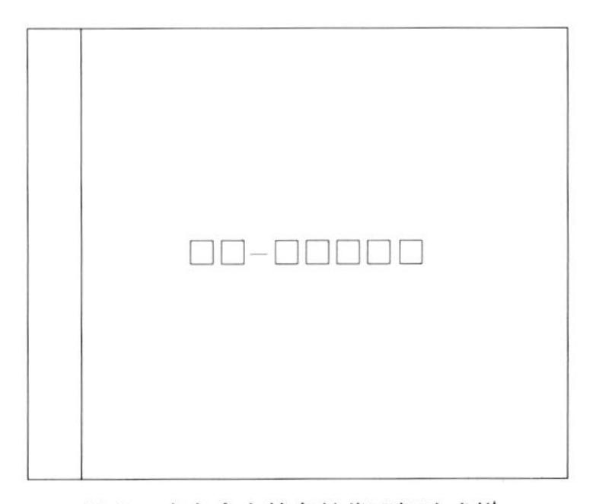

图 3　光盘盒上的身份代码标注式样

11.2.4　登记表

登记表应装于 A4 纸张（210mm × 297mm）大小的白色信封内，信封上注明参赛方的个人身份代码。

■ 登记表用 PDF 或 MS Word 文档格式打印，内容全部使用英文，可以手写或用 MS Word 输入。签名请用钢笔书写。

■ 所有参赛方都要签署登记表，作为接受竞赛条款和相关规则的凭证。

□□-□□□□□

图 4　A4 信封上个人身份代码标记示意

11.3　提交方式

11.3.1　提交日期与方式

■ 提交日期：2005 年 10 月 18 日上午 10 点至 10 月 25 日下午 5 点（GMT+09）

• 通过国际邮政或快递方式提交作品的邮戳日期不得迟于 2005 年 10 月 25 日，并于 2005 年 10 月 31 日下午 5 点（GMT+09）之前寄到竞赛办公室方为有效。

■ 提交方式：送交或邮寄

■ 提交地点

• 地址：韩国京畿道安阳市东安区官阳洞1591-6　韩国国土研究院（KRIHS），“国际竞赛组”　邮编 431-712

• 电话：+82-31-380-0555

• 传真：+82-31-380-0554

• 电邮：macc@krihs.re.kr

11.3.2　提交作品的其他问题

■ 为避免作品在经过韩国海关时缴纳进口关税或产生不必要的耽搁，所有国外作品都应注明“非商业用途”。

■ 主办方对作品在提交至指定地点之前发生的遗失或损坏现象概不负责。

11.4　匿名方式

11.4.1　个人身份代码（PIN）

■ 为满足作品匿名要求，所有参赛方须在登记表信封、3# 展板右上方（不被方案内容覆盖处）以及 CD 盒三个位置标明个人身份代码。参赛方可以选取一组由任意 2 个英文字母和 5 位阿拉伯数字组成的字符串作为个人身份代码。

例如：AB—12345

■ 宣布竞赛获奖名单之前参赛作品将一直保持匿名状态。

■ 个人身份代码需用黑色钢笔书写，不得有任何其他符号或标记。有关个人身份代码的详细信息请参照附录一。

11.4.2　个人身份代码管理

■ 作品提交后所有参赛方的个人身份代码将被提交编号取代。如果有两个或多个参赛方选用了相同的个人身份代码，主办方将有权单方面更换重复的个人身份代码。内含登记表的信封将被装入一个经重新编号的信封内，并且在竞赛结束之前一直保持密封状态。

12. 奖项设置

■ 为入围的获胜方案设置有以下奖项：

• 一等奖（1 名）：奖金 20 万美元

• 二等奖（2 名）：每名奖金 10 万美元

• 三等奖（3 名）：每名奖金 5 万美元

• 荣誉奖（4 名）：每名奖金 2 万美元

■ 作为一次概念性设计竞赛，此轮获奖者不代表由此获得新复合型行政中心城市后续实施方案的设计权。

■ 如有必要，主办方会与获奖者签订聘用合同，委托其担任新复合型行政中心城市总体规划咨询师。

■ 一、二和三等奖的获得者可能有机会参与新复合型行政中心城市的深化设计竞赛。

■ 所有参赛团体或个人均可获得纪念品一份和内含获奖方案的 CD 光盘一张。

■ 获奖者在领取奖金的同时也代表将其方案的版权提供给了竞赛主办方。

■ 所有奖金都将依照韩国税法进行征税。

13. 版权

■ 所有参赛团体和个人提交的成果都应是原创作品，不得侵犯他人的著作权。

■ 竞赛主办方保留直接或间接的使用或修改获奖方案的创意并应用

于新复合型行政中心城市规划设计中的权利。参赛者在签署登记表时即被认为同意了以上协议内容。

- 主办方拥有提交作品的所有权，并且不再返还给参赛方。
- 主办方保留将作品中的部分或全部内容、照片、图形和设计用于影印、出版、展览和商业等用途的权利。

14. 申诉

- 此次竞赛将严格遵守韩国的法律与法规，任何针对赛事提出的申诉将由依据韩国版权条例或韩国商务仲裁委员会的裁决规则成立的版权委员会(http://www.copyright, or. kr)通过协商和调解后进行裁决，未决的申诉将由首尔地方法院进行调解或仲裁。

第3章
项目概况

1. 项目背景

- 韩国始于20世纪60年代的快速城市化、工业化浪潮和高速的经济增长使首尔都市圈（SMA）承受了因行政、经济、人口和产业等的过度集中而带来的巨大压力。
- 当首尔都市圈因为人口和产业的过度集中而引发了诸如住房短缺、环境污染、交通拥堵和城市无序蔓延等大城市病时，其他地区的发展却异常缓慢，甚至停滞不前。这种国内发展的不均衡性已经开始影响国家的凝聚力和竞争力。
- 韩国政府自20世纪70年代早期以来则采取了一系列的措施防止首尔都市圈的过度集中并促进其他地区的发展，但成效甚微，首尔都市圈过度集中的发展状态一直持续到现在。
- 任由这样的势态延续下去将会严重阻碍国家的发展进程。因此，韩国政府紧急采取了力度更大的措施来解决因过度集中引发的社会问题，进而促进整个地区的协调发展。这些措施包括将公共机构转移到地方，营造整个地区内有机、分散的布局形式和建设新复合型行政中心城市等几个方面。
- 建设新复合型行政中心城市是以上措施中的关键所在。将目前位于首尔地区的大部分公共机构迁往忠清南道地区，以建成国家的另一个行政中心。通过此举可以在一定程度上改变韩国的空间结构体系。从根本上来说，这项计划的目标是通过改变国家的空间规划来创造新的国土管理模式。
- 韩国政府和民众热切期望新行政中心能够建成一座极富吸引力和自豪感的国际城市，并能在协调国内发展、提高国家竞争力和引领21世纪城市发展模式等方面起到积极作用。

2. 项目宗旨

新行政中心城市工程是要推动韩国在21世纪新兴的东北亚地区里的不断创新与发展，这也是实现国家“协调发展和跨越式增长”目标的一种有效途径，同时象征着韩国即将步入一个崭新的时代。韩国在经历了过去50年的政治和经济发展后，要在全球化和信息化的今天继续保持领先地位就必须明确新行政中心城市建设的以下四点宗旨。

2.1 在韩国中部建设一座具有“全球都市区”特点的“核心城市”。

- 新行政中心城市将成为韩国中部的一处“全球都市区”。包括新行政中心城市及其周边城镇在内的整个地区将激发出新的发展动力，并且承担起中央政府的大部分职能，例如国际交流、研发、新型都市产业和教育业等。基于以上背景，这一崭新的全球都市区将成为未来导向型的国家管理中心。

2.2 建设一座可以提高国内其他“城市地区”全球化水平的“催化型城市”。

- 新行政中心作为一座“催化型城市”将推动与国内其他城市地区的协调发展，共同组成两小时城市圈。新行政中心城市的开发既能给首尔都市圈和其他城市地区带来巩固自身优势和扩大发展空间的机遇，又可以提升它们在全球化背景下的综合竞争实力。

2.3 营造一座能够促进公共和私营部门协同合作的“复合型功能城市”。

■ 设有中央行政机关和相关公共机构的新行政中心城市可以使政府的办公效率提高到一个全新的水平，在克服僵化的官僚体制运作模式的同时，也可以推动经济活动的活跃发展。显然，新知识经济、教育、研发、信息产业和国际贸易等领域的多功能复合对新行政中心城市的成功发展起着至关重要的作用，同时也要意识到公共和私营部门的创新与协作所蕴含的巨大效力。

2.4 作为一个检验和宣传各项城市创新理论的“试验型城市”。

■ 新的行政中心城市将引发行政管理、信息产业、交通运输、商业交往、国际贸易、住房供给、资源保护、城市发展、基础教育、文化传统和社会服务等所有领域里的“城市化创新”。把创新的城市发展构想付诸于实践可以解决过去产生的一些失误，新行政中心城市在这里扮演着一个向其他城市宣传创新成果的“试验型城市”的角色，这在一定程度上可以改善城市未来的环境品质。

3. 城市景观

基于以上四点宗旨，我们对城市景观的要求可以概括为以下四个方面，它们将共同反映出社会、政治、科技和文化等因素在未来城市里的演变，我们希望通过提议富有创造性和想象力的设计理念来实现城市结构、形态、空间、环境、意象和各种活动等的良好视觉效果，并在文化、教育、社会服务、住房、运输、通信、商业和基础设施等领域也能满足同样的要求。

3.1 营造一座可以展示韩国在信息领域里适应性与灵活性等领先水平的“普适型城市”。

■ 新行政中心城市将确保通过完善的“普适型城市”理念在韩国城际之间处于领先地位，这不仅意味着是一种科技上的进步，而且是城市品质的根本性改善。一个具有适应性、开放性、灵活性和24小时可达性等特点的高速网络化城市需要一种全新的设计理念来促进新一轮的科技进步。我们相信这一革命性的变革将在新行政中心城市中得以充分体现。

3.2 营造一座让不同价值取向的市民和世界游客都能向往并认可的“人性化城市”。

城市内的居民和世界各地的游客都将体会到这座新城带给他们的喜悦以及对他们不同生活方式的尊重，例如接受他们在民主性、自发性、自主性、自由性、创造性、构想性、探索性、人际交往和文化习俗等方面的不同价值取向。新行政中心城市就是将这种“城市感受”以人文关怀的形式表现出来，进而提升市民们的归属感和对新城生活的认同感。我们相信，这种永恒的人文精神会在新城中得以完美展现。

3.3 发展成一座人与自然结合城市功能，达到可持续动态平衡的“进化型城市”。

■ 这将是一座保持城市文明可持续发展的“进化型城市”，城市的功能在任何发展阶段和成长周期里都可以保持平衡，促进自然环境与人工环境、城市与居民、用地与开发以及人类与动植物之间形成亲切宜人的关系。我们希望通过富有创新性和想象力的设计手法来实现进化型的生态城市价值理念。

3.4 通过对历史与未来富有创造性的阐释来展现一座在东亚“富有韩国城市特色的文化名城”。

■ 在21世纪初这一特定的时间，地处东亚的韩国建设这座新行政中心城市涉及到了“全球化背景下的韩国特色城市”这一文化议题。就目前来看，东亚和韩国在历史上有着极富哲学性和思想性的城市营造理念。我们希望将“韩国特色城市”的理念通过创造性、感悟性和文化性的方式融入到新行政中心城市的建设中来，而不是沿用常规或老套的做法。我们由衷地希望新行政中心城市能够丰富世界的城市文化历史。

4. 项目地点和范围

- 地点：韩国忠清南道燕岐郡（南面、金南面和东面）公州市（长岐面）附近地区
- 用地面积：73.14km1(7314hm^2)
- 人口指标：500000
- 为防止城市无计划地占用山林和农田，新行政中心周边 4 ～ 5km 范围内的 224km^2 土地将通过政府规划进行严格的控制。
- 城市职能：
  - 主要职能：国家行政管理。
  - 辅助职能：国家政策研究、国际文化交流、高等教育、科技创新、观光旅游、休闲度假及其他一些基本的城市功能（例如商业、办公和居住等）。

5. 进度安排

- 新复合型行政中心城市工程进度表：
  - 2005 年 3 月 18 日：颁布复合型行政中心城市特别法案（具有法律效力）；
  - 2005 年 5 月 24 日：确定并公布项目选址；
  - 2005 年 5 月 27 日：发布国际城市概念竞赛信息；
  - 2006 年上半年：确定新行政中心城市的总体规划方案；
  - 2006 年下半年：新行政中心城市功能发展规划；
  - 2007 年上半年：确定新行政中心城市实施方案；
  - 2007 年下半年：开工建设；
  - 2012 年：国家行政中心机构搬迁启动。
- 新行政中心城市将分阶段发展，并于 2030 年前后发展为稳定在大约 500 万人口的城市规模。

6. 城市职能

6.1 主要职能：国家行政中心。

- 新行政中心城市将容纳 12 个政府部门、6 个政府相关机构以及一些国家保障部门。预计这些机构将有大约 10000 名政府职员。
- 迁往新行政中心城市的政府机构名单请参照附录 II 。

6.2 辅助职能：国际公共政策研究、国际文化交流、科技创新和观光旅游等。

- 与国家行政部门密切相关的公共研究机构连同其几千名雇员将搬迁到新行政中心城市。
- 会议中心、国际机构和配套的文化设施（展览馆、剧院、美术馆等）可以完善城市的主要行政职能。
- 包括研究所和四年制大学在内的多层次教育机构。
- 采用新技术工艺的无污染工业可为城市提供新的就业机会。
- 新行政中心城市将为游客提供丰富的游览胜地及配套服务设施。

6.3 基本职能

- 与城市发展阶段和人口规模变化相适应的居住、商业、办公、教育、街区商业、公园以及其他城市公共服务设施等的规划建设。

7. 尊重基地现状环境

- 项目界线内的国道、地方道路、高压走廊及其他基础设施均可迁址、移动或重建。
- 交通运输网可以重新组织或连接到现有网络上。
- 参赛者可以对现有的住宅、工厂和其他设施提出搬迁或保留的建议。
- 项目界线内的历史遗迹和文物除了已注明“必须保护”的以外，均可搬迁或异地保护。
- “必须保护”条目以外的山地、丘陵、平原和河流等自然要素在经过参赛方的环境评估后可作适当调整。

附录 I
个人身份代码（PIN）格式

■ PIN 码由 2 位英文字母和 5 位阿拉伯数字组成。

■ 填写 PIN 码的表格线条宽度为 0.5mm。

■ PIN 码的格式如下所示：

1）注册信封和项目说明书

- 注册信封和项目说明书文本上的 PIN 码要填写在一个尺寸为 10mm × 80mm 的矩形表格内。

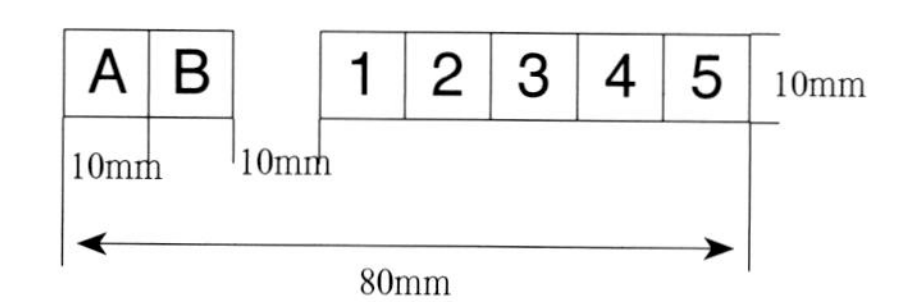

2）展板和包裹

- 展板上的 PIN 码要填写在一个尺寸为 20mm × 160mm 的矩形表格内。

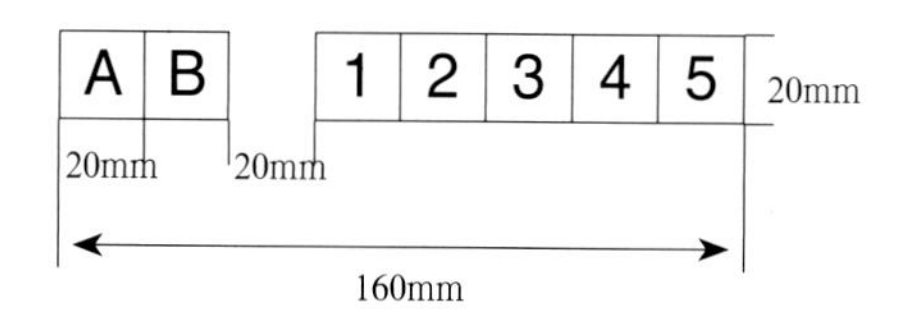

3）CD 光盘盒

- CD 光盘盒上的 PIN 码要填写在一个尺寸为 5mm × 40mm 的矩形表格内。

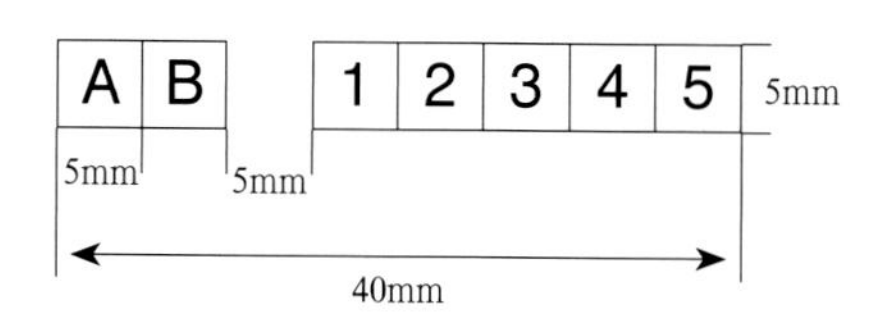

附录 II
迁往新行政中心城市的国家行政机构名单

类 别	机构名称	主要职能	雇员人数
总计			10234
总统委员会	中央人事委员会	管理国家公务员人力资源	243
	中小企业总统委员会	研究和协调中小型企业政策	29
	韩国反腐败独立委员会	预防政治腐败	171
总理下属机构	国务协调室	评估和协调政府政策	218
	总理秘书处	协助总理	82
	应急委员会(*)	编制全国性的突发事件处理计划	84
	公平贸易委员会(*)	制订竞争政策和公平贸易章程	360
	韩国检察院(*)	管理针对政府机关的投诉	92
	国家青少年委员会(*)	制订与实施青少年保护政策	54
	计划预算处(*)	制订国家预算政策	338
	政府立法院(*)	管理和协调立法计划	163
	政府信息处(*)	拓展在国内和国际中的政府管理	302
	爱国人士和退伍军人事务部(*)	制订爱国人士和退伍军人的政策	253
国家政府机构及下属部门	财政经济部	制订国家经济政策	727
	教育与人力资源部	制订教育和人力资源政策	495
	科技部	制订科学与技术进步政策	366
	文化观光部	制订文化与旅游政策	435
	农林部	制订农业与林业政策	514
	产业资源部	制订商业、工业和能源政策	747
	情报通信部	制订信息与通信政策	707
	健康福利部	制订公共健康和福利政策	477
	环境部	制订自然和环境保护政策	462
	劳动部	制订工人和老年人福利政策	479
	建设交通部	研究全国开发与住宅政策	934
	海洋水产部	制订海事、港口和渔业政策	516
	国税厅	征收税款	718
	国家应急管理委员会	管理全国突发事件	268

注：雇员人数统计数据截止到2004年12月。
(*)这些总理委任的机构由政府组织编制，但拥有独立职权。

1. 宣传 /Promotion

■ 国际竞赛传单 / International Competition Leaflet

■ 展览会传单 / Exhibition Leaflet

海报及相关图片 / Posters and Related Images

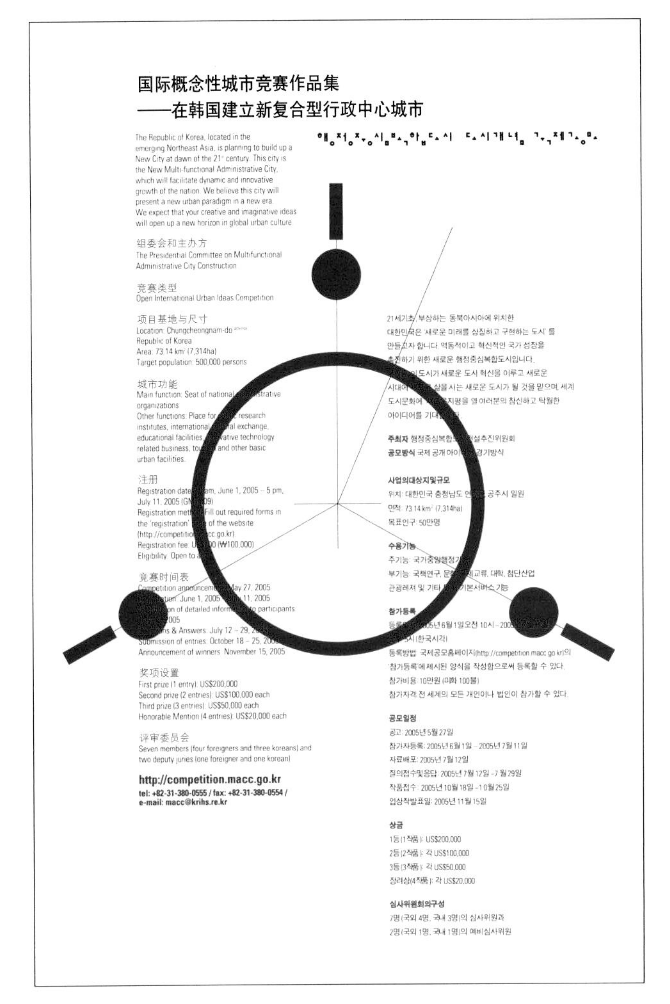

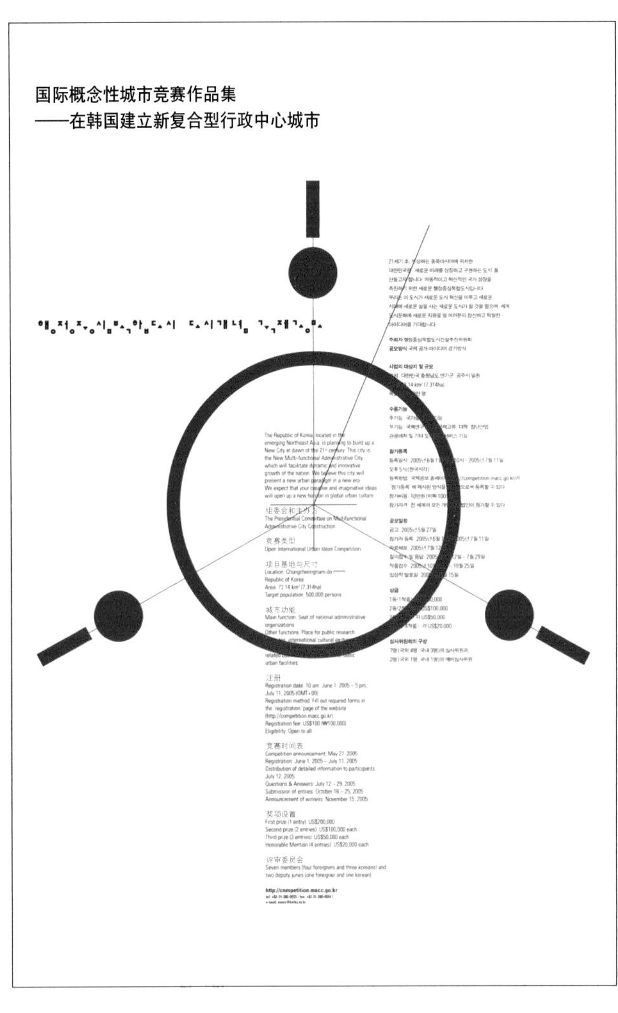

国际概念性城市竞赛作品集
——在韩国建立新复合型行政中心城市

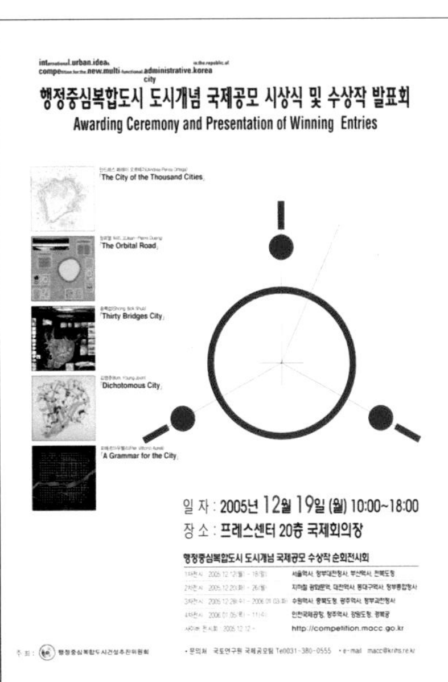

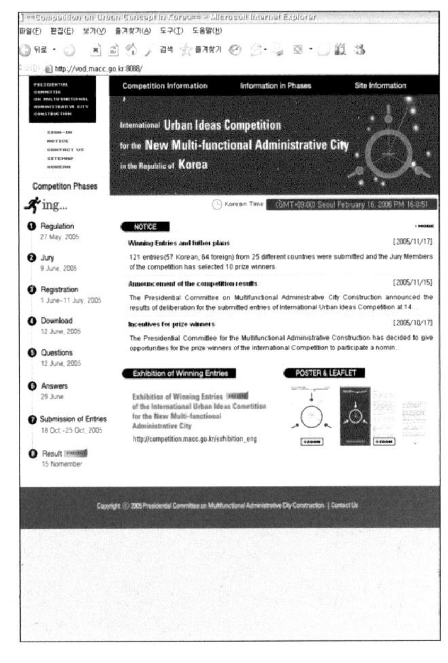

2. 注册数据 /Registration Data

■ 职业统计表 /Statistics per Occupation

	韩国（Korean）	外国（Foreign）	总计（Total）	(%)
建筑师（Architect）	90	134	**224**	63.8
城市规划师（Urban Planner）	10	30	**40**	11.4
教授（Professor）	21	7	**28**	8.0
学生（Student）	31	9	**40**	11.4
其他（Others）	17	2	**19**	5.4
总计（Total）	169	182	**351**	100

■ 年龄统计表 /Statistics per Age

	韩国（Korean）	外国（Foreign）	总计（Total）	(%)
20's	43	28	**71**	20
30's	45	62	**107**	30
40's	49	46	**95**	27
50's	23	23	**46**	13
60's	9	19	**28**	8
70's	0	4	**4**	1
总计（Total）	169	182	**351**	100

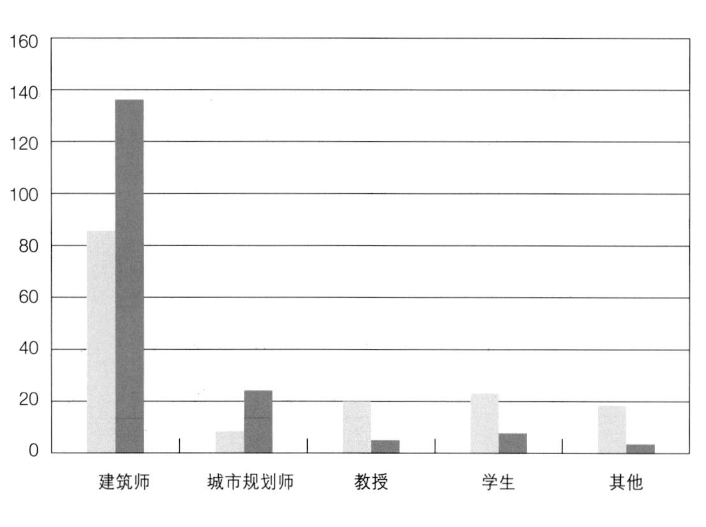

职业统计表

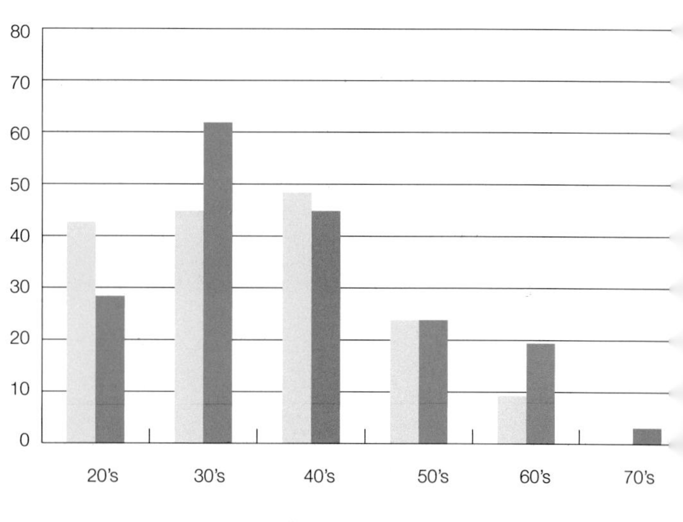

年龄统计表

3. 详细信息发布 /Distribution of Detailed Information

■ 基本信息 /Basic Information

1. **底图**

   CAD 格式的基地底图（比例尺 1:25000）

2. **其他地图**

   韩国地图、忠清道区域图、周边地区地图及项目地区影像档案

3. **项目地区图片**

   航拍图、模型图及不同位置的人眼视角照片

4. **项目地区视频剪辑**

5. **基地信息**

   – 区域信息：通达或环绕项目地区的交通（高速路、国道、高速铁路、铁路、国际机场）、基础设施主干线和关键的景观要素；

   – 自然环境：气候（季节特征、风、温度、湿度、降水量）、地形、植物种类、地质特点和水系；

   – 人文环境：人口、经济区和历史遗迹等。

6. **登记表（随作品一起提交）**

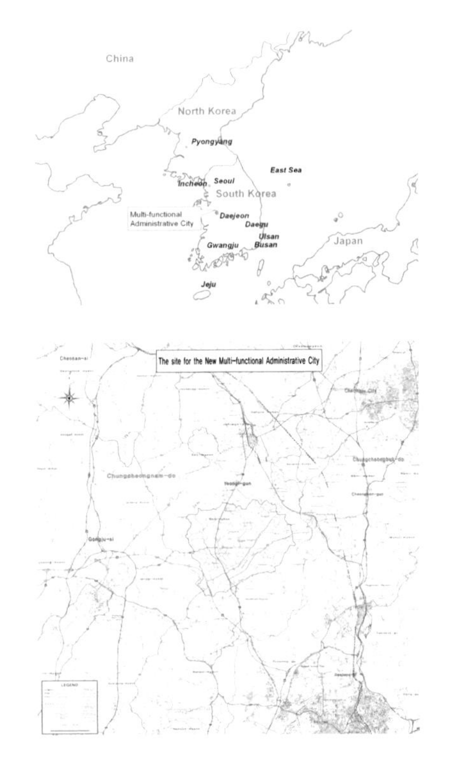

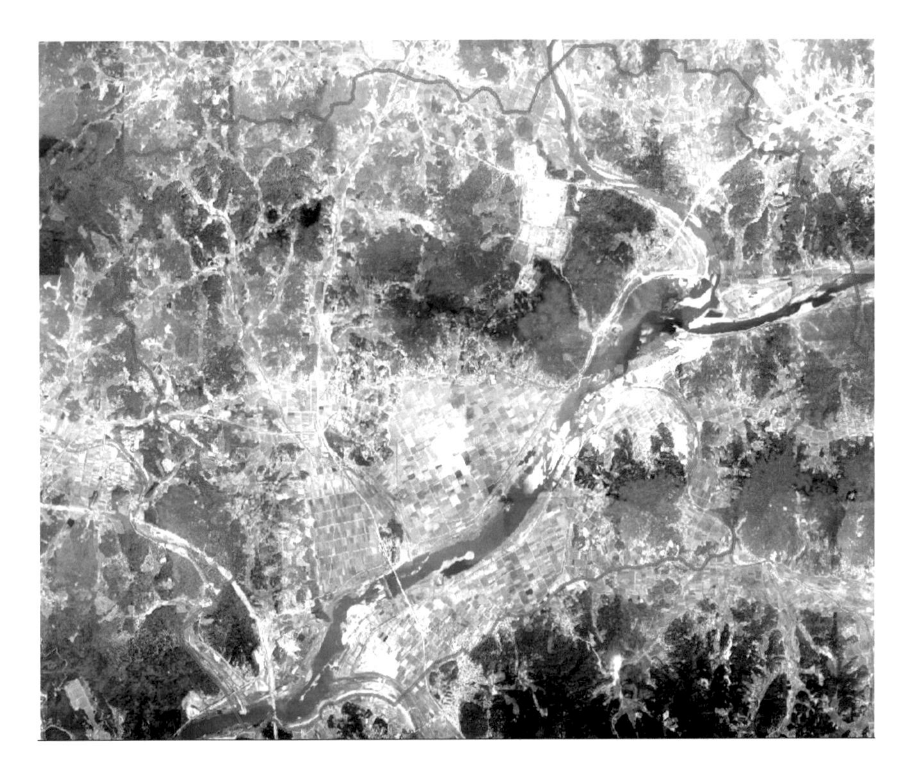

■ 附加说明 / Additional Information

1. **咨询与解答**

2. **底图**

   (6. 附带底图)

3. **项目地点及其周边地区地图**

   (4. 项目地点及其周边地区地图)

4. **其他地图与信息**

   (1. 交通图 /2. 高速路 /3. 忠清道区域图 /5. 等高线图 /7. 地图格式与字体 /8. 场地断面 (金江) /9. 场地照片 /10. 周边地区信息)

○ 补发资料清单 (2005.7.29)

연번	구분		내용	파일명
1	도로망현황 그림	수정	분석자료 <Figure 1> 수정 북유성인터체인지와 동공주인터체인지 표시, 북유성 IC→JC로 수정, 대상지 외곽라인표시	도로망현황 그림.jpg
2	고속도로현황 표 (분석자료 1페이지)	수정	대전당진 고속도로의 대상지접근 I.C 수정(북유성인터체인지 → 유성인터체인지, 동공주인터체인지)	고속도로현황 표 (분석자료 1페이지).pdf
3	충청권 지도	수정	Map of Chungcheong Region.jpg 수정 (반경 50km, 100km 표시선 추가)	충청권 지도.jpg
4	대상지 및 주변지역지도	수정	Map of periphery area and project site.jpg 수정 (Exit(Interchange)위치 추가)	대상지 및 주변지역지도.jpg
5	대상지 주변의 제방높이 분석자료 <표 1>	수정	영문표기 수정 The altitude of the Geum Riverbed is 0.06~6.73m → The waterdepth of the Geum Riverbed is 0.06~6.73m	대상지 주변의 제방높이 분석자료 <표 1>.pdf
6	Base Map	수정	CAD파일에 GIS자료 포함 (주요 레이어 : GIS-Conservation-region : 산성, 보호수목 GIS-Site-area : 예정지역, 주변지역 경계 GIS-Village-hamlet : 기존취락지구 GIS-Ecologic_1 : 자연생태 1등급)	Add_Base Map(2000 Kor).dwg
7	Plot styles & Font	신규	CAD용 CTB파일과 FONT	Plot styles & Font.zip
8	대상지 단면도(금강주변)	신규	금강주변에 대한 대지단면도와 키맵 (세군데)	대상지 단면도 (금강주변).jpg 대상지 단면도 (금강주변) 키맵.jpg
9	대상지사진	신규	대상지 사진 추가	대상지사진.zip
10	주변지역정보	신규	충청북도, 충청남도, 대전시 엑셀파일 정보	주변지역정보.zip

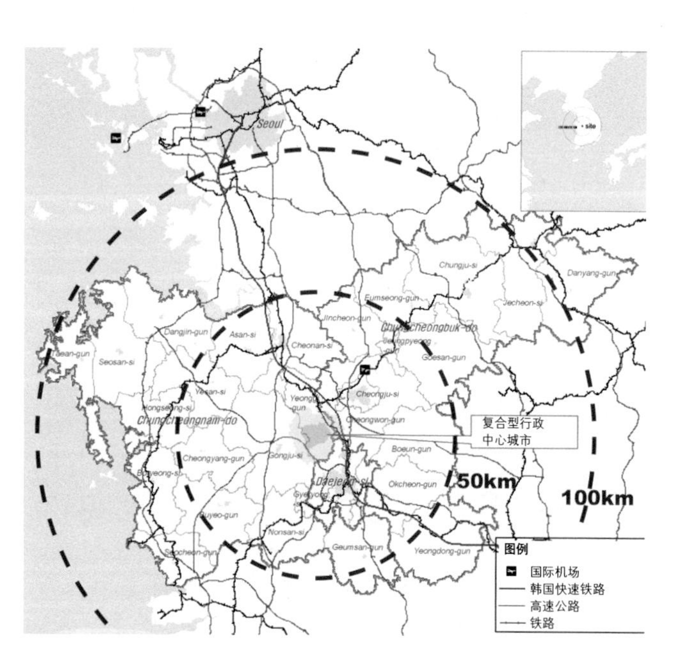

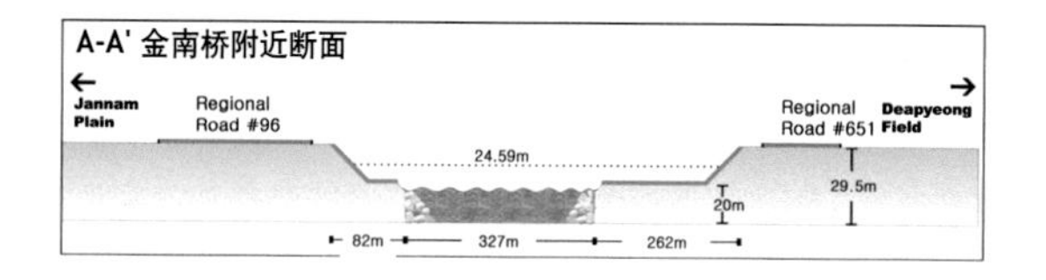

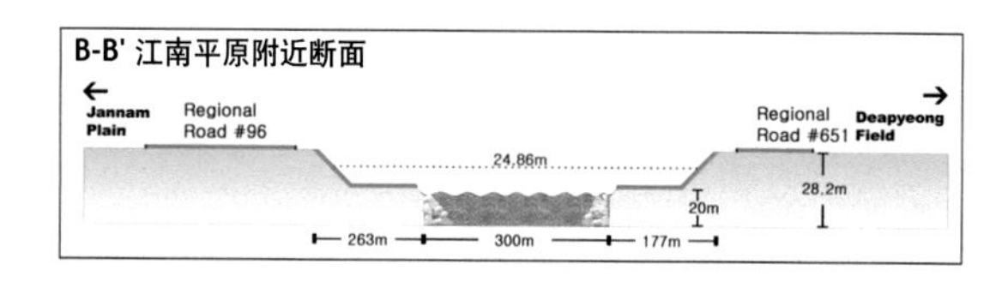

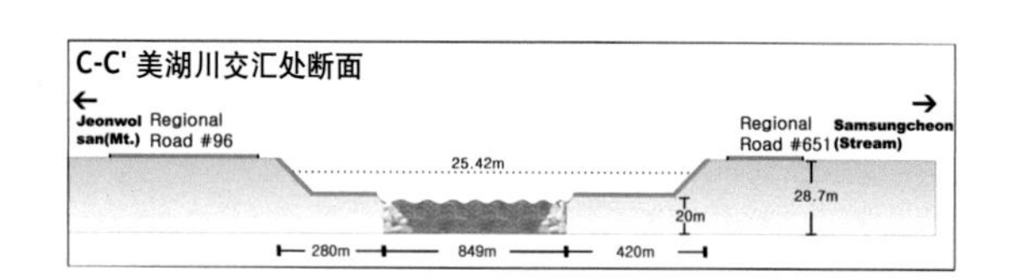

4. 现场调研 / Site Trip

■ 第一次现场调研：2005. 7. 29（24 人参与）

■ 第二次现场调研：2005. 8. 12（32 人参与）

■ 第三次现场调研：2005. 8. 27（51 人参与）

■ 行程安排 / Time Schedule

09:30 ~ 12:20：从国土研究院出发前往大田站（玉山停歇站：午餐）

12:30 ~ 13:20：从大田站前往金南小学

13:20 ~ 13:40：测量数据并介绍复合型行政中心城市

13:40 ~ 14:00：抵达元帅山入口处
参与者可在 A 组和 B 组之间任意选择

14:00 ~ 16:00：A 组—攀登元帅山，介绍基地情况
B 组—乘坐大巴考察基地

16:00 ~ 17:00：A 组和 B 组汇合，进行基地（金江与美湖川交汇处）考察与介绍

17:00 ~ 18:00：返回大田站

18:00 ~ 20:00：从大田站返回国土研究院（晚餐）

5. 方案提交与评审 / Submission of Entries and Deliberation

10月31日

■ 方案提交与确认 / Submission and Confirmation of Entries

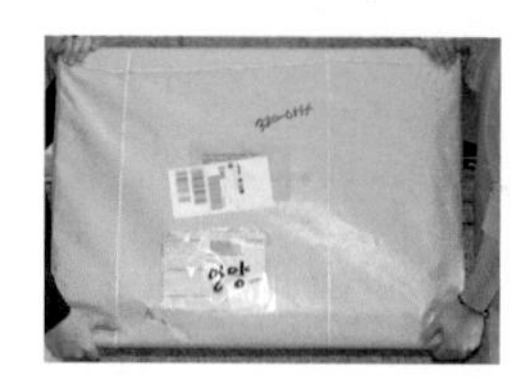

11月1日

■ 排列展板 / Panel Work and Arranging the Entries

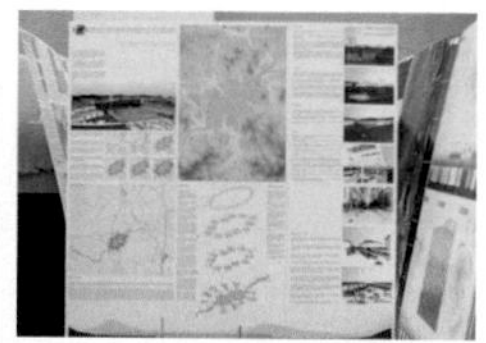

11月11日

■ 首次招待会 / First Reception

11月12日

■ 现场调研 / Site Trip

■ PA 介绍 / PA Briefing

11月12日

■ 评审第一天 / First day of Deliberation

11月13日

■ 评审第二天 / Second day of Deliberation

■ 总理视察 / Prime Minister's Visit

11月14日

■ 评审第三天 / Third day of Deliberation

■ 第二次招待会 / Second Reception

11月15日

■ 宣布获奖方案及召开记者招待会 / Announcement of Prize Winners and Press Conference

12月19日

■ 颁奖典礼 / Awarding Ceremony

■ 获奖方案介绍 / Presentation of Winning Entries

■ 获奖者招待会 / Reception for Prize Winners

6. 展览 / Exhibition

■ 海报 / Promotion

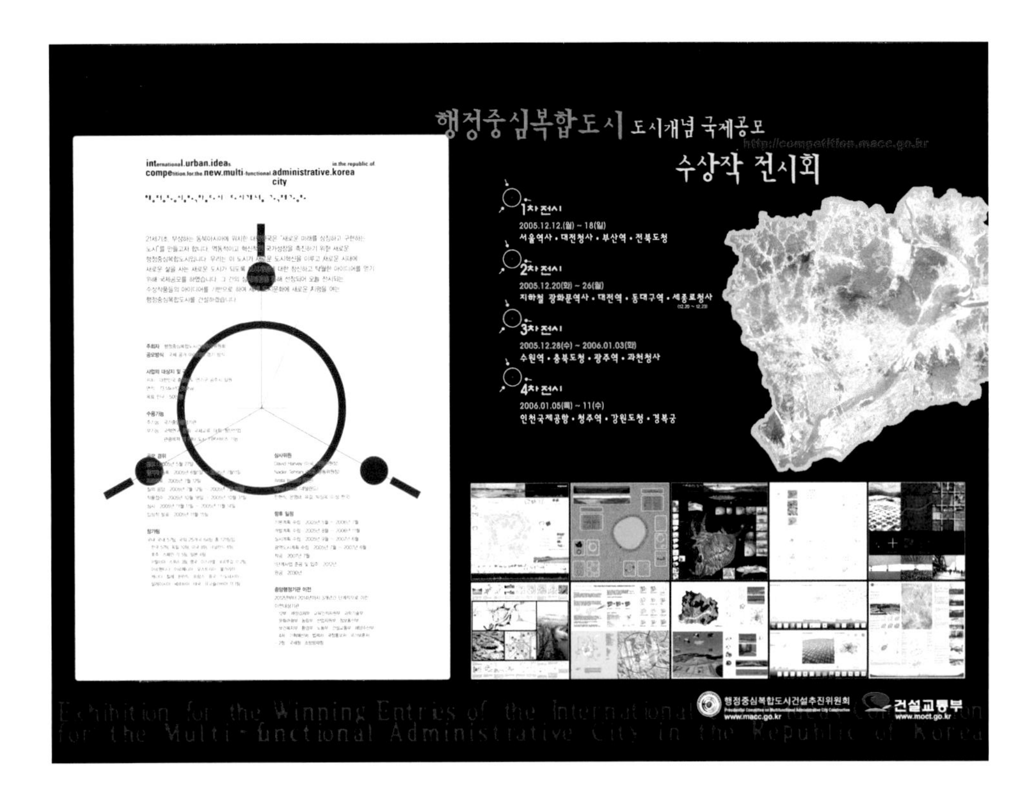

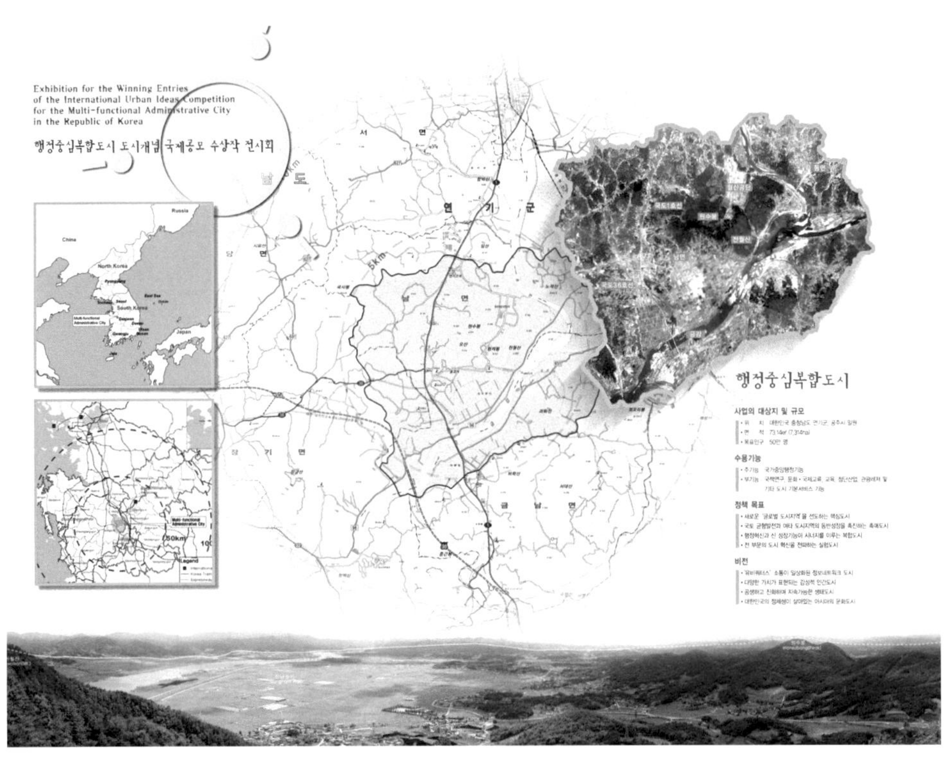

■ 首次展览：2005.12.12（周一）~ 18（周日）

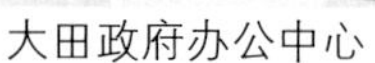
大田政府办公中心

釜山站

首尔站

金罗北道政府办公楼

■ 第二次展览：2005.12.20（周二）~ 26（周一）

大邱东站

光化门地铁站

大田站

中央政府办公中心

■ 第三次展览：2005.12.28（周三）~ 2006.01.03（周二）

水原站

光州站

果川政府办公中心

忠清北道政府办公楼

■ 第四次展览：2006 年 1 月 5 日 ~ 11 日

仁川国际机场

江原道政府办公楼

景福宫

清州站

■ 韩国首尔迁都规划竞赛获奖作品展示 / 主页

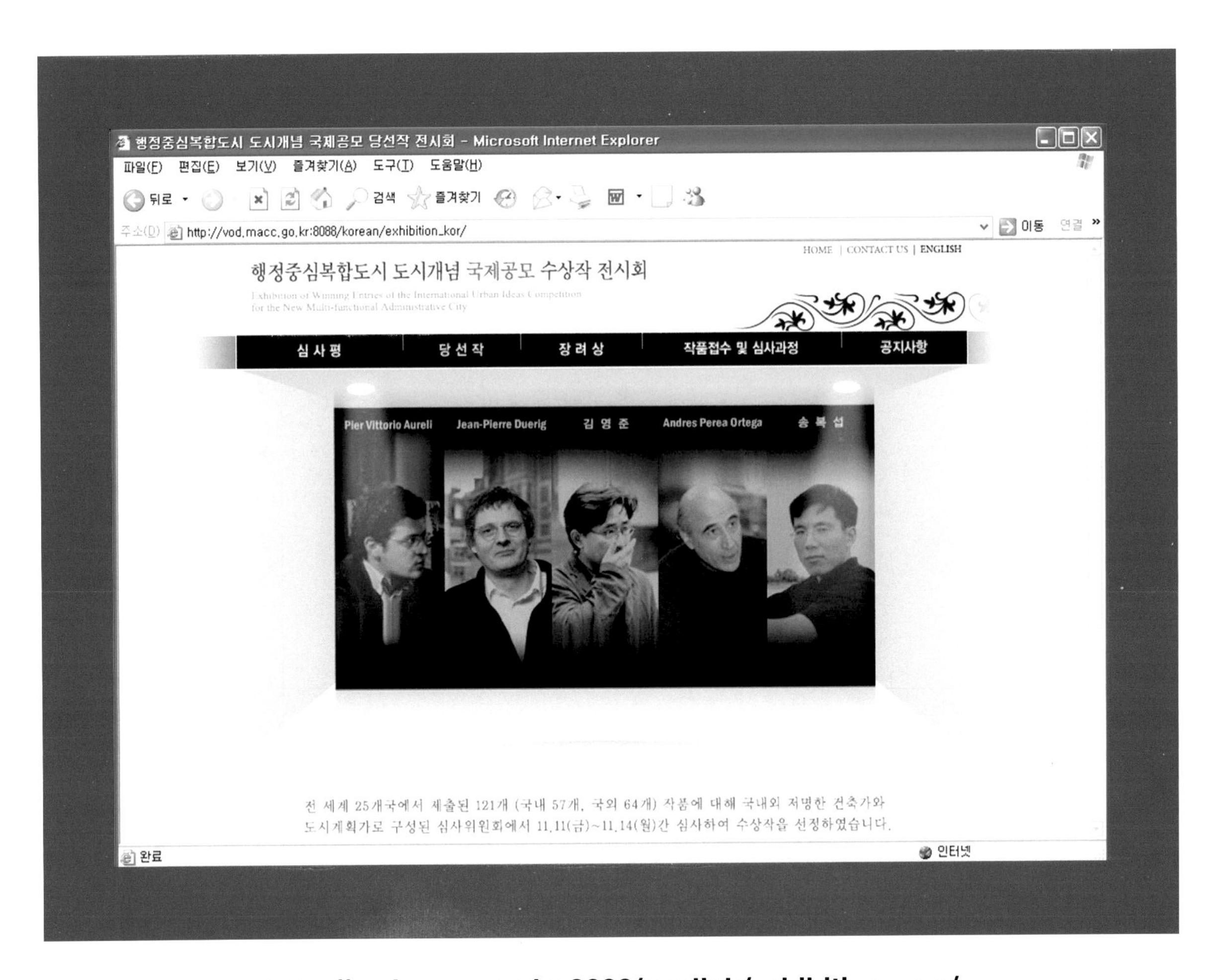

http://vod.macc.go.kr:8088/english/exhibition_eng/

http://vod.macc.go.kr:8088/korea/exhibition_kor/index_kor.asp

译后记 (Afterword to The Translation)

中韩两国是一衣带水的邻邦，自古以来文化交融共生，交流源远流长。韩国新复合型行政中心城市的参赛方案中包含了当今世界上诸多优秀设计师的精心之作。能有机会翻译此书，也是希望将其中优秀的规划理念和设计方法介绍给国内的读者，藉以用一种全新的目光审视我们的城市在发展中所面临的问题与抉择，而书中的很多方案为我们提供了不错的答案和思考的空间。

本书的原著为韩文，此次翻译是在韩文和英文对照版的基础上进行的，因其语言表达习惯和语法语义的微妙差别，在本书的翻译过程中译者力求忠于原著，不惜多方请教韩语和英语的相关专业人士，共同反复推敲论证，力争无误。对于译文中出现的有关人名、地名以及韩国行政机构名称等词汇的描述由于同音多译的缘故，可能与其他译法存在差异，如给读者的阅读造成不便敬请谅解。

本书的翻译是集体合作的结晶，凝结着译者的心血和劳动，在此衷心感谢配合本书翻译的北京工业大学的傅博和冯辽，同时也真诚地感谢中国建筑工业出版社的白玉美主任和戚琳琳编辑，是她们的热情和努力促成了本书的出版。由于时间仓促，加之译者水平有限，若有翻译不当之处，敬请读者批评斧正。

北京工业大学　武凤文

2009 年 11 月

著作权合同登记图字：01－2009－7860 号

图书在版编目（CIP）数据

韩国首尔迁都规划竞赛作品集／（韩）复合型行政中心城市建设推进委员会，复合型行政中心城市建设厅编；武风文等译．—北京：中国建筑工业出版社，2009
ISBN 978－7－112－11007－0

Ⅰ．韩… Ⅱ．①复…②复…③武… Ⅲ．城市规划－建筑设计－作品集－韩国－现代
Ⅳ．TU984.312.6

中国版本图书馆 CIP 数据核字（2009）第 085258 号

责任编辑：白玉美 戚琳琳
责任设计：郑秋菊
责任校对：陈 波 关 健

韩国首尔迁都规划竞赛作品集

[韩] 复合型行政中心城市建设推进委员会 / 复合型行政中心城市建设厅 编
武风文 傅博 冯辽 译

*

中国建筑工业出版社出版、发行（北京西郊百万庄）
各地新华书店、建筑书店经销
北京嘉泰利德公司制版
北京画中画印刷有限公司印刷

*

开本：880×1230 毫米 1/16 印张：$16^3/_4$ 字数：670 千字
2010 年 1 月第一版 2010 年 1 月第一次印刷
定价：136.00 元
ISBN 978－7－112－11007－0
(18257)